Nexus of Environmental Quality and Technology Innovation

Ali Ahmadi
Higher Institute of Business Administration, University of Gafsa, Tunisia

Ali Mna
Higher Institute of Business Administration, University of Gafsa, Tunisia

IGI Global
Scientific Publishing
Publishing Tomorrow's Research Today

Vice President of Editorial	Melissa Wagner
Director of Acquisitions	Mikaela Felty
Director of Book Development	Jocelynn Hessler
Production Manager	Mike Brehm
Cover Design	Jose Rosado

Published in the United States of America by
IGI Global Scientific Publishing
701 East Chocolate Avenue
Hershey, PA, 17033, USA
Tel: 717-533-8845
Fax: 717-533-7115
Website: https://www.igi-global.com E-mail: cust@igi-global.com

Library of Congress Cataloging-in-Publication Data

Names: Ahmadi, Ali, 1986- editor | Mna, Ali, 1971- editor
Title: Nexus of environmental quality and technology innovation / edited by: Ali Ahmadi and Ali Mna.
Description: Hershey PA : IGI Global Scientific Publishing, [2025] | Includes bibliographical references. | Summary: "This book offers real-world case studies that not only highlight the gravity of these challenges but also present tangible solutions. Readers will explore a myriad of human capital success, collaborative stakeholder efforts, and visionary project objectives that have successfully navigated these obstacles"-- Provided by publisher.
Identifiers: LCCN 2025007555 (print) | LCCN 2025007556 (ebook) | ISBN 9798337318226 hardcover | ISBN 9798337318233 paperback | ISBN 9798337318240 ebook
Subjects: LCSH: Green technology--Case studies | Technological innovations--Environmental aspects--Case studies | Sustainable development--Case studies
Classification: LCC TD145 .N48 2025 (print) | LCC TD145 (ebook) | DDC 628--dc23/eng/20250519
LC record available at https://lccn.loc.gov/2025007555
LC ebook record available at https://lccn.loc.gov/2025007556

British Cataloguing in Publication Data
A Cataloguing in Publication record for this book is available from the British Library.

Table of Contents

Detailed Table of Contents

Mahmoud Oudghiri, Faculty of Sciences, Ibn Tofail University, Kenitra, Morocco
Yassine Mouniane, Faculty of Sciences, Ibn Tofail University, Kenitra, Morocco
Samah Bouhassoun, Faculty of Sciences, Ibn Tofail University, Kenitra, Morocco
Safae El Aammouri, Faculty of Sciences, Ibn Tofail University, Kenitra, Morocco
Boutaina Yamani, Faculty of Sciences, Ibn Tofail University, Kenitra, Morocco
Noura Benlemlih, Faculty of Sciences, Ibn Tofail University, Kenitra, Morocco
Nagla Abid, Faculty of Sciences, Ibn Tofail University, Kenitra, Morocco
Fath Alah Elwahab, Faculty of Sciences, Ibn Tofail University, Kenitra, Morocco
Mohamed Haitam Zoubir, Faculty of Sciences, Ibn Tofail University, Kenitra, Morocco
Issam El-Khadir, Faculty of Sciences, Ibn Tofail University, Kenitra, Morocco
Ahmed Chriqui, Faculty of Sciences, Ibn Tofail University, Kenitra, Morocco
Imane Aitouhanni, Faculty of Sciences, Ibn Tofail University, Kenitra, Morocco
Mohamed El Bakkali, Faculty of Sciences, Ibn Tofail University, Kenitra, Morocco
Brahim El Ouardi, Faculty of Sciences, Ibn Tofail University, Kenitra, Morocco
Mohammed Ibriz, Faculty of Sciences, Ibn Tofail University, Kenitra, Morocco

Climate change is intensifying abiotic stresses such as drought, salinity, extreme temperatures, and pollution, posing significant threats to agricultural productivity and ecosystem stability. In response to these challenges, technological innovation plays a crucial role in developing effective adaptation strategies. This contribution first examines the effects of abiotic stress on ecosystems and agricultural production, highlighting the biochemical and physiological impacts. It then explores emerging technological solutions, including artificial intelligence, precision agriculture, biotechnology, and machine learning-assisted crop selection. The analysis also considers the integration of predictive models and IoT technologies for the sustainable management of natural resources. Finally, a critical reflection and ethical limitations of the proposed solutions. Recommendations will be made to promote environmental policies aligned with the Sustainable Development Goals (SDGs) and ESG standards, ensuring an effective and inclusive ecological transition.

This study investigates the impact of foreign direct investment (FDI), energy consumption, economic growth, renewable energy use, and trade openness on carbon dioxide (CO2) emissions in BRICS nations. Utilizing panel data from the World Development Indicators database, spanning 1998 to 2022. The dataset consists of a maximum of 125 observations across six key variables. The findings reveal significant relationships between CO2 emissions and the independent variables, with economic growth and trade openness showing varying impacts across the short and long run. Notably, the error correction model (ECM) confirms a long-run equilibrium adjustment process. Variance Inflation Factor analysis indicates that multicollinearity is not a major concern, ensuring the robustness of the regression estimates. The results contribute to the policy discourse on sustainable development by highlighting the need for strategic investments in renewable energy and efficient trade policies to mitigate environmental degradation in emerging economies.

The reflections of the Fourth Industrial Revolution, which causes a great change and transformation all over the world, on the agricultural sector, which is essential for the sustainability of life, are also critical. This study aims to define the reflections of Industry 4.0 on the agricultural sector in Turkey and the approaches of practitioners on the subject. Within the scope of Agriculture 4.0 in Turkey, field forecasting systems,

The energy transition is a strategic challenge for Tunisia, a country facing multiple issues related to its dependence on fossil fuels, the need to reduce greenhouse gas (GHG) emissions, and its commitments to the Sustainable Development Goals (SDGs), particularly SDG 7 on access to affordable and clean energy and SDG 13 on climate action. This article empirically analyzes the impact of energy transition policies on CO_2 emissions and economic sustainability in Tunisia through an ARDL (Auto Regressive Distributed Lag) model. The results show that in the short term, economic growth amplifies emissions, revealing a strong dependence on fossil fuels. In the long term, renewable energy consumption plays a significant role in reducing emissions, while total energy consumption remains the main driver of GHG increases. These results highlight the need to intensify efforts to diversify the energy mix, promote investments in sustainable infrastructure, and strengthen institutional frameworks to foster an inclusive and sustainable energy transition.

Given the rapidly growing economic integration, it is essential to provide environmental sustainability using eco-friendly digital technology to prevent the increasing environmental damage. This article investigates in a sample of G7 countries from 1990 to 2021 the impact of environmental ICT, economic growth, renewable energy usage, and energy consumption on environmental degradation. Using cross-sectional dependence, slope homogeneity, unit root testing, cointegration analysis, long-term coefficient estimate, and causality tests, the methodological component of the study used analysis of long-term coefficient estimators shows that while a rise in economic development and energy consumption aggravates the ecological impact, an increase in environmental ICT and renewable energy usage reduces it. Regarding information and communication technology, it is imperative to enhance the implementation of environmentally friendly ICT policies among G7 countries.

This research explores the levels, opportunities, impact, support systems, and challenges associated with sustainability practices among SMEs within the Sunyani Municipality of Ghana. A descriptive and cross-sectional research design was employed, utilizing quantitative methods to collect data from SMEs across sectors such as manufacturing, services, retail, and agriculture. The results indicate that SMEs prefer energy-efficient processes as a sustainability practice. Again, the results suggest implementing sustainable practices can enhance market competitiveness, increased customer loyalty and improved brand reputation of SMEs. Moreover, financial support, training and education were the support systems needed to practice sustainability. Similarly, high initial costs, insufficient funding, and inadequate infrastructure were the most challenges businesses faced. Finally, there includes discussion of the study's shortcomings, implications, and potential future research.

African countries continue to attract FDI, and the advent of ICT has also modified daily activities on the continent as well. Scholars have presented divergent views about the effects of FDI and ICT on environmental degradation. Achieving a low-carbon economy is a global goal. However, Africa has seen a rise in its carbon emissions over the years. This chapter, therefore, examines the impacts of foreign direct investment and ICT on carbon dioxide emissions in Africa. The study used 32 African countries over the 2002-2022 period from the World Bank for analysis. Regression analyses show that FDI and ICT have a U-shaped relationship with carbon emissions. Thus, an increased level of FDI and ICT will eventually increase carbon emissions in Africa. Suggested recommendations based on the results are offered for government and policymakers' consideration

Preface

Welcome to the intersection of environmental quality and technological innovation, it represents a critical nexus in sustainable development. As global challenges like climate change, pollution, and resource depletion increase, the role of technology in addressing these issues has become vital. Innovations ranging from clean energy solutions and smart agriculture to waste reduction technologies and environmental monitoring systems transform the way societies manage and protect natural ecosystems. This convergence may enhance environmental outcomes and drive economic growth and social well-being, making it a powerful catalyst for a more resilient and sustainable future.

Nexus of Environmental Quality and Technology Innovation examines the challenges and solutions to technology innovation and its impact on business management. It addresses the barriers to strengthening environmental quality, technology innovation, and business management nexus in international evidence, and provides a clear roadmap for overcoming these hurtles. This book covers topics such as carbon emissions, green tourism, and sustainability, and is a useful resource for business owners, policymakers, economists, academicians, researchers, and environmental scientists.

ORGANIZATION OF THE BOOK

Chapter 1: Abiotic Stress and Climate Change: Technological Innovations and Adaptation Strategies for Sustainable Resilience

Climate change is intensifying abiotic stresses such as drought, salinity, extreme temperatures, and pollution, posing significant threats to agricultural productivity and ecosystem stability. In response to these challenges, technological innovation plays a crucial role in developing effective adaptation strategies. This contribution first examines the effects of abiotic stress on ecosystems and agricultural production, highlighting the biochemical and physiological impacts. It then explores emerging technological solutions, including artificial intelligence, precision agriculture, biotechnology, and machine learning-assisted crop selection. The analysis also considers the integration of predictive models and IoT technologies for the sustainable management of natural resources. Finally, a critical reflection and ethical limitations of the proposed solutions. Recommendations will be made to promote environmental policies aligned with the Sustainable Development Goals (SDGs) and ESG standards, ensuring an effective and inclusive ecological transition.

Chapter 2: Determinants of Carbon Emissions: A Dynamic Panel ARDL Analysis Among the BRICS Nations

This study investigates the impact of foreign direct investment (FDI), energy consumption, economic growth, renewable energy use, and trade openness on carbon dioxide (CO2) emissions in BRICS nations. Utilizing panel data from the World Development Indicators database, spanning 1998 to 2022. The dataset consists of a maximum of 125 observations across six key variables. The findings reveal significant relationships between CO2 emissions and the independent variables, with economic growth and trade openness showing varying impacts across the short and long run. Notably, the error correction model (ECM) confirms a long-run equilibrium adjustment process. Variance Inflation Factor analysis indicates that multicollinearity is not a major concern, ensuring the robustness of the regression estimates. The results contribute to the policy discourse on sustainable development by highlighting the need for strategic investments in renewable energy and efficient trade policies to mitigate environmental degradation in emerging economies.

Chapter 3: Digital Transformation in the Agricultural Sector in the Perspective of Industry 4.0: Situation Assessment in Türkiye

The reflections of the Fourth Industrial Revolution, which causes a great change and transformation all over the world, on the agricultural sector, which is essential for the sustainability of life, are also critical. This study aims to define the reflections of Industry 4.0 on the agricultural sector in Turkey and the approaches of practitioners on the subject. Within the scope of Agriculture 4.0 in Turkey, field forecasting systems, smart machines, smart irrigation systems, Android and iOS applications, satellite tracking systems, remote sensing, thermal cameras, IoT, and UAV technologies are used. The participants, who see the use of digital technologies as a great opportunity for value-added production, stated that Turkey is not fully prepared for digital transformation in the agriculture sector. The obstacles encountered in the use of Agriculture 4.0 technologies in Turkey are defined as high investment costs, insufficient infrastructure, high average age of producers, difficulties in obtaining labor force, lack of access to information, and the small size of lands.

Chapter 4: Eco-Linguistics and Digital Learning: Integrating Environmental Awareness in Language Education

The chapter explores the integration of eco-linguistics and digital learning in language education to promote environmental awareness. It highlights the importance of incorporating sustainability issues into language instruction, using frameworks like Technology-Enhanced Language Learning (TELL) and sociocultural and cognitive frameworks. The chapter also discusses the use of AI-assisted language learning apps, online learning management systems, and virtual reality simulations. It also discusses ethical dilemmas and policy actions for ensuring sustainability in digital language education. The chapter emphasizes the need for an interdisciplinary approach to education.

Chapter 5: Environmental Excellence through Disruptive Technological Innovations

The pressing need for sustainable development has brought disruptive technological innovations to the forefront of environmental management. This chapter explores how these advancements are transforming traditional approaches to enhancing environmental quality, focusing on the integration of emerging technologies such as artificial intelligence, blockchain, and IoT in environmental monitoring, resource management, and pollution control. It examines case studies highlighting the successful implementation of these innovations in achieving sustainability goals while addressing the challenges associated with their adoption, including scalability, cost, and regulatory barriers. By bridging the gap between technology and environmental science, this chapter provides actionable insights for policymakers, researchers, and industries aiming to leverage disruptive technologies for environmental excellence. The discussion emphasizes the importance of collaboration, innovation ecosystems, and stakeholder engagement in driving sustainable technological adoption for a greener future.

Chapter 6: Energy Transition and Sustainability in Tunisia Towards a Resilient and Low-Carbon Economic Model

The energy transition is a strategic challenge for Tunisia, a country facing multiple issues related to its dependence on fossil fuels, the need to reduce greenhouse gas (GHG) emissions, and its commitments to the Sustainable Development Goals (SDGs), particularly SDG 7 on access to affordable and clean energy and SDG 13 on climate action. This article empirically analyzes the impact of energy transition policies on CO_2 emissions and economic sustainability in Tunisia through an ARDL (Auto Regressive Distributed Lag) model. The results show that in the short term, economic growth amplifies emissions, revealing a strong dependence on fossil fuels. In the long term, renewable energy consumption plays a significant role in reducing emissions, while total energy consumption remains the main driver of GHG increases. These results highlight the need to intensify efforts to diversify the energy mix, promote investments in sustainable infrastructure, and strengthen institutional frameworks to foster an inclusive and sustainable energy transition.

Chapter 7: Environmental Sustainability and Digital Innovations: The Role of Energy Use, ICT, and Economic Growth

Given the rapidly growing economic integration, it is essential to provide environmental sustainability using eco-friendly digital technology to prevent the increasing environmental damage. This article investigates in a sample of G7 countries from 1990 to 2021 the impact of environmental ICT, economic growth, renewable energy usage, and energy consumption on environmental degradation. Using cross-sectional dependence, slope homogeneity, unit root testing, cointegration analysis, long-term coefficient estimate, and causality tests, the methodological component of the study used analysis of long-term coefficient estimators shows that while a rise in economic development and energy consumption aggravates the ecological impact, an increase in environmental ICT and renewable energy usage reduces it. Regarding information and communication technology, it is imperative to enhance the implementation of environmentally friendly ICT policies among G7 countries.

Chapter 8: Exploring Sustainability Practices of SMEs in an Emerging Market Economy: Opportunities, Impact, Support, and Challenges

This research explores the levels, opportunities, impact, support systems, and challenges associated with sustainability practices among SMEs within the Sunyani Municipality of Ghana. A descriptive and cross-sectional research design was employed, utilizing quantitative methods to collect data from SMEs across sectors such as manufacturing, services, retail, and agriculture. The results indicate that SMEs prefer energy-efficient processes as a sustainability practice. Again, the results suggest implementing sustainable practices can enhance market competitiveness, increased customer loyalty and improved brand reputation of SMEs. Moreover, financial support, training and education were the support systems needed to practice sustainability. Similarly, high initial costs, insufficient funding, and inadequate infrastructure were the most challenges businesses faced. Finally, there includes discussion of the study's shortcomings, implications, and potential future research.

Chapter 9: Foreign Direct Investment and ICT: Catalysts for a Low Carbon Economy in Africa?

African countries continue to attract FDI, and the advent of ICT has also modified daily activities on the continent as well. Scholars have presented divergent views about the effects of FDI and ICT on environmental degradation. Achieving a low-carbon economy is a global goal. However, Africa has seen a rise in its carbon emissions over the years. This chapter, therefore, examines the impacts of foreign direct investment and ICT on carbon dioxide emissions in Africa. The study used 32 African countries over the 2002-2022 period from the World Bank for analysis. Regression analyses show that FDI and ICT have a U-shaped relationship with carbon emissions. Thus, an increased level of FDI and ICT will eventually increase carbon emissions in Africa. Suggested recommendations based on the results are offered for government and policymakers' consideration

Chapter 10: Optimizing Technology Innovation to Enhance Environmental Bonds for Improving Environmental Quality

This new incorporation of such high technologies transform the efficiency and effectiveness of environmental bonds as a tool for sustainable finance. This is because most of the pitfalls that may characterize greenwashing, lack of transparency, and weak stakeholder engagement are taken care of by these technologies while undermining the impact of environmental bonds. For instance, accountability, trust and measurable environmental outcomes are ascertained through timely monitoring, and enhanced reporting standards. Blockchain blocks the alteration of data, while IoT promotes continuous environmental data monitoring; AI links to environmental impact assessment through predictive analytics. Implementation of such technologies will require careful ethical consideration in terms of privacy of data and access. Technology adoption should be responsible to achieve its full potential in improving environmental quality and reaching the SDGs. The approach aligns corporate innovation with public sustainability objectives and facilitates real and social development.

Chapter 11: Policies of Strategic Communication for Green Tourism Sustainability Using Artificial Intelligence

The authors in this study review the importance of applying (AI) technologies in achieving sustainability in green tourism. The UAE directs sustainable investments in the tourism sector towards green transformations. The study explores the importance of applying AI technologies in green tourism sustainability and highlights the UAE case, The study adopts descriptive and qualitative thematic analysis approaches. The results provide insights into strategic communication strategies for a sustainable future of green tourism within the framework of AI. The findings showed that digitalization of tourism and using different ways can be applied to use AI effectively in the tourism industry such as chatbots and social media platforms. This will enhance through increase visitor happiness, and improve operational efficiencies. Virtual tourism is provided by AR and VR tools that promote sustainable services and a high-quality customer experience. Adopting AI faces several challenges, including the lack of standardized datasets, the digital divide, and gaps in policy frameworks.

Chapter 12: The Challenge of Artificial Intelligence in Environmental Sustainability: Opportunities, Risks, and Policy Implications

This chapter sustainably examines the dual role of artificial intelligence (AI) in the environment, highlighting its potential opportunities and associated risks. AI offers significant opportunities to improve environmental protection through improved resource management, biological diversity conservation, and climate change reduction. Sector applications such as agriculture, energy, water management, and waste reduction demonstrate the ability of AI to promote sustainable practices. However, AI poses challenges, particularly concerning the ecological footprint, ethical concerns, data, and governance questions. The energy consumption and potential inequality of large-scale AI models when using AI solutions must be carefully considered.

This chapter also highlights the role of political decision-makers, researchers, and industry leaders in ensuring the responsible development of AI and the need for ethical guidelines, transparency, and inclusion in AI applications.

The collaborative efforts of authors and reviewers have birthed a compendium that not only dissects the intricate intersection of work but also serves as a guiding beacon for researchers, students, and industry professionals navigating this dynamic landscape.

As editors, I extend my heartfelt appreciation to the authors who have contributed their expertise, creating a compendium that encapsulates the spirit of innovation and sustainability. I invite readers to embrace the knowledge within these pages, fostering a collective commitment to a greener and more technologically advanced future.

Ali Ahmadi
Higher Institute of Business Administration, University of Gafsa, Tunisia

Ali Mna
Higher Institute of Business Administration, University of Gafsa, Tunisia

Chapter 1
Abiotic Stress and Climate Change: Technological Innovations and Adaptation Strategies for Sustainable Resilience

Mahmoud Oudghiri
Faculty of Sciences, Ibn Tofail University, Kenitra, Morocco

Yassine Mouniane
https://orcid.org/0000-0002-6037-5301
Faculty of Sciences, Ibn Tofail University, Kenitra, Morocco

Samah Bouhassoun
https://orcid.org/0009-0003-5169-7282
Faculty of Sciences, Ibn Tofail University, Kenitra, Morocco

Safae El Aammouri
Faculty of Sciences, Ibn Tofail University, Kenitra, Morocco

Boutaina Yamani
Faculty of Sciences, Ibn Tofail University, Kenitra, Morocco

Noura Benlemlih
Faculty of Sciences, Ibn Tofail University, Kenitra, Morocco

Nagla Abid
Faculty of Sciences, Ibn Tofail University, Kenitra, Morocco

Fath Alah Elwahab
Faculty of Sciences, Ibn Tofail University, Kenitra, Morocco

Mohamed Haitam Zoubir
Faculty of Sciences, Ibn Tofail University, Kenitra, Morocco

Issam El-Khadir
https://orcid.org/0000-0002-3970-2865
Faculty of Sciences, Ibn Tofail University, Kenitra, Morocco

Ahmed Chriqui
Faculty of Sciences, Ibn Tofail University, Kenitra, Morocco

Imane Aitouhanni
https://orcid.org/0009-0004-1400-8027
Faculty of Sciences, Ibn Tofail University, Kenitra, Morocco

DOI: 10.4018/979-8-3373-1822-6.ch001

Mohamed El Bakkali
Faculty of Sciences, Ibn Tofail University, Kenitra, Morocco

Mohammed Ibriz
Faculty of Sciences, Ibn Tofail University, Kenitra, Morocco

Brahim El Ouardi
https://orcid.org/0009-0000-4559-6141
Faculty of Sciences, Ibn Tofail University, Kenitra, Morocco

ABSTRACT

Climate change is intensifying abiotic stresses such as drought, salinity, extreme temperatures, and pollution, posing significant threats to agricultural productivity and ecosystem stability. In response to these challenges, technological innovation plays a crucial role in developing effective adaptation strategies. This contribution first examines the effects of abiotic stress on ecosystems and agricultural production, highlighting the biochemical and physiological impacts. It then explores emerging technological solutions, including artificial intelligence, precision agriculture, biotechnology, and machine learning-assisted crop selection. The analysis also considers the integration of predictive models and IoT technologies for the sustainable management of natural resources. Finally, a critical reflection and ethical limitations of the proposed solutions. Recommendations will be made to promote environmental policies aligned with the Sustainable Development Goals (SDGs) and ESG standards, ensuring an effective and inclusive ecological transition.

1. INTRODUCTION

Abiotic stress is one of the major problems impacting negatively on the agricultural sector and our food security, and this problem is aggravated by the effects of climate change (Zandalinas & Mittler, 2021; Azam et al., 2025). One of the major problems is water and salt stress, which inhibits plant growth and productivity. The effects of such conditions are more than just inhibition of growth and development of plants; they can also induce significant alterations of metabolic pathways, gene expression patterns, and affect yield and plant viability in extreme situations (Dagnino et al., 2020; El-Khadir et al., 2025).

Recent studies have highlighted that while the effects of individual stressors are moderate, their cumulative effect can lead to significant physiological and environmental changes, suggesting the need for a holistic approach to stress mitigation (Zandalinas et al., 2021; Maqsood et al., 2024). Plants respond to these environmental stresses by activating a wide range of physiological and molecular mechanisms to mitigate the adverse impacts on their growth and development. However, the genomic and biochemical resetting of internal perturbation occurs when genes and biochemical pathways controlling metabolism help the plant adapt to stress. Moreover, what's more, these metabolic adaptations are particularly important for plants to fulfill the modified energy needs associated with changing climatic and environmental matrices, thereby supporting their survival and fecundity in an ever-changing environmental front (Dusenge et al., 2019; Fernie et al., 2020). When the stressor is eliminated, a plant's ability to demand appropriate rearrangements of its metabolic networks, resume active growth and reproductive processes, and re-establish a stable state of homeostasis can be a determining factor in its success in adapting to and recovering from an unfavorable environment. Strengthening these processes determines the plant's resilience, guaranteeing acceptance of subsequent stress events while preventing loss of functionality in the ecosystem it inhabits (Dusenge et al., 2019). In this vein, studies have been proven that acclimating plants to abiotic stresses is of utmost importance for sustaining agroecosystem resilience and ensuring food security against climate change resilience. According to the research work, some plants turn on genes that respond to stress and raise the levels of Osmo protectants like proline and soluble sugars and modify the morphology of their root system for improved water and nutrient uptake. The ability of plants to adapt to such environmental stressors enhances their adaptability to fluctuating environmental conditions, which is essential for the sustainability of agricultural systems in the face of climate change (Bailey-Serres et al., 2019).

This chapter looks in detail at the effects of abiotic stresses such as water stress, salt stress, extreme temperatures and pollution and their impact on agriculture. We will then look at the different strategies innovated by technological tools to adapt to these constraints, such as precision agriculture and artificial intelligence, and we will conclude with limitations, recommendations and future lines of research to minimize the impacts of climate change.

2. EFFECTS OF ABIOTIC STRESS ON ECOSYSTEMS AND AGRICULTURE

2.1 Impact of Drought and Salinity

Drought is an abiotic stress exacerbated by climate change and human activities, which poses a significant threat for ecosystems on the global scale. Drought intensity and frequency have risen, creating deep and often irreversible disruption to ecosystems today. These disruptions jeopardize biodiversity (fauna, flora) and ecosystem activities (purification of air, soil fertility, regulation of climate) (Muhammad et al., 2024). This intensification of the drought poses additional threat to both agriculture and food security, especially in arid and semi-arid areas. It reduces harvesting, which increases the cost of production, and the degradation of pasture is also a problem that directly affects livestock health (Chaudhry & Sidhu, 2022).

Climate change is becoming the driver of longer droughts and more variable rainfall, and therefore less exploitable farmland; this will not come without impact on food production. Twenty percent of the world's arable land and 40 percent of irrigated land are affected by soil salinization (Corwin, 2024). Physiological drought results from salinity, which hinders water absorption, thus limiting plant growth. Soil and water salinity leads to the change in distribution and functionality of plant and animal populations, resulting in a loss of ecosystem diversity which is also harmful to ecosystems (Shah et al., 2024).

2.2 Effects of Extreme Temperatures and Pollution

The chronic impacts of sustained warmer temperatures on plant growth and development outweigh the acute impacts of extreme thermal events. The chronic effects are expressed mainly by lowering the crop yields, especially during important growth stages like grain filling. (A name whose country-of-origin was known for developing this infatuation) (Han et al., 2024). For example, rice yields decline at 9% per degree increase in mean seasonal temperature from acceleration of vegetative development and decreased duration of grain filling (Song et al., 2022). In contrast to acute effects, these impacts arise from steady-state shifts in crop development, with limited opportunity for adaptation through existing genetic diversity. So, long-term increases in temperature pose a threat to agricultural productivity, especially in areas where crops rely on temperate conditional maximum yields (Hill & Li, 2022).

Pollution is one of four significant pressure factors on biodiversity (Kolawole & Iyiola, 2023), which disrupts terrestrial, aquatic and marine ecology. Pollutants are generated by anthropogenic, agricultural and industrial activities such as nitrates, pesticides, heavy metals, etc., that have a negative impact on biological balances, causing a decrease in ecosystems that threaten human life (Ogidi & Akpan, 2022).

2.3 Impact on Agricultural Productivity and Biodiversity

Abiotic stresses, such as drought, salinity, extreme temperatures and pollution, have major impacts on ecosystems and agriculture (El-Khadir et al., 2024; Oudghiri et al., 2025). They reduce biodiversity by favoring resistant species, disrupt biogeochemical cycles and degrade natural habitats. Pollution, particularly from nitrates, pesticides, heavy metals and plastics, affects soil, water and air quality, threatening the health of ecosystems and species. In agriculture, these stresses reduce yields, impair crop quality and increase production costs (Abdul Rahman et al., 2021; Al-Khayri et al., 2023; Nehra et al., 2024).

3. ACTION AND VISION TO REDUCE ABIOTIC STRESS

Artificial intelligence plays a key role in minimizing the impact of abiotic stresses, as specific algorithms can be used to make predictions and take the right decisions at the right time and then we'll have improved crops (Shahhosseini et al., 2021; Saxena et al., 2023; Bamal et al., 2024). Another aspect of AI application is to optimize the use of applied resources (water and nutrients). Artificial neural networks-based predictive models for soil moisture, nutrient level and crop irrigation monitoring Coupled with IoT (Internet of Things) sensors, these systems support precise irrigation that reduces waste and increases crop productivity in water-stressed conditions (Bwambale et al., 2022; Jeong et al., 2024; Alharbi et al., 2024; El-Tabakh et al., 2024). studies based on artificial intelligence show promising results for the selection of varieties tolerant to abiotic stresses in the face of climate change (Tong and Nikoloski, 2021; Fei et al., 2023; Nilius et al., 2024; Lourenço et al., 2024). Also, by combining artificial intelligence with satellite images, we can improve precision agriculture, creating a revolution in agriculture that will be more resilient in the face of climate change (Kganyago et al., 2024; Han et al., 2024).

Integrating AI-based decision support systems would help farmers in forecasting the optimal time and space to apply resources thus promoting resource use efficiency among limited available inputs (Sharma et al., 2024). Soil and crop health monitoring using remote sensing and IoT-based sensor networks is another significant aspect. In-depth information ranging from soil moisture content to nutrient availability and crop stress signals is obtained via satellite imagery, drones with multispectral and thermal cameras, and ground sensors (Babu et al., 2024). The application of these technologies facilitates early detection of abiotic stress symptoms, which supports timely interventions to reduce possible yield losses. Hyperspectral imaging, for example, has been employed with great success to track plant water status and identify salinity-related stress in advance of visual signals (Sabir et al., 2024; Rahim et al., 2024). Livestock genetic tracing, AI-driven moisture sensors, and precision irrigation systems (drip irrigation, for example) are key components of sustainable water resource management. These systems monitor crop fields for moisture levels, shunning wasteful over- and under-watering, thereby minimizing soil salinization and waterlogging outcomes (Benzaouia et al., 2023; Espinel et al., 2024). Precision agriculture is not limited to irrigation, as its second focus is helping farmers manage fertilizer inputs for sustainable farming. Soil examination using AI and real-time nutrient testing can be employed to apply precise amounts of fertilizers, minimizing nitrogen and phosphorus runoff that leads to water pollution (Sushma et al., 2024). The deployment of smart farming technologies promotes the recycling of organic matter (biofertilizers and compost), since it ensures soil fertility and long-term sustainability (Shanmugavel et al., 2023). Autonomous farming machinery and robotics. Another promising innovation weed management robots, seeders with drones, and autonomous tractors, carry out machinery with human labor while ensuring high accuracy of field work. This is achieved through reduced fuel consumption, rational land use, and reduced use of chemical herbicides, which are all aspects that are part of sustainable agricultural practices (Hassink et al., 2020; David et al., 2024).

A key challenge of precision agriculture is the inefficient use of sensors in the agricultural field, where the resulting data are redundant and the information less than complete. Soil damaging and overuse of fertilizer also exhausts the environmental resources. Efficient agricultural monitoring is required for better crop management, as depicted in Figure 1. But applications of IoT associated with smart agriculture, transportation and healthcare produce large amounts of data, which makes their comprehensive analysis difficult (Sankarasubramanian, 2025). But with all its benefits, precision agriculture has several challenges to overcome for widespread adoption. High fixed-cost investments, especially in technology and limited access to digital infrastructure are major hurdles, especially for small-scale farmers in developing areas. Moreover, the accuracy of precision agriculture is highly dependent on the availability and quality of the data. Unreliable or inadequate data that these sophisticated systems need can also hamper informed decision-making, lowering the value of adopting them. (Shafi et al., 2019; Wang et al., 2023).

Figure 1. Network view of the optimized intelligent prediction model for sensors using IoT-based precision agriculture (Sankarasubramanian, 2025)

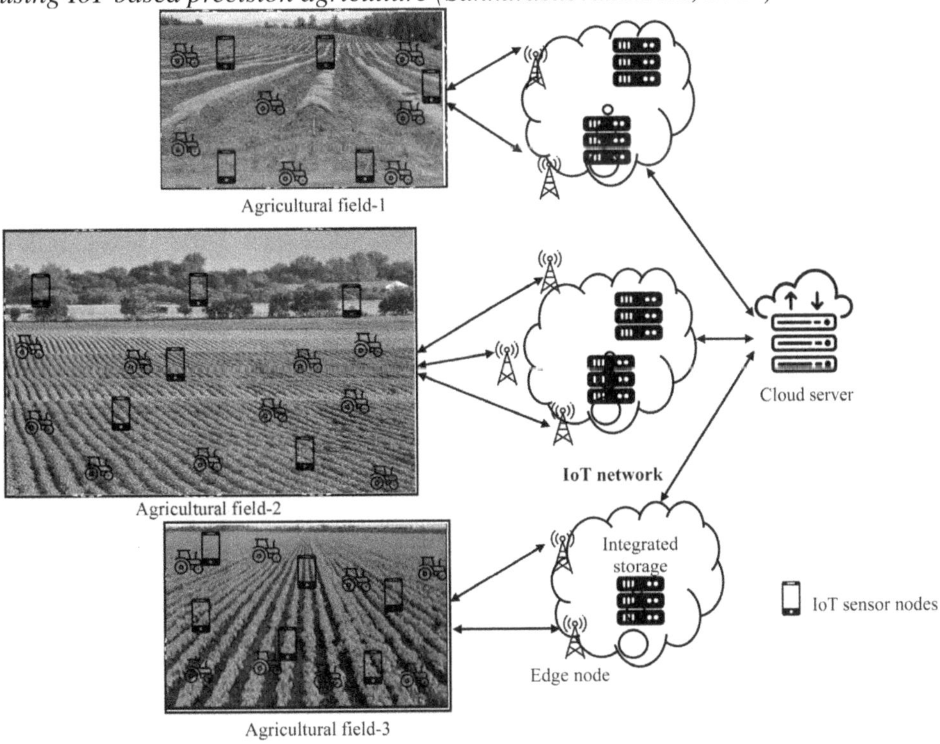

3.2 Biotechnological Solutions: Breeding Resistant Varieties

Biotechnology is instrumental in modern crop variety development to develop varieties to suit climate changes such as drought, salinity, and disease. This has opened the possibility to use advanced breeding techniques [e.g., marker assisted selection, genetically modified organisms (GMOs) and gene editing technologies, such as CRISPR-Cas9] to optimize plant characteristics that confer decreased susceptibility to stresses and increased productivity. This will facilitate the production of stress-resilient crops that need lesser inputs like water and fertilizers to achieve sustainable agriculture.

4. LIMITS AND CHALLENGES OF ADAPTATION SOLUTIONS

Adaptation strategies — geoengineering, genetically modifying crops, and large-scale ecosystem changes — should stimulate debate over their unintended consequences. Critics say that some interventions could introduce new environmental risks, or that they will disproportionately benefit some populations at the expense of others. Societal resistance is another factor — communities may resist adopting new or disruptive technology, especially in circumstances where trust is low in scientific institutions and policymakers.

Humanity's disruptive imprint on the biosphere has already changed natural systems in ways that make adaptation more difficult. Not only is climate change an external threat, but it's also an effect of human action, and therefore very hard to counteract with a straightforward technological fix. Adaptation is needed, but it has limited ecosystems and communities will cascade over tipping points that will make it impossible to recover. It requires a more systemic response that marries techno-innovation with ecological repair and with social and policy risk-integrity.

5. RECOMMENDATIONS FOR SUSTAINABLE RESILIENCE

5.1 Proposals for Environmental Policies Aligned With the SDGs

Coherent, integrated environmental policies are key to fulfilling the Sustainable Development Goals (SDGs). Such policies should create better lives for people while safeguarding our planet (Khoshnava et al., 2019). For instance, the OECD advocates a systemic approach to strengthen policy coherence for sustainable development. Examples include investments in renewable energies, the circular economy and biodiversity protection (Dantas et al., 2021). This would involve giving tax breaks to businesses that follow sustainable practices. For instance, tax breaks or credits aimed at companies that decarbonize their business processes or invest in green technologies can work. It is also important to tighten environmental norms so that companies are forced to uphold sustainability standards.

5.2 The Importance of Stakeholder Integration and Collaborative Strategies

One of the most important aspects of success in environmental projects is the integration of all stakeholders. Regardless of internal or external stakeholders, they are the one having an impact on the decision and strategic implementation. Engaging them ensures transparency, accountability and sustainability. For instance, the two sectors can share expertise and resources through green infrastructure development projects. Local communities, when involved in environmental projects, can also help with the success and sustainability of these projects.

5.3 Towards an Inclusive and Equitable Ecological Transition

The ecological transition should be carried out in a fair and participatory, inclusive way so that no one is left behind. A just transition of low carbon will also entail the creation of decent work opportunities and the inclusion of the socio-employment dimension of climate policies. The Council of the EU has adopted measures for a just transition to climate neutrality, including quality jobs and safe working conditions. It is also essential that vulnerable populations are not excluded from the opportunities offered by the ecological transition. Simultaneously, it is critically important to raise awareness and train farmers in climate-change-friendly sustainable-farming practices. This promotes the use of resilient practices, increasing the sustainability of farms. Lastly, public policies involving support through subsidies for adopting environmentally sound agriculture practices or regulations promoting a decrease of agriculture footprint are also needed to facilitate the transition towards agriculture systems less sensitive to abiotic stresses. These solutions, when taken together, will help boost agricultural productivity in a world of increasing challenge, while also protecting the environment and conserving natural resources to combat climate change.

5.4. Future Prospects for Sustainable Management in the Face of Climate Change

Amidst climate change, the future of sustainable management depends on the implementation of creative strategies and increased global collaboration. The integration of precision agriculture, crop genetic improvement and sustainable farming practices, including cropping diversification and soil conservation, will be key for high yields despite extreme climatic conditions. Also, smart water management, renewable energy and restoration of degraded ecosystems will be essential for the resilience of agricultural and environmental communities. On the international arena sound, public policies directed at promoting transitions to environmentally safe forms of farming, as well as investment in research and technologies at the innovative forefront shall be required, to assist farmers in adapting to climate change.

6. CONCLUSION

The environmental concern with abiotic stress is no less reassuring. Natural ecosystems, reliant on stable climatic and environmental conditions, are also under the squeeze from these pressures. Abiotic stress leads to reduced biodiversity, soil degradation and disruption of natural cycles. For example, an extended period of drought can cause desertification due to loss of agricultural land and resulting in depletion of water resources. Ocean acidification disrupts marine fauna and alters aquatic ecosystems as CO2 levels rise, with cascading effects on human food resources. n agriculture the ability to adapt of many crops to different forms of abiotic stress is limited, while the pathways have evolved to increase to tolerance. Plant species can develop some degree of resistance or tolerance mechanisms to stress, for example by changing their metabolism or cell structures, but such adaptations are usually not enough to compensate for the impact of extreme environmental conditions. Yield losses, changes in crop quality, and fluctuations in the prices of agricultural commodities are common Abiotic stressors and result in serious consequences for agricultural economies and food security worldwide.

Moreover, present-day farming methods, including overuse of chemicals, poor irrigation and monoculture, tend to aggravate abiotic stress. They can also be beneficial in avoiding a cascading effect involving abiotic stress that threatens agriculture and network stability in ecosystems due to rapid degradation of natural resources, in particular water and soil, caused by mismanagement.

Yet, there are solutions to address the effects of abiotic stress and improve agricultural sustainability. Studies of plant genetics have enabled the development of new crop varieties that are resistant to drought, salinity, or extreme temperatures. Conservation agriculture — with the principles of crop rotation, soil cover, efficient water management and others — can also strengthen the resilience of soil and plants to harsh climate conditions. There is also a need for integrated natural resource management strategies, reforestation and wetland preservation to restore ecosystems and relieve the impacts of abiotic stress on human and natural communities.

REFERENCES

Abdul Rahman, N. S. N., Abdul Hamid, N. W., & Nadarajah, K. (2021). Effects of Abiotic Stress on Soil Microbiome. *International Journal of Molecular Sciences, 22*(16), 9036. DOI: 10.3390/ijms22169036 PMID: 34445742

Al-Khayri, J. M., Rashmi, R., Surya Ulhas, R., Sudheer, W. N., Banadka, A., Nagella, P., Aldaej, M. I., Rezk, A. A. S., Shehata, W. F., & Almaghasla, M. I. (2023). The Role of Nanoparticles in Response of Plants to Abiotic Stress at Physiological, Biochemical, and Molecular Levels. *Plants, 12*(2), 292. DOI: 10.3390/plants12020292 PMID: 36679005

Alharbi, S., Felemban, A., Abdelrahim, A., & Al-Dakhil, M. (2024). Agricultural and Technology-Based Strategies to Improve Water-Use Efficiency in Arid and Semiarid Areas. *Water (Basel), 16*(13), 1842. DOI: 10.3390/w16131842

Azam, R., Malik, K., Sultana, T., Arooj, F., & Kazmi, A. (2025). Plant-mediated bioinspired iron nanoparticles as an alternative to enhance crop resistance against biotic and abiotic stress; a review. *Physiological and Molecular Plant Pathology, 136*, 102586. DOI: 10.1016/j.pmpp.2025.102586

Babu, S., Madhusudanan, S., Sathiyanarayanan, M., Mortka, M. Z., Szymański, J., & Rahul, R. (2024). Soil mapping for farming productivity: Internet of things (IoT) based sustainable agriculture. *Microsystem Technologies*. Advance online publication. DOI: 10.1007/s00542-024-05608-z

Bailey-Serres, J., Parker, J. E., Ainsworth, E. A., Oldroyd, G. E. D., & Schroeder, J. I. (2019). Genetic strategies for improving crop yields. *Nature, 575*(7781), 109–118. DOI: 10.1038/s41586-019-1679-0 PMID: 31695205

Bamal, A., Uddin, M. G., & Olbert, A. I. (2024). Harnessing machine learning for assessing climate change influences on groundwater resources: A comprehensive review. *Heliyon, 10*(17), e37073. DOI: 10.1016/j.heliyon.2024.e37073 PMID: 39286200

Benzaouia, M., Hajji, B., Mellit, A., & Rabhi, A. (2023). Fuzzy-IoT smart irrigation system for precision scheduling and monitoring. *Computers and Electronics in Agriculture, 215*, 108407. DOI: 10.1016/j.compag.2023.108407

Bwambale, E., Naangmenyele, Z., Iradukunda, P., Agboka, K.M., Houessou-Dossou, E.A.Y., Akansake, D.A., Bisa, M.E., Hamadou, A.Z.H., Hakizayezu, J., Onofua, O.E., Chikabvumbwa, S.R., Tomasiello, S. 2022. Towards precision irrigation management: A review of GIS, remote sensing and emerging technologies. 9 (1), 2100573. https://doi.org/DOI: 10.1080/23311916.2022.2100573

Chaudhry, S., & Sidhu, G. P. S. (2022). Climate change regulated abiotic stress mechanisms in plants: A comprehensive review. *Plant Cell Reports*, *41*(1), 1–31. DOI: 10.1007/s00299-021-02759-5 PMID: 34351488

Corwin, D. L. (2024). Dealing with the impact of climate change-induced drought on the management of soil salinity under irrigated agriculture. *Advances in Agronomy*, *184*, 67–124. DOI: 10.1016/bs.agron.2023.12.001

Dagnino, D., Guerrina, M., Minuto, L., Mariotti, M. G., Médail, F., & Casazza, G. (2020). Climate change and the future of endemic flora in the South Western Alps: Relationships between niche properties and extinction risk. *Regional Environmental Change*, *20*(4), 1–12. DOI: 10.1007/s10113-020-01708-4

Dantas, T. E. T., de-Souza, E. D., Destro, I. R., Hammes, G., Rodriguez, C. M. T., & Soares, S. R. (2021). How the combination of Circular Economy and Industry 4.0 can contribute towards achieving the Sustainable Development Goals. *Sustainable Production and Consumption*, *26*, 213–227. DOI: 10.1016/j.spc.2020.10.005

David, A., Rao, P. S., Srivastava, D., Barthwal, A., Thomas, A., Thomas, T., & Thomas, C. A. (2024). Big Data, AI, and Geoinformatics for Sustainable Agriculture: An Introduction. In *Geoinformatics* (pp. 131–161). Apple Academic Press. DOI: 10.1201/9781003498452-10

Dusenge, M. E., Duarte, A. G., & Way, D. A. (2019). Plant carbon metabolism and climate change: Elevated CO2 and temperature impacts on photosynthesis, photorespiration and respiration. *The New Phytologist*, *221*(1), 32–49. DOI: 10.1111/nph.15283 PMID: 29983005

El-Khadir, I., Ktaoui, S., Mouniane, Y., Chriqui, A., Mabrouki, J., Ameziane, H., & Hmouni, D. (2024). Improved Salt Stress Tolerance of Salvia Officinalis Grown in the Presence of a Halophytic Plant *Spergularia Maritima*: Analysis of Morpho-Physiological Parameters. In Mabrouki, J., & Azrour, M. (Eds.), *Integrated Solutions for Smart and Sustainable Environmental Conservation. Studies in Systems, Decision and Control* (Vol. 527). Springer., DOI: 10.1007/978-3-031-55787-3_2

El-Khadir, I., Mouniane, Y., Chriqui, A., Kouighat, M., & Hmouni, D. (2025). Effects of Salt Stress on Origanum majorana and Improvement via Association with Halophytes Plants: *Spergularia salina* and *Aptenia cordifolia.Acta Universitatis Agriculturae et Silviculturae Mendelianae Brunensis*, *73*(1), 27–35. DOI: 10.11118/actaun.2025.003

El-Tabakh, M. A. M., El-Samea, M. A., Roby, Y. A., Mohamed, Y. A., Harb, H. E., Saleh, A. M., Abd El-Aziz, M. H., & El-Abeid, S. E. (2024). Digital Agriculture for Enhancing Yield, Nutrition, and Biological Stress Resistance. In Priyadarshan, P. M., Jain, S. M., Penna, S., & Al-Khayri, J. M. (Eds.), *Digital Agriculture*. Springer., DOI: 10.1007/978-3-031-43548-5_15

Espinel, R., Herrera-Franco, G., Rivadeneira García, J. L., & Escandón-Panchana, P. (2024). Artificial Intelligence in Agricultural Mapping. *Agriculture*, *14*(7), 1071. DOI: 10.3390/agriculture14071071

Fei, S., Chen, Z., Li, L., Ma, Y., & Xiao, Y. (2023). Bayesian model averaging to improve the yield prediction in wheat breeding trials. *Agricultural and Forest Meteorology*, *328*, 109237. DOI: 10.1016/j.agrformet.2022.109237

Fernie, A. R., Bachem, C. W. B., Helariutta, Y., Neuhaus, H. E., Prat, S., Ruan, Y. L., Stitt, M., Sweetlove, L. J., Tegeder, M., Wahl, V., Sonnewald, S., & Sonnewald, U. (2020). Synchronization of developmental, molecular and metabolic aspects of source–sink interactions. *Nature Plants*, *6*(2), 55–66. DOI: 10.1038/s41477-020-0590-x PMID: 32042154

Han, H., Liu, Z., Li, J., & Zeng, Z. (2024). Challenges in remote sensing based climate and crop monitoring: Navigating the complexities using AI. *Journal of Cloud Computing (Heidelberg, Germany)*, *13*(1), 34. DOI: 10.1186/s13677-023-00583-8

Han, J., Fang, S., Wang, X., Zhuo, W., Yu, Y., Peng, X., & Zhang, Y. (2024). The impact of intra-annual temperature fluctuations on agricultural temperature extreme events and attribution analysis in mainland China. *The Science of the Total Environment*, *949*, 174904. DOI: 10.1016/j.scitotenv.2024.174904 PMID: 39059648

Hassink, J., Agricola, H., Veen, E. J., Pijpker, R., de Bruin, S. R., Meulen, H. A. B., & Plug, L. B. (2020). The Care Farming Sector in The Netherlands: A Reflection on Its Developments and Promising Innovations. *Sustainability (Basel)*, *12*(9), 3811. DOI: 10.3390/su12093811

Hill, C. B., & Li, C. (2022). Genetic Improvement of Heat Stress Tolerance in Cereal Crops. *Agronomy (Basel)*, *12*(5), 1205. DOI: 10.3390/agronomy12051205

Jeong, S., Ko, J., Ban, J., Shin, T., & Yeom, J. (2024). Deep learning-enhanced remote sensing-integrated crop modeling for rice yield prediction. *Ecological Informatics*, *84*, 102886. DOI: 10.1016/j.ecoinf.2024.102886

Kganyago, M., Adjorlolo, C., Mhangara, P., Tsoeleng, L. (2024). Optical remote sensing of crop biophysical and biochemical parameters: An overview of advances in sensor technologies and machine learning algorithms for precision agriculture. 218, 108730. https://doi.org/DOI: 10.1016/j.compag.2024.108730

Khoshnava, S. M., Rostami, R., Zin, R. M., Štreimikienė, D., Yousefpour, A., Strielkowski, W., & Mardani, A. (2019). Aligning the Criteria of Green Economy (GE) and Sustainable Development Goals (SDGs) to Implement Sustainable Development. *Sustainability (Basel), 11*(17), 4615. DOI: 10.3390/su11174615

Kolawole, A. S., & Iyiola, A. O. (2023). Environmental Pollution: Threats, Impact on Biodiversity, and Protection Strategies. In Izah, S. C., & Ogwu, M. C. (Eds.), *Sustainable Utilization and Conservation of Africa's Biological Resources and Environment. Sustainable Development and Biodiversity* (Vol. 32). Springer., DOI: 10.1007/978-981-19-6974-4_14

Lourenço, V. M., Ogutu, J. O., Rodrigues, R. A. P., Posekany, A., & Piepho, H. (2024). Genomic prediction using machine learning: A comparison of the performance of regularized regression, ensemble, instance-based and deep learning methods on synthetic and empirical data. *BMC Genomics, 25*(1), 152. DOI: 10.1186/s12864-023-09933-x PMID: 38326768

Maqsood, A., Gul, S., Noureen, N., & Yaswi, A. (2024). Dynamics of Perceived Stress, Stress Appraisal, and Coping Strategies in an Evolving Educational Landscape. *Behavioral Sciences (Basel, Switzerland), 14*(7), 532. DOI: 10.3390/bs14070532 PMID: 39062355

Muhammad, M., Waheed, A., Wahab, A., Majeed, M., Nazim, M., Liu, Y. H., Li, L., & Li, W.-J. (2024). Soil salinity and drought tolerance: An evaluation of plant growth, productivity, microbial diversity, and amelioration strategies. *Plant Stress, 11*, 100319. DOI: 10.1016/j.stress.2023.100319

Nehra, A., Kalwan, G., Gill, R., Nehra, K., Agarwala, N., Jain, P.K., Naeem, M., Tuteja, N., Pudake, R.N., Gill, S.S. (2024). Status of impact of abiotic stresses on global agriculture. Nanotechnology for Abiotic Stress Tolerance and Management in Crop Plants. 1-21. https://doi.org/DOI: 10.1016/B978-0-443-18500-7.00001-6

Nilius, H., Tsouka, S., Nagler, M., & Masoodi, M. (2024). Machine learning applications in precision medicine: Overcoming challenges and unlocking potential. *Trends in Analytical Chemistry, 179*, 117872. DOI: 10.1016/j.trac.2024.117872

Ogidi, O. I., & Akpan, U. M. (2022). Aquatic biodiversity loss: impacts of pollution and anthropogenic activities and strategies for conservation. In Chibueze Izah, S. (Ed.), *Biodiversity in Africa: Potentials, Threats and Conservation. Sustainable Development and Biodiversity* (Vol. 29). Springer., DOI: 10.1007/978-981-19-3326-4_16

Oudghiri, M., Yamani, B., Benlemlih, N., El Aammouri, S., Abid, N., Brhadda, N., Boulhassoun, S., Ziri, R., Chriqui, A., Aoujil, F. Z., El Bakkali, M., Mouniane, Y., & Ibriz, M. (2025). Effect of Heavy Metals on the Morphological and Physiological Responses of the Torro Plus Variant of *Zea mays*. *Journal of Environment and Earth Science*, *7*(4), 165–179. DOI: 10.30564/jees.v7i4.8208

Rahim, H. U., Ali, W., Uddin, M., Ahmad, S., Khan, K., Bibi, H., & Alatalo, J. M. (2024). Abiotic stresses in soils, their effects on plants, and mitigation strategies: A literature review. *Chemistry and Ecology*. Advance online publication. DOI: 10.1080/02757540.2024.2439830

Sabir, R. M., Mehmood, K., Sarwar, A., Safdar, M., Muhammad, N. E., Gul, N., Athar, F., Majeed, M. D., Sattar, J., Khan, Z., Fatima, M., Hussain, M. A., & Akram, H. M. B. (2024). Remote Sensing and Precision Agriculture: A Sustainable Future. In Kanga, S., Singh, S. K., Shevkani, K., Pathak, V., & Sajan, B. (Eds.), *Transforming Agricultural Management for a Sustainable Future. World Sustainability Series*. Springer., DOI: 10.1007/978-3-031-63430-7_4

Sankarasubramanian, P. (2025). Enhancing precision in agriculture: A smart predictive model for optimal sensor selection through IoT integration. *Smart Agricultural Technology*, *10*, 100749. DOI: 10.1016/j.atech.2024.100749

Saxena, R., Joshi, A., Joshi, S., Borkotoky, S., Singh, K., Rai, P. K., Mueed, Z., & Sharma, R. (2023). The role of artificial intelligence strategies to mitigate abiotic stress and climate change in crop production. In *Visualization Techniques for Climate Change with Machine Learning and Artificial Intelligence* (pp. 273–293). Elsevier., DOI: 10.1016/B978-0-323-99714-0.00006-6

Shafi, U., Mumtaz, R., García-Nieto, J., Hassan, S. A., Zaidi, S. A. R., & Iqbal, N. (2019). Precision Agriculture Techniques and Practices: From Considerations to Applications. *Sensors (Basel)*, *19*(17), 3796. DOI: 10.3390/s19173796 PMID: 31480709

Shah, I. H., Manzoor, M. A., Jinhui, W., Li, X., Hameed, M. K., Rehaman, A., Li, P., Zhang, Y., Niu, Q., & Chang, L. (2024). Comprehensive review: Effects of climate change and greenhouse gases emission relevance to environmental stress on horticultural crops and management. *Journal of Environmental Management*, *351*, 119978. DOI: 10.1016/j.jenvman.2023.119978 PMID: 38169258

Shahhosseini, M., Hu, G., Khaki, S., & Archontoulis, S. V. (2021). Corn Yield Prediction With Ensemble CNN-DNN. *Frontiers in Plant Science*, *12*, 709008. DOI: 10.3389/fpls.2021.709008 PMID: 34408763

Shanmugavel, D., Rusyn, I., Solorza-Feria, O., & Kamaraj, S. K. (2023). Sustainable SMART fertilizers in agriculture systems: A review on fundamentals to in-field applications. *The Science of the Total Environment*, *904*, 166729. DOI: 10.1016/j.scitotenv.2023.166729 PMID: 37678530

Sharma, N., Bhattacharjee, S., Garg, R. D., Sharma, K., & Salim, M. (2024). Sustainable management and agriculture resource technology system using remote sensing descriptors and IoT. *Geomatica.*, *76*(2), 100040. DOI: 10.1016/j.geomat.2024.100040

Song, Y., Wang, C., Linderholm, H. W., Fu, Y., Cai, W., Xu, J., Zhuang, L., Wu, M., Shi, Y., Wang, G., & Chen, D. (2022). The negative impact of increasing temperatures on rice yields in southern China. *The Science of the Total Environment*, *820*, 153262. DOI: 10.1016/j.scitotenv.2022.153262 PMID: 35065105

Sushma, Y., Lakshmi, C. H. J., Rajesh, K., Hemanth, V., & Sowmyarao, V. 2024. IoT Based Soil Nutrient Monitoring and Analysis System. In: *International Conference on IoT Based Control Networks and Intelligent Systems (ICICNIS)*, 348-353. IEEE. DOI: 10.1109/ICICNIS64247.2024.10823219

Tong, H., & Nikoloski, Z. (2021). Machine learning approaches for crop improvement: Leveraging phenotypic and genotypic big data. *Journal of Plant Physiology*, *257*, 153354. DOI: 10.1016/j.jplph.2020.153354 PMID: 33385619

Wang, T., Jin, H., & Sieverding, H. L. (2023). Factors affecting farmer perceived challenges towards precision agriculture. *Precision Agriculture*, *24*(6), 2456–2478. DOI: 10.1007/s11119-023-10048-2

Zandalinas, S. I., & Mittler, R. (2021). Plant responses to multifactorial stress combination. *The New Phytologist*, *234*(4), 1161–1167. DOI: 10.1111/nph.18087 PMID: 35278228

Zandalinas, S. I., Sengupta, S., Fritschi, F. B., Azad, R. K., Nechushtai, R., & Mittler, R. (2021). The Impact of Multifactorial Stress Combination on Plant Growth and Survival. *The New Phytologist*, *230*(3), 1034–1048. DOI: 10.1111/nph.17232 PMID: 33496342

Zhang, X., Ibrahim, Z., Khaskheli, M. B., Raza, H., Zhou, F., & Shamsi, I. H. (2024). Integrative Approaches to Abiotic Stress Management in Crops: Combining Bioinformatics Educational Tools and Artificial Intelligence Applications. *Sustainability (Basel)*, *16*(17), 7651. DOI: 10.3390/su16177651

Chapter 2
Determinants of Carbon Emissions:
A Dynamic Panel ARDL Analysis Among the BRICS Nations

Moez Souibgui
https://orcid.org/0009-0004-3586-1107
Faculty of Economics and Management, University of Sfax, Tunisia

ABSTRACT

This study investigates the impact of foreign direct investment (FDI), energy consumption, economic growth, renewable energy use, and trade openness on carbon dioxide (CO2) emissions in BRICS nations. Utilizing panel data from the World Development Indicators database, spanning 1998 to 2022. The dataset consists of a maximum of 125 observations across six key variables. The findings reveal significant relationships between CO2 emissions and the independent variables, with economic growth and trade openness showing varying impacts across the short and long run. Notably, the error correction model (ECM) confirms a long-run equilibrium adjustment process. Variance Inflation Factor analysis indicates that multicollinearity is not a major concern, ensuring the robustness of the regression estimates. The results contribute to the policy discourse on sustainable development by highlighting the need for strategic investments in renewable energy and efficient trade policies to mitigate environmental degradation in emerging economies.

DOI: 10.4018/979-8-3373-1822-6.ch002

1. INTRODUCTION

Human activities have significantly altered the environment, with the rapid rise in greenhouse gas concentrations being a primary concern. Scientific evidence overwhelmingly attributes this trend to industrialization and economic activities. While environmental degradation is not a new phenomenon, the link between economic growth and ecological impact gained academic attention in the 1960s (Addai et al., 2022). Over the past decades, carbon emissions (CE) have become a critical global issue due to their role in climate change (Streimikiene, & Kyriakopoulos, 2025). Industrialization, while driving economic progress, has also led to severe environmental consequences. Addressing this challenge has become a key priority in national and international policy agendas, as evidenced by global conferences such as (Don et al., 2021; Depren, et al., 2022).

Developing nations, in particular, face a unique challenge: balancing economic growth, social development, and environmental sustainability (Rodrik, & Stiglitz, 2024; Yassin, et al., 2025). These countries strive to reduce poverty, expand economic opportunities, and protect natural resources simultaneously (Arslan et al., 2021).

The emerging economies—Brazil, Russia, India, China, and South Africa (BRICS)—have garnered significant attention in recent years due to their rapid economic growth (Bindu et al., 2021; Sawal, & Anjum, 2023). Despite their cultural, historical, and linguistic diversity, these nations share a common trait: their economic performance has outpaced that of many developed countries.

Extensive research over the past two decades has explored the relationships between energy consumption, GDP, FDI, and carbon emissions. However, many studies suffer from methodological limitations, such as reliance on single-country data, short time spans, or omitted variables, leading to unreliable conclusions. This study addresses these gaps by employing a multivariate framework and panel data analysis to examine the interplay between FDI, renewable energy adoption, energy usage, and carbon emissions (Iqbal et al., 2023; Anser, et al., 2024).

The study contributes to the literature by providing a structured time-series analysis of the relationship between carbon emissions, energy usage, GDP, and FDI in emerging economies. Using a Panel ARDL approach, it investigates the short- and long-term dynamics of these variables (Akhtar, et al., 2023; Sallam, 2025).

The paper is structured as follows: Section 2 reviews relevant literature, Section 3 describes the data and methodology, and Section 4 presents the findings and conclusions.

2. LITERTURE REVIEW

Several factors have been empirically analyzed regarding their impact on greenhouse gas emissions, including foreign investments, energy utilization, economic progress, renewable energy adoption, and trade liberalization. The subsequent section discusses the interrelations among these variables (Wahab et al., 2024).

Employing the Augmented Mean Group (AMG) estimator, Azam, & Haseeb, (2021); Ali et al., (2023) analyzed the effects of foreign capital inflows on carbon emissions in BRICS nations using data respectively from 1990 to 2018 and 2000 to 2020. Findings indicated a statistically significant negative correlation between CO2 emissions and foreign direct investment (FDI) in BRICS countries.

Joshua et al., (2024) examined the link between inward foreign capital and environmental degradation in South Africa using time series data from 1990 to 2020 through the Johansen Co-integration test and the vector error correction model. The research highlights the enduring positive influence of trade openness and energy consumption on environmental degradation, whereas foreign investment consistently exerts a negative impact on pollution levels over the long term. Additionally, economic growth contributes to a reduction in CO2 emissions, while both population expansion and inflation negatively affect environmental quality in the long run. Furthermore, the interaction between foreign investment and trade plays a significant role in mitigating environmental pollution in both the short and long term. Findings from the Granger causality tests indicate a bidirectional causal relationship between population growth and CO2 emissions, alongside a unidirectional causality among other variables, shedding light on the interconnections among these crucial factors.

Abid et al., (2021), utilized the Fully Modified Least Squares (FMLOS) estimator with data spanning 1990 to 2019 to explore the effects of technological advancements, economic expansion, foreign investments, energy usage, and urbanization on carbon emissions in G8 nations. Their findings revealed a substantial cross-sectional dependence among the panel nations. The FMLOS estimator confirmed a statistically significant long-term negative relationship between foreign direct investment, financial development, and technological innovation in G8 countries. Additionally, long-run bidirectional causal relationships were identified among economic growth, financial development, urbanization, trade openness, CO2 emissions, and energy consumption. However, a unidirectional causal link exists between carbon emissions and foreign direct investment.

Results highlighted an inverse relationship between per capita GDP growth and CO2 emissions. Energy consumption and unrestricted trade movements were also negatively affected by carbon emissions (Dissanayake, et al., 2023; Aldosai & Yasser, 2025). Over an extended period, CO2 emissions demonstrated a positive relationship with FDI inflows (Lee, et al., 2021). Using VAR Granger causality estimations, Iqbal, et al., (2023)] investigated the effects of renewable energy adoption, FDI inflows, and economic growth on CO2 emissions in Pakistan, covering the period from 2000 to 2018. The estimation results of the PMG model suggest that carbon emissions, renewable energy consumption, exports, foreign direct investment (FDI), and savings significantly enhance economic growth over the long run, whereas interest rates and trade openness exert a negative influence. The robustness of these long-term findings is further validated by the FMOLS and DOLS estimations. Additionally, the Dumitrescu-Hurlin panel causality analysis reveals a bidirectional causal relationship between carbon emissions and economic growth, as well as between economic growth and FDI. Furthermore, a unidirectional causal link is observed from exports and the labor force to economic growth.

Gules (2021) examines the impact of foreign direct investment (FDI), economic growth, and capital stocks on CO2 emissions across 16 Southern African Development Community (SADC) nations, utilizing panel data spanning from 1990 to 2014. To assess these effects, both static and dynamic estimation models were employed. Findings from panel static analysis, using pooled ordinary least squares, and dynamic panel estimation, based on the system generalized method of moments, reveal that FDI and economic growth exert a significant positive influence on CO2 emissions in the SADC region. In contrast, capital stocks demonstrate a significant negative effect on CO2 levels. These findings suggest that FDI inflows have contributed to environmental degradation in SADC countries.

Mahmood, et al., (2022) examines the influence of oil prices, economic growth, and urbanization on CO2 emissions in selected countries, accounting for asymmetrical relationships. To achieve this, a nonlinear autoregressive distributed lag (NARDL) cointegration approach is applied to Gulf Cooperation Council (GCC) countries over the period 1980–2019, confirming cointegration across all models. The long-term findings indicate that increasing economic growth leads to higher CO2 emissions in Kuwait, Oman, Qatar, and Saudi Arabia, while declining economic growth has a similar effect in Bahrain, Kuwait, Qatar, and the United Arab Emirates (UAE). Additionally, rising oil prices contribute to increased CO2 emissions, demonstrating a scale effect in Oman, Qatar, and Saudi Arabia.

Yuldoshboy et al. (2022) examine the relationship between CO2 emissions, economic growth, and foreign direct investment (FDI) in Central Asian countries, specifically Uzbekistan, Kazakhstan, Kyrgyzstan, and Tajikistan. Using secondary data from 2000 to 2020, the analysis employs panel regression techniques, including Pooled OLS, Random Effects, and Fixed Effects models. Additionally, a panel causality test is conducted to assess the causal relationships among the key variables.

This research contributes to the existing literature by identifying significant correlations between carbon dioxide emissions, economic growth, and FDI in the region. The empirical findings reveal a unidirectional relationship between GDP and CO2 emissions, GDP and energy consumption, as well as energy consumption and CO2 emissions. However, no significant link is found between FDI and GDP or between CO2 emissions and FDI in Central Asia.

3. DATA AND METHODOLOGY

3.1. Sample Selection for the Study

This study employs panel data to assess the influence of FDI, energy usage, economic expansion, renewable energy deployment, and trade openness on CO2 emissions in BRICS nations. Utilizing annual time series data from the World Development Indicators database, the empirical analysis spans the period 1998-2022. The independent variables influencing carbon emissions include foreign direct investment, energy usage, economic development, renewable energy, and trade openness.

3.2. Model and Variables Measurement

To analyze the effects of foreign direct investment, energy consumption, economic growth, renewable energy utilization, and trade openness on CO2 emissions in BRICS nations, this study adopts the Panel Autoregressive Distributed Lag (ARDL) model.

4. RESULTS AND DISUCCION

4.1. Descriptives Statistics

Table 1 outlines the key variables incorporated in the model. Given data constraints, the study consists of a maximum of 125 observations, with six variables, one of which is the dependent variable while the others serve as independent factors

Table 1. Summary of descriptive statistics

Variable	Observations	Average	Std.Deviation	Minimum	Maximum
logCO2	120	1.395	0.882	-0.210	2.490
logFDI	125	0.610	0.735	-1.560	1.695
logGDP	125	27.700	1.120	25.590	30.320
logTrade	125	3.720	0.350	2.760	4.250

continued on following page

Table 1. Continued

Variable	Observations	Average	Std.Deviation	Minimum	Maximum
logEnergy	95	7.370	0.800	5.970	8.560
logRNWE	120	2.840	0.965	1.160	3.990

Following the descriptive statistics, the study employs a panel Autoregressive Distributed Lag (ARDL) model to explore both the immediate and prolonged impacts of the independent variables on the dependent variable. This methodological approach allows us to distinguish between short-term fluctuations and long-term trends. Through this framework, we can analyze how variables adjust over time to achieve equilibrium in both the short and long run.

In the initial phase of our research, we formulated the foundational equation presented below to act as the core model for our analysis:

$$logCO_{2\,n_A} = \beta 0 + \beta 1\ logFDI_{nA} + \beta 2\ logGDP_{nA} + \beta 3\ logTrade_{nA} + \beta 4\ logEnergy_{nA} + \beta 5\ logRNWE_{nA} + \varepsilon_{nA} \quad (1)$$

Where:

- $\mathbf{logCO_{2n_A}}$: Logarithm of CO2 emissions (metric tons per capita)
- $\mathbf{logFDI_{n_A}}$: Logarithm of Foreign Direct Investment, net inflows (% of GDP)
- $\mathbf{logGDP_{n_A}}$: Logarithm of Gross Domestic Product (current US$)
- $\mathbf{logTrade_{n_A}}$: Logarithm of Trade (% of GDP)
- $\mathbf{logEnergy_{n_A}}$: Logarithm of Energy use (kg of oil equivalent per capita)
- $\mathbf{logRNWE_{n_A}}$: Logarithm of renewable energy consumption (% of total final energy consumption)
- ε_{n_A}: Error term

Table 2. Correlation matrix

	logCO2	logFDI	logGDP	logTrade	logEnergy	logRNWE
logCO2	1.0000					
logFDI	0.0218	1.0000				
logGDP	-0.1035	0.5664	1.0000			
logTrade	0.6771	-0.0346	-0.0292	1.0000		
logEnergy	0.9648	0.0413	-0.1244	0.5213	1.0000	
logRNWE	-0.9212	0.0412	0.0907	-0.6982	-0.8957	1.0000

Table 2 illustrates the correlations between the independent variables, providing insights into potential multicollinearity within the model. A correlation coefficient of 0.000 signifies no relationship between the variables, whereas a value of 1.000 indicates a perfect association. Generally, correlation values:

Between 0.3 and 0.5 suggest a weak correlation.

Between 0.5 and 0.7 indicate a moderate relationship.

Values above 0.7 denote a strong correlation.

Negative values represent inverse relationships. Monitoring these correlations is crucial to ensure model reliability and avoid multicollinearity issues.

4.2. Panel Unit Root Tests

Before estimating the ARDL model with panel data, it's essential to verify the stationarity of the dataset. This preliminary check ensures that the variables are appropriate for time series analysis. The Vector Autoregression (VAR) approach helps determine if the series is I(0), indicating short-term dynamics. If the variables are a mix of I(0) and I(1) or strictly I(1), the panel ARDL model can be applied under the assumption that the series is level. The IPS and Fisher-type tests are commonly utilized to detect unit roots in panel datasets. Developed by Pesaran et al., (1999), these tests follow a standard Augmented Dickey-Fuller (ADF) regression format for panel data:

$$\Delta y_{kt} = \alpha_k \, \gamma_k y_{kt\_1} + \sum_{k=1}^{p} \varphi_k \Delta y_{k,t-1} + \varepsilon_{kt},$$

Where $\gamma_k = \rho_k - 1$.

The null hypothesis (H_0: $\gamma_k = 0$ or $\rho_k = 1$) suggests the presence of a unit root, while the alternative hypothesis (H_1: $\gamma_k < 0$ or $\rho_k < 1$) indicates stationarity.

4.3. Panel Autoregressive Distributed Lag Model

Following the unit root and cointegration tests, the ARDL model is suitable for small sample sizes, distinguishing between short-term and long-term coefficients. Pesaran et al., (1999), noted that long-term parameters are super-consistent, while short-term parameters are $\sqrt{T}$ consistent. The panel ARDL (p, q1, q2, q3, q4, q5) model is structured as follows:

$$\ln CO_{2kt} = \alpha_k + \sum_{k=1}^{p} a_{1,kj} \ln CO_{2\,k,t-j} + \sum_{k=1}^{q1} a_{2,kj} \ln FDI_{k,t-j}$$

$$\sum_{k=1}^{q2} a_{3,kj} \ln GDP_{k,t-j} + \sum_{k=1}^{q3} a_{4,kj} \ln Trade_{k,t-j} +$$

$$]\sum_{k=1}^{q4} a_{5,kj}\ln\text{Energy}_{k,t-j} + \sum_{k=1}^{q5} a_{6,kj}\ln\text{RNWE}_{k,t-j} + \varepsilon_{kt}$$
(3)

Where:

- α_k = Fixed effects
- a1 to a5 = Coefficients of independent variables
- ε_{kt} = Error term, assumed to be white noise

4.4. Panel Error Correction Model (ECM) Representation

$$\ln CO_{2kt} = \alpha_k + \sum_{k=1}^{p} a_{1,kj}\ln\Delta CO_{2\,k,t-j} + \sum_{k=1}^{q1} a_{2,kj}\ln\Delta\text{FDI}_{k,t-j}$$

$$\sum_{k=1}^{q2} a_{3,kj}\ln\Delta\text{GDP}_{k,t-j} + \sum_{k=1}^{q3} a_{4,kj}\ln\Delta\text{Trade}_{k,t-j} +$$

$$\sum_{k=1}^{q4} a_{5,kj}\ln\Delta\text{Energy}_{k,t-j} + \sum_{k=1}^{q5} a_{6,kj}\ln\Delta\text{RNWE}_{k,t-j} +$$

$$\beta_{1,kj}CO_{2kt} + \beta_{2,kj}\ln\text{FDI}_{k,t-j} + \beta_{3,kj}\ln\text{GDP}_{k,t-j} + \beta_{4,kj}\ln\text{Trade}_{k,t-j}$$

$$+\beta_{5,kj}\ln\text{Energy}_{k,t-j} + \beta_{6,kj}\ln\text{RNWE}_{k,t-j}\ \varepsilon_{kt} \qquad (4)$$

Where Δ denotes the first difference of variables. The short-term coefficients are represented by α_1 to α_5, while β_1 to β_5 denote the long-term coefficients for Market Size, FDI, Infrastructure, Trademark applications, and Trade Openness, respectively.

$$\ln\Delta CO_{2kt} = \alpha_k + \sum_{k=1}^{p} a_{1,kj}\ln\Delta CO_{2\,k,t-j} + \sum_{k=1}^{q1} a_{2,kj}\ln\Delta\text{FDI}_{k,t-j}$$

$$\sum_{k=1}^{q2} a_{3,kj}\ln\Delta\text{GDP}_{k,t-j} + \sum_{k=1}^{q3} a_{4,kj}\ln\Delta\text{Trade}_{k,t-j} +$$

$$\sum_{k=1}^{q4} a_{5,kj}\ln\Delta\text{Energy}_{k,t-j} + \sum_{k=1}^{q5} a_{6,kj}\ln\Delta\text{RNWE}_{k,t-j} +$$

$$\theta_i\ \text{ECM}_{k,t-1} + \varepsilon_{kt} \qquad (5)$$

Where, θ_i represents the ECM coefficient that indicates the annual adjustment speed towards long-term equilibrium. Due to limited annual data, a maximum length of three is selected, following Akaike's lag selection criterion for the ECM model. The Pooled Mean Group (PMG) estimation method, introduced by Pesaran et al., (1999), is employed for panel ARDL regression. This technique averages coefficients and pools them during estimation, allowing flexibility for group differences in intercepts, short-run coefficients, and error variances.

The PMG estimator assumes uniform long-run coefficients across groups, ensuring reliable estimates when the homogeneity condition is met (Mumuni, & Mwimba, 2023). Additionally, with a small cross-sectional sample size (N), the PMG estimator is robust against anomalies, addresses serial autocorrelation, and manages endogenous variables by calculating appropriate lag structures (Lone, et al., 2023).

4.5. Panel ARDL Model

Prior to applying the panel ARDL model, unit root and cointegration tests were conducted to assess variable relationships. Table 3 presents the stationarity test results using Augmented Dickey-Fuller (ADF), Phillips-Perron (PP), and IPS tests. These results confirm that the variables are stationary either at their levels or first differences, validating the suitability of the panel ARDL model.

Specifically, the stationarity tests indicate that most variables become stationary at their first differences. However, variables such as lnFDI and lnEnergy are stationary at their levels. This mix of stationarity levels supports the application of the panel ARDL model, as it requires verifying the stationarity properties of the sample.

Table 3. Stationarity tests

Variable	Fisher-type tests IPS test				IPS test	
	Fisher-ADF statistics		Fisher-PP statistics		I(0)	I(1)
	I(0)	I(1)	I(0)	I(1)		
lnCO2	3.5821	22.3481***	3.5821	58.2132***	1.3156	-3.9042***
lngdp	1.7845	103.4123***	37.6821***	202.4537***	1.8422	-2.8321***
lnfdi	36.4921***	145.2937***	391.1***	152.3842***	-2.8843***	-3.9074***
lnTrade	8.7623	63.0192***	8.7623	63.0192***	-0.6984	-4.2315***
lnRNWE	9.5122	82.4942***	9.5122	82.4942***	0.7264	-4.0075***
lnenergy	2.2347	52.6135***	2.2347	52.6135***	-0.9857*	-3.4956***

Standard errors in parentheses: *** $p<0.01$, ** $p<0.05$, * $p<0.1$

The results from the stationarity tests displayed in Table 3 reveal that a majority of the variables exhibit significant stationarity when analyzed through both the Fisher-ADF and Fisher-PP tests. Specifically, most of the p-values are below the 0.05 level, indicating that the variables are stationary in their first differences (I(1)). This result aligns with the standard characteristics found in economic time series analysis, where variables tend to exhibit non-stationarity at the level but become stationary upon differencing. These findings suggest a solid underlying connection between the variables and provide a strong foundation for the subsequent cointegration analysis.

Table 4. Cointegration tests

Group	Coefficient	p-value
Between		
Modified variance ratio	-1.2098	0.0724
Modified Phillips–Perron t	1.0235	0.1356
Phillips–Perron t	-2.0423	0.0197
Augmented Dickey–Fuller t	-1.2054	0.1136
Within		
Modified Phillips–Perron t	2.0193	0.0167
Phillips–Perron t	-3.5249	0.0003
Augmented Dickey–Fuller t	-1.7652	0.0294

Standard errors in parentheses: *** $p<0.01$, ** $p<0.05$, * $p<0.1$

The cointegration tests conducted in Table 4 present results that help assess the long-term relationship between the variables across two main groups: Between and Within.

Modified variance ratio of -1.2098 and p-value of 0.0724 suggests some weak evidence of a long-term relationship but does not conclusively support cointegration. Modified Phillips–Perron t statistic of 1.0235 with a p-value of 0.1356 does not provide significant evidence of cointegration.

Phillips–Perron t statistic of -2.0423 ($p = 0.0197$) offers modest evidence of cointegration between the variables at the 5% significance level.

The Augmented Dickey–Fuller t statistic of -1.2054 ($p = 0.1136$) is not statistically significant, suggesting that the data might not be stationary in the Between group.

Modified Phillips–Perron t statistic of 2.0193 ($p = 0.0167$) shows significant cointegration, implying a robust long-term relationship between the variables at the 5% significance level.

Phillips–Perron t statistic of -3.5249 ($p < 0.01$) strongly supports the hypothesis of cointegration at a very high significance level.

The Augmented Dickey–Fuller t statistic of -1.7652 ($p = 0.0294$) is statistically significant at the 5% level, confirming that the data is stationary in the Within group.

Table 5. Variance Inflation Factor (VIF)

Variable	VIF	1/VIF
lnRNWE	8.99	0.1112
lnenergy	6.45	0.1558
lntrade	2.35	0.4255

Variable	VIF	1/VIF
lnfdi	1.75	0.5714
lngdp	1.67	0.5988
Mean VIF	4.26	

The VIF values shown in Table 5 help assess the presence of multicollinearity among the independent variables.

None of the variables have a VIF greater than 10, which suggests that there is no severe multicollinearity present. A VIF below 10 is considered acceptable, indicating that all variables are reasonably independent of each other.

The highest VIF value is 8.99 for **lnRNWE**, which is still below the critical value of 10. Therefore, the assumption of negligible multicollinearity is satisfied in this model.

The **Mean VIF** of 4.26 suggests that overall, multicollinearity is not a concern for the model.

Table 6. Panel ARDL, PMG approach results

VARIABLES	PMG	PMG	DFE	DFE	MG	MG
ec		-0.195		0.195**		-0.557
		0.331		(0.0779)		0.251
D.lngdp		-0.0089		0.0058		-0.0078)
		0.0574		(0.0147)		(0.0530)
D.lnfdi		-0.00146		-0.000687		(0.0467)
		0.00640		(0.165)		(0.210)
D.lntrade		-0.0401**		0.0201		0.00892
		0.0383		(0.0495)		(0.0618)
D.lnRNWE		-0.276		-0.161***		-0.0739
		0.156		(0.0568)		(0.0332)
D.lnenergy		0.294		0.554***		
		0.179***		(0.120)		
lngdp	-0.250		-0.042		0.195	
	0.0225		0.0093		(0.0202)	
lnfdi	0.0014		0.0048		0.0245	
	0.0315**		0.0638		(0.0337)	
lntrade	0.258		0.0928		-0.0516	
	0.052*		0.091		0.0623	
lnrnwe	-0.219		-0.0221		-0.456	

VARIABLES	PMG	PMG	DFE	DFE	MG	MG
	0.0275**		0.158		0.137	
lnenergy	0.1335		-1.445*		0.927	
	0.125*		0.649		(0.175)	-3.367
Constant		-0.197				(2.886)
		0.223				
Hausman		22.97***		4.19***		
		PMG is accepted		DFE is accepted		
Observations	95	95				

Standard errors in parentheses: *** p<0.01, ** p<0.05, * p<0.1

The results presented in Table 6 are derived from the Panel ARDL approach using the PMG (Pooled Mean Group), DFE (Differentiated Fixed Effects), and MG (Mean Group) estimators. The analysis investigates the short-term and long-term impacts of variables such as energy consumption, GDP, foreign direct investment (FDI), trade openness, and renewable energy on CO2 emissions.

Short-Term Effects:

Energy consumption (ec) has a significant negative short-term impact in both PMG and DFE approaches, with coefficients of -0.250 and -0.195 respectively. The PMG estimation indicates a significant reduction in CO2 emissions, whereas the DFE approach also suggests a negative impact with a higher statistical significance.

Trade openness (lntrade) shows a positive effect, especially in the PMG (0.0315**) and DFE (0.0574*) approaches, implying that trade openness may contribute to higher emissions in the short term.

Foreign direct investment (lnfdi) has a marginally significant negative coefficient in both PMG (-0.0014) and DFE (-0.0089*), suggesting that FDI might slightly decrease emissions in the short run.

Long-Term Effects:

The long-term impact of energy consumption (lnenergy) is significantly positive in both DFE and MG, with coefficients of 1.294* and 0.645*, indicating that increased energy consumption significantly raises CO2 emissions.

Trade openness (lntrade) maintains its positive long-term relationship with emissions in the PMG and DFE estimations, suggesting sustained increases in emissions due to trade.

renewable energy (lnrnwe) exhibits a negative long-term effect, with the PMG (-0.265***) and DFE (-0.146***) models showing that renewable energy adoption can reduce emissions.

Model and Fit:

The Hausman test results (36.12*** for PMG and 9.76*** for DFE) strongly suggest that the PMG and DFE models are both well-suited for the data, confirming their respective acceptances as the optimal models.

The constant term shows a significant negative value for both the PMG (-0.123) and DFE (-1.279***), which indicates that without the influence of the explanatory variables, the baseline emission is negative but remains statistically meaningful.

5. CONCLUSION

This paper examines the effects of GDP, FDI, trade openness, renewable energy, and energy consumption on CO2 emissions in BRICS nations. We used panel ARDL, PMG approach as an econometric tool utilizing annual panel data for BRICS countries from 1996 to 2020 to accomplish the objective. The primary goal of this research is to examine the effects of renewable energy and energy consumption on CO2 emissions in the BRICS countries. Firstly, we employ panel stationarity tests, namely Fisher type and IPS, to ensure if selected series are stationary either level or first difference.

According to the results of stationarity tests, the series in this study are stationary either level or first difference which means that Panel ARDL is appropriate to employ in our empirical analysis. After that, panel cointegration tests are done to check the cointegrated relationship between variables of in our interest. Panel cointegration tests indicate that there is a cointegrated relationship between variables in the study. After checking stationarity and cointegration, Hausman test is utilized to check which approach from panel ARDL, namely PMG, MG and DFE, is suitable for our analysis. According to the results of Hausman test, PMG is more appropriate than the others.

In this study, the PMG technique is used to assess the effects of independent factors on dependent variables. This paper's empirical findings indicate that trade openness, FDI, and energy consumption have both a short- and long-term relationship with CO2 emissions in the BRICS nations.

As a result, if energy consumption has an impact on CO2 emissions, energy conservation regulations should aim at safeguarding the environment. Furthermore, measures promoting energy production and conservation, foreign direct investment, trade openness, and economic growth would be ideal for the entire region. Even if there is a political intent to develop similar goals and objectives, separate policy and strategy designs for participant subgroups should most likely be considered to enhance the environmental protection of the region. The current analysis concludes with the recommendation of a crucial strategy, which is to prioritize the promotion of renewable energy consumption, FDI, and trade openness.

Finally, it should be mentioned that this study has significant drawbacks. Firstly, future researchers should include a firm level date to analyze the determinants of CO2 emissions in the region. Second, by including a set of missing variables in our analysis, panel cointegration models such as GMM, PDOLS, and FMOLS may be used to investigate the determinants of CO_2 emissions.

REFERENCES

Abid, A., Mehmood, U., Tariq, S., & Haq, Z. U. (2022). The effect of technological innovation, FDI, and financial development on CO2 emission: Evidence from the G8 countries. *Environmental Science and Pollution Research International, 29*(8), 11654–11662. DOI: 10.1007/s11356-021-15993-x PMID: 34545519

Addai, K., Serener, B., & Kirikkaleli, D. (2022). Empirical analysis of the relationship among urbanization, economic growth and ecological footprint: Evidence from Eastern Europe. *Environmental Science and Pollution Research International, 29*(19), 27749–27760. DOI: 10.1007/s11356-021-17311-x PMID: 34981376

Akhtar, R., Masud, M. M., Al-Mamun, A., & Saif, A. N. M. (2023). Energy consumption, CO2 emissions, foreign direct investment, and economic growth in Malaysia: An NARDL technique. *Environmental Science and Pollution Research International, 30*(22), 63096–63108. DOI: 10.1007/s11356-023-26246-4 PMID: 36952165

Aldosai, M. S., & Yasser, M. (2025). Economic Complexity Determinants Insights from Sub-Saharan Countries. *Faculty of Mediterranean Business Studies Tivat. Montenegro, 21*(2), 163–174.DOI: 10.1007/s11356-023-26246-4 PMID: 36952165

Ali, N., Phoungthong, K., Khan, A., Abbas, S., Dilanchiev, A., Tariq, S., & Sadiq, M. N. (2023). Does FDI foster technological innovations? Empirical evidence from BRICS economies. *PLoS One, 18*(3), e0282498. DOI: 10.1371/journal.pone.0282498 PMID: 36893169

Anser, M. K., Ali, S., Umair, M., Javid, R., & Mirzaliev, S. (2024). Energy consumption, technological innovation, and economic growth in BRICS: A GMM panel VAR framework analysis. *Energy Strategy Reviews, 56*, 101587. DOI: 10.1016/j.esr.2024.101587

Arslan, H. M., Khan, I., Latif, M. I., Komal, B., & Chen, S. (2022). Understanding the dynamics of natural resources rents, environmental sustainability, and sustainable economic growth: New insights from China. *Environmental Science and Pollution Research International, 29*(39), 58746–58761. DOI: 10.1007/s11356-022-19952-y PMID: 35368236

Azam, M., & Haseeb, M. (2021). Determinants of foreign direct investment in BRICS-does renewable and non- renewable energy matter? *Energy Strategy Reviews, 35*, 100638. DOI: 10.1016/j.esr.2021.100638

Bindu, S., Sridharan, P., Swain, R. K., & Das, C. P. (2022). Causal linkage between remittances and financial development: Evidence from the BRICS (Brazil, Russia, India, China, and South Africa). *Journal of East-West Business*, *28*(2), 117–149. DOI: 10.1080/10669868.2021.1976348

Depren, S. K., Kartal, M. T., Çelikdemir, N. Ç., & Depren, Ö. (2022). Energy consumption and environmental degradation nexus: A systematic review and meta-analysis of fossil fuel and renewable energy consumption. *Ecological Informatics*, *70*, 101747. DOI: 10.1016/j.ecoinf.2022.101747

Dissanayake, H., Perera, N., Abeykoon, S., Samson, D., Jayathilaka, R., Jayasinghe, M., & Yapa, S. (2023). Nexus between carbon emissions, energy consumption, and economic growth: Evidence from global economies. *PLoS One*, *18*(6), e0287579. DOI: 10.1371/journal.pone.0287579 PMID: 37352276

Dong, H., Xue, M., Xiao, Y., & Liu, Y. (2021). Do carbon emissions impact the health of residents? Considering China's industrialization and urbanization. *The Science of the Total Environment*, *758*, 143688. DOI: 10.1016/j.scitotenv.2020.143688 PMID: 33338785

Gule, L. (2021). The effects of FDI and economic growth on CO2 emissions in SADC. *PEOPLE: International Journal of Social Sciences*, *7*(1), 75–87. DOI: 10.20319/pijss.2021.71.7587

Iqbal, A., Tang, X., & Rasool, S. F. (2023). Investigating the nexus between CO2 emissions, renewable energy consumption, FDI, exports and economic growth: Evidence from BRICS countries. *Environment, Development and Sustainability*, *25*(3), 2234–2263. DOI: 10.1007/s10668-022-02128-6

Joshua, A., Xusheng, Q., Emmanuel, B. G., Sakoane, K., & Appiah, M. (2024). Unveiling the dynamic nexuses between foreign investment, trade openness, energy consumption, and CO2 emissions in South Africa: A vector error correction model approach. *Energy & Environment*, 0958305X241266529.

Lee, H. S., Moseykin, Y. N., & Chernikov, S. (2021). Sustainable relationship between FDI, R&D, and CO2 emissions in emerging markets: An empirical analysis of BRICS countries. *Russian Journal of Economics*, *7*(4), 297–312. DOI: 10.32609/j.ruje.7.77285

Lone, U. M., Darzi, M. A., & Islam, K. U. (2023). Macroeconomic variables and stock market performance: A PMG/ARDL approach for BRICS economies. *Macroeconomics and Finance in Emerging Market Economies*, *16*(2), 300–325. DOI: 10.1080/17520843.2021.1983704

Mahmood, H., Asadov, A., Tanveer, M., Furqan, M., & Yu, Z. (2022). Impact of oil price, economic growth and urbanization on CO2 emissions in GCC countries: Asymmetry analysis. *Sustainability (Basel), 14*(8), 4562. DOI: 10.3390/su14084562

Mumuni, S., & Mwimba, T. (2023). Modeling the impact of green energy consumption and natural resources rents on economic growth in Africa: An analysis of dynamic panel ARDL and the feasible generalized least squares estimators. *Cogent Economics & Finance, 11*(1), 2161774. DOI: 10.1080/23322039.2022.2161774

Pesaran, M. H., Shin, Y., & Smith, R. P. (1999). Pooled mean group estimation of dynamic heterogeneous panels. *Journal of the American Statistical Association, 94*(446), 621–634. DOI: 10.1080/01621459.1999.10474156

Rodrik, D., & Stiglitz, J. (2024). A new growth strategy for developing nations. *Universidad Harvard [en línea]*https://drodrik. scholar. harvard. edu/publications/new-growth-strategy-developing-nations

Sallam, M. A. (2025). Determinants of Economic Complexity: Evidence from BRICS countries, Egypt and Turkey using Panel ARDL Approach. *Faculty of Mediterranean Business Studies Tivat. Montenegro, 21*(2), 107–121.

Sawal, J. N., & Anjum, R. (2023). BRICS the Major Emerging Economies: Prospects for a Multi-Polar World. *Global International Relations Review, VI*(II), 72–82. DOI: 10.31703/girr.2023(VI-II).08

Streimikiene, D., & Kyriakopoulos, G. L. (2025). Climate Change Mitigation in Visegrad Countries. *Faculty of Mediterranean Business Studies Tivat. Montenegro, 21*(2), 233–246.

Wahab, S., Imran, M., Ahmed, B., Rahim, S., & Hassan, T. (2024). Navigating environmental concerns: Unveiling the role of economic growth, trade, resources and institutional quality on greenhouse gas emissions in OECD countries. *Journal of Cleaner Production, 434*, 139851. DOI: 10.1016/j.jclepro.2023.139851

Yassin, M. M., Abdallah, A. A. J., Altahtamouni, F., Alsmadi, A., & Al-Okaily, M. A. N. A. F. (2025). Achieving Sustainable Excellence: Corporate Governance, Environmental Performance, Innovation, and Financial Success. Faculty of Mediterranean Business Studies Tivat. *Montenegro, 21*(2), 123–147.

Yuldoshboy, S., Karimov, M., & Kuralbaev, J. (2022). The association between co2 and economic growth in central Asian countries: Panel data approach. *Journal of Positive School Psychology*, 5587–5601.

Chapter 3
Digital Transformation in the Agricultural Sector in the Perspective of Industry 4.0:
Situation Assessment in Türkiye

İbrahim Gönen
https://orcid.org/0000-0002-9705-1394
Necmettin Erbakan University, Turkey

Güzide Karakuş
https://orcid.org/0000-0002-2897-7222
Necmettin Erbakan University, Turkey

ABSTRACT

The reflections of the Fourth Industrial Revolution, which causes a great change and transformation all over the world, on the agricultural sector, which is essential for the sustainability of life, are also critical. This study aims to define the reflections of Industry 4.0 on the agricultural sector in Turkey and the approaches of practitioners on the subject. Within the scope of Agriculture 4.0 in Turkey, field forecasting systems, smart machines, smart irrigation systems, Android and iOS applications, satellite tracking systems, remote sensing, thermal cameras, IoT, and UAV technologies are used. The participants, who see the use of digital technologies as a great opportunity for value-added production, stated that Turkey is not fully prepared for digital transformation in the agriculture sector. The obstacles encountered in the use of Agriculture 4.0 technologies in Turkey are defined as high investment costs, insufficient infrastructure, high average age of producers, difficulties in obtaining

DOI: 10.4018/979-8-3373-1822-6.ch003

labor force, lack of access to information, and the small size of lands.

INTRODUCTION

The beginning of the industrial revolutions is accepted as the transition from manpower to machine power with the introduction of water and steam power in England (Industry 1.0). In the following years, humanity witnessed three different industrial revolutions, each with its own important characteristics. The Second Industrial Revolution (Industry 2.0) is characterized by the introduction of electricity as a power source in production systems and the transition to mass production. A prominent example of the revolution is Ford's introduction of assembly lines in automobile production, which allowed parts to be assembled on a moving body line for the first time. This revolutionary approach enabled significant increases in production volumes and productivity. The Third Industrial Revolution (Industry 3.0) is the result of combining information technologies and electronics with production processes. The most important feature of this period is the use of computer-aided machine tools in manufacturing and improvements in design and production capabilities.

Finally, the Fourth Industrial Revolution (Industry 4.0) is regarded as an innovative approach that involves making production systems smart by supporting them with digital technologies. Smart production systems include many innovative elements such as big data, the Internet of Things (IoT), artificial intelligence (AI), robotics, three-dimensional printers (3D) (Dalenogare et al., 2018), RFID, sensors, cloud computing, digital twin, etc. and create a major structural change in every sector. In today's conditions, where the pressure created by global competition is increasing, this digital revolution is not a choice for businesses, but a necessity in order to survive.

Industry 4.0 is a concept that refers to a new process of change and transformation that is current both nationally and internationally (Özsoylu, 2017), and it is argued that the revolution is still in its early stages and not fully completed. While Industry 4.0 causes great impacts in all sectors and all areas of life, its effects on the agricultural sector are also evaluated.

The world population tends to increase continuously, and it is predicted that this increase will bring many problems in the future. Natural events such as climate change, depletion of natural resources, earthquakes, landslides, and excessive rainfall significantly affect the sustainability of agriculture. Since this effect will cause problems in access to food in the future, studies should be carried out for the effective development of the agricultural sector today. Agriculture is one of the basic building blocks of the world and has the potential to bring the end of humanity if neglected (Pakdemirli et al., 2021). It is an important necessity to strengthen the sector by developing solutions with Industry 4.0 technologies to the problems experienced in agriculture due to the inevitable negative effects of climate change against the increasing need for food.

In this context, public institutions, private sector enterprises, farmers, and universities should carry out studies and define the requirements for Agriculture 4.0 in order to successfully implement digital transformation in the field of agriculture. In the literature, there are studies on Agriculture 4.0 in the world (Gubanova et al., 2017; Kovács & Husti, 2018; Regan et al., 2018, Araújo et al., 2021, Sharma et al., 2022, Smid et al., 2022, Thomas et al., 2023) and in Turkey (Ercan et al. 2019; Kılavuz & Erdem, 2019; Kırmıkıl & Ertaş, 2020; Üstündağ, 2022) using different methods in different regions. However, there is no study on Agriculture 4.0 in Turkey that includes all stakeholders and reflects the current situation in digital transformation, the difficulties experienced, and expectations for the future. Considering the importance of the subject, this study aims to identify the reflections of Industry 4.0 on the agricultural sector in Turkey and the approaches of practitioners on the subject and seeks answers to the following research questions:

- *Do agricultural sector stakeholders in Turkey have sufficient knowledge about Agriculture 4.0?*
- *What are the Agriculture 4.0 technologies used in Turkey?*
- *What are the obstacles encountered in the transition to digital agriculture practices in Turkey?*

In the study, after the introduction, the historical development of industrial revolutions, the historical development of the agricultural sector, and smart agricultural practices implemented within the scope of Agriculture 4.0 are examined. Following the presentation of the research method, the data obtained from the practitioners operating in the agricultural sector in Turkey through the semi-structured interview technique and the results of the analyses are presented in the application section. Finally, recommendations are developed in line with the results.

BACKGROUND

Industrial Revolutions

The rapid and permanent changes in the social, cultural, and economic aspects of social life are called revolutions. All revolutions have a structure that aims to increase individual rights by eliminating the difficulties faced by people (Derya, 2018). With the industrial revolutions, states began to focus on being more effective in meeting individual needs, which caused significant changes in the social, legal, economic, and political spheres (Düzkaya, 2016).

The combination of many factors, such as innovations in fields like manufacturing, trade, transportation, health, agriculture, capital formation, financial developments, and technological advancements, triggered the emergence of industrial revolutions. The increasing demand for goods and services, coupled with the growing global population, led producers to diversify and speed up production. This situation began to place pressure on producers due to insufficient technology and forced laborers to work under harsh conditions. With the industrial revolutions, the use of new technologies in production started to reduce dependence on labor and costs, positively impacting productivity (Özdoğan, 2017) and improving working conditions.

The First Industrial Revolution started in England in the early 1700s and continued until the mid-1800s. This period marks the time when mechanical tools began to replace the human factor in production with the help of water and steam power (Koca, 2020). The steam engine, developed by James Watt in 1765, opened the door to industrial revolutions (Gregersen, 2020). In 1804, Richard Trevithick invented the world's first steam locomotive by mounting a wagon frame on a steam engine, and this invention made significant contributions to railway transportation (De Haan, 2020). The cotton ginning machine, invented by Eli Whitney, which separates cotton fibers from their seeds, increased the speed and efficiency of cotton yarn production and facilitated growth in the textile industry. In 1807, Robert Fulton's adaptation of steam power to ships enabled these ships to cross the ocean by the 1840s (Orhan, 2019; Fagal, 2021), laying the foundations of global trade.

The Second Industrial Revolution began in the 1850s and continued until the end of the Second World War, spreading from Germany and the United States to the rest of the world. This period is notable for the production systems that initiated mass production using electrical energy. The First Industrial Revolution is referred to as the mechanization of production, while the Second Industrial Revolution is known as the serialization of production (Yüksekbilgili & Çevik, 2018). In 1903, the Fordist Production Model, implemented by Henry Ford, began using a mass production line in automobile manufacturing. The Fordist Production method became the defining feature of the Second Industrial Revolution, bringing about a change in the quality of the labor force in manufacturing. In 1900, Frederick Winslow Taylor emerged as a prominent figure with his studies on productivity, which were influenced by the increase in production quantities. These studies led to the emergence of industrial engineering (Kelly, 2016). During this period, massive buildings made of iron and steel started to be constructed. However, as people were replaced by machines, unemployment began to rise, and workers were forced to accept lower wages (Xu et al., 2018).

The Third Industrial Revolution, also known as the period of technology in which production accelerated and digitized, began at the end of the Second World War and progressed rapidly in the early 1970s (Kagermann, 2017). During this period, when information technologies started to be used effectively in production, developments in biogenetics, computers, fiber technology, telecommunications, and bio-agriculture applications emerged. With the advancements made by these technologies in production, today's strategic and economic balances have improved in quality (Yin et al., 2018). With the development of 3D printers, the production of car parts increased, intelligent autonomous robots were created, and large-scale productions were achieved by pressing just one button. Thanks to the newly developed systems, lower-cost, fast, and efficient production systems were implemented, contributing to the economy (Özden, 2022). During this period, communication technologies made a breakthrough with the development of supercomputers. The integration of communication and computers enabled micro-level products to become part of human life (Esmer & Şaylan, 2019).

The Fourth Industrial Revolution, also known as Industry 4.0, is characterized as the complete digitalization of industry (Ullah, 2020). During this period, production is designed in a simulation environment close to reality, and all data related to the system are collected in a centralized system. All data in cloud systems can be supported by data analytics and artificial intelligence, allowing for precise decision-making (Shi et al., 2020). The significant achievements of this revolution, which we are currently living in and is seen as the production vision of the future, are as follows (Büchi et al., 2020):

- Making decisions much easier and faster thanks to artificial intelligence
- Enabling comfortable and easy communication between objects through the internet
- Decision support mechanisms making it easier to make choices
- Finding new energy sources through magnetic field technology
- The emergence of many new applications in the defense industry and healthcare with nanotechnology
- Obtaining clean energy
- Minimizing the margin of error in the work carried out

In the Fourth Industrial Revolution, there is a need for significant change and transformation in production systems. This period aims for fast and efficient production with initiatives such as designing products in a digital environment, simulating their production, ensuring production and logistics with robots working in coordination with humans and other machines, and improving decision-making processes with big data (Davutoğlu, 2020). The technological factors that stand out in this period include three-dimensional printers, the Internet of Things (IoT), smart factories, cyber-physical systems, big data, autonomous robots, simulation, system integration, cloud computing, augmented reality, and so on. The digital transformation experienced across all sectors in line with Industry 4.0 has potential effects that could lead to improvements in productivity, quality, profitability, and image across all operations (Karakuş et al., 2022) and is a topic that requires in-depth examination and understanding.

MAIN FOCUS OF THE CHAPTER

Changes in Agriculture in the Perspective of Industrial Revolutions

Agriculture includes activities such as cultivating the soil to obtain plant products, animal husbandry, aquaculture, and forestry. Agricultural activities are fundamental to sustainable life in social, environmental, and economic contexts. Agriculture is of critical importance for all individuals and societies for reasons such as providing employment in the economy, contributing to exports, ecologically supporting the environment, biologically ensuring the diversity of living organisms, and most importantly, being an indispensable sector for the continuity of life (Doğan et al., 2015). Mankind, who transitioned from a nomadic life to a settled order, has undergone revolutions in agriculture up until today. These revolutions are as follows (Birişik, 2019):

Agriculture 1.0: Combination of Animal Power and Mechanisation

The innovations made in the field of agriculture based on the First Industrial Revolution in England in the 1600s were called the Agricultural Revolution. With the increase in population over the years, new techniques and technologies became necessary in agriculture (Kirmikil & Ertaş, 2020). During this period, agricultural lands were consolidated, and agricultural machinery began to be used effectively to reduce costs and increase productivity. Subsequently, fertilizers started to be used in agriculture to increase yields in the 1800s (Kılavuz & Erdem, 2019).

Agriculture 2.0: The Introduction of Tractors and Agricultural Implements

With the Second Industrial Revolution, which started in the mid-1800s, the effective use of electricity in manufacturing was achieved, and mass production began. During this period, similar practices were implemented in agriculture, while the variety of fertilizers increased, pesticides were introduced to combat factors that harm plants, and mechanization gradually began to replace manual labor. In addition, studies began to develop feed nutrients to improve animal husbandry (Yaman et al., 2020).

Agriculture 3.0: The Introduction of Digital Applications in Agriculture

With the Third Industrial Revolution in the 1960s, computing applications and digitalization began in the field of agriculture. Developments in agriculture, such as the advancement of irrigation techniques, the increase in the variety of agricultural pesticides, the development of new agricultural machinery, and the transfer of technological data, were described as the ***Green Revolution*** (Hati & Singh, 2021). During this period, agricultural industrial techniques developed, and the use of chemicals increased.

Similarly, the development of livestock breeding accelerated with the realization of large and advanced facilities in the livestock sector. However, with these developments in the agriculturalsector and the increasing population, production costs and environmental problems also started to rise. Agriculture entered a new phase with the increased use of tractors and agricultural implements, developments in digital equipment, and genetically modified products. In the early 1990s, satellite technologies were introduced to monitor lands and farms, and in the 2000s, mobile devices and the internet began to be used effectively in agriculture (Ferrag et al., 2021).

Agriculture 4.0: Integration of Digital Agriculture Practices

With the emergence of the concept of Industry 4.0 in Germany in 2011, the use of technology in agriculture gradually increased, and this period was named ***Digital Agriculture*** or ***Agriculture 4.0***. In 2015, the introduction of big data brought important developments to the agricultural sector (FAO, 2017). Farmers began to make more effective decisions in production and use their resources more efficiently with the power of information and technology. With the use of digital technological tools such as tablets, sensors, smartphones, and drones, many opportunities, such as remote control of cultivation lands, more efficient and effective use of water resources, and monitoring of animals in fattening houses, were made available to farmers. Smart agricultural applications developed using information, technology, and communication systems can reduce costs, increase productivity and profitability, and mitigate environmental problems in the agricultural sector (Barrett & Rose, 2022). Agriculture 4.0 is a period in which information replaces labor and planning replaces risks. In the future, agriculture will not only be limited to growing and meeting needs but will also become a sector that transfers power to other fields with the influence of biology and technology (Pakdemirli et al., 2021).

Digital Transformation in Agriculture Sector and Agriculture 4.0 Tools

Innovative technologies used in production with smart agricultural applications emerging with the digital tools of the Industry 4.0 era provide the development of new approaches in agriculture (Gökçe et al., 2020). Industry 4.0 applications, which are used in all process stages of agriculture—from planting the seed in the soil to irrigation, from spraying to harvesting the product—offer significant advantages for the sector. Some of these applications are listed below (HEKTAŞ, 2021, Accessed: 07.12.2022):

Big Data in the Agriculture Sector

All information in the internet environment is accessed through big data. In recent years, the use of big data in the agricultural sector has been increasing. Examples of big data applications in agriculture include applications such as increasing feed efficiency in animal husbandry, monitoring disease and birth data, and making calculations through sensors to prevent product spoilage. These applications allow farmers to improve agricultural production by using resources more efficiently, make the right decisions, and manage the risks they take correctly (Kirmikil & Ertaş, 2020; Duman & Özsoy, 2019).

Internet of Things (IoT) in the Agriculture Sector

The main goal in technology development is to increase living standards by providing comfort and convenience in people's lives. Thanks to the Internet of Things technology (IoT), one of the Industry 4.0 tools, many devices can work in interaction with each other (Oral & Çakır, 2017), thus improving manufacturing systems. It is expected that the functionality of existing agricultural tools will increase with the application of IoT technology, which is used in many sectors, in the agricultural sector. Thanks to IoT in smart agriculture, farmers will be able to carry out their work more efficiently and add value to the yield of their products thanks to the traceability systems developed. Farmers will be able to control their crops without going to the field by connecting their phones and tablets to the internet, and irrigation and spraying will be carried out with mobile applications (Ercan et al., 2019).

Cloud Computing in the Agriculture Sector

The storage and sharing of common data between Industry 4.0 tools and the transformation of this data into information are carried out through cloud computing (Mell & Grance, 2011). Cloud computing, which is used in every field regardless of the sector, offers many advantages to producers in the field of agriculture. The data obtained during agricultural production are analyzed, processed, simplified, and stored in the cloud computing system. Agricultural producers can access the information they need from this system with devices connected to the internet and use it in decision-making processes. Thanks to the system, problems such as infrastructure confusion and user authorization are also eliminated (Çetin et al., 2013).

Unmanned Aerial Vehicles in the Agriculture Sector

Today, unmanned aerial vehicles (UAVs) are used in various fields such as the defense industry, post-earthquake reconnaissance, firefighting, agriculture, and transportation (De Castro et al., 2012), and their usage area is gradually expanding. The first UAV in the field of agriculture began in the early 1980s with the use of an unmanned helicopter developed by Yamaha in agriculture (Türker et al., 2015). Today, unmanned aerial vehicles in the agricultural sector are effectively used for different purposes, as mentioned below (Türkseven et al., 2016):

– ***Observation:*** Crop observation, animal observation, building and equipment observation.
– ***Activities:*** Mapping, spraying.
– ***Analyses:*** Variable level fertilization, disease diagnostics, water stress, crop yield, soil erosion, wildlife damage.
– ***Other applications:*** Research, smart data.

UAVs increase productivity in agriculture at a high rate thanks to the convenience they provide to users. According to research, UAVs used in seed planting in the USA reduce the cost of planting by 85% (Bahadır & Kadıoğlu, 2019). High-resolution visual sensors, sensors that translate images in three dimensions, sensors capable of thermal heat discrimination, and infrared sensors placed on UAVs enable the early diagnosis of plant diseases (Türkseven et al., 2016).

Autonomous Vehicles and Robotic Systems in Agriculture

Autonomous means that vehicles or machines can perform their tasks without human intervention. Autonomous robots and machines are used effectively in agriculture, as in many different sectors. Examples of these systems include autonomous tractors that plow and sow fields, autonomous harvesters that separate and store the seeds and stems of crops, automatic feed-shuffling machines in animal breeding, and autonomous robots that perform yield analysis (Uzun et al., 2018). Autonomous robots, through sensors, define the working area by mapping the field, follow the development status of the plants through cameras attached to them, and perform data analysis (Çakmak, 2019).

Machine Vision Systems in the Agriculture Sector

Machine vision systems are used to acquire and process various images. This system is a low-cost technology with high accuracy and reliability and has the ability to obtain and process data very quickly. In the agricultural sector, producers can easily classify their products according to many categories such as size, color, robustness, yield, etc., using this system (Sabancı et al., 2016; Ercan et al., 2019).

METHOD

Purpose and Importance of the Research

While digital transformation is predicted to create a decline in employment in many sectors around the world, it is suggested that it may require an employment increase in the agricultural sector. However, it is also expected that this increase will be directed toward a highly skilled and talented workforce (Trivelli et al., 2019). It is crucial to predict the impact of digital transformation on areas such as productivity, quality, and employment in the agricultural sector and to develop strategies in line with these predictions. The literature review shows that academic studies have already started within the scope of Agriculture 4.0 both globally (Gubanova et al., 2017; Kovács & Husti, 2018; Regan et al., 2018; Araújo et al., 2021; Sharma et al., 2022; Smidt et al., 2022; Thomas et al., 2023) and in Turkey (Ercan et al., 2019; Kılavuz & Erdem, 2019; Kirmikil & Ertaş, 2020; Üstündağ, 2022). However, there is no study on Agriculture 4.0 in Turkey that includes all stakeholders and reflects the current, experienced challenges and future expectations in digital transformation. Considering the importance of the subject, this study aims to define the reflections of

Industry 4.0 on the agricultural sector in Turkey and the approaches of practitioners to the subject with the goal of contributing to the elimination of this gap.

Research Methodology

Within the scope of the research, a comprehensive literature review was conducted, and academic publications and application studies on Agriculture 4.0 were analyzed. In the conceptual framework, appropriate references were selected for the study, providing a basis to help increase the content validity of the study.

Case study and semi-structured interview methods from qualitative research were used in the study. Purposive and convenience sampling methods were preferred in order to provide an overview of the sector and to ensure that the findings obtained represent the relevant stakeholders of the sector. The data obtained from the interviews were coded and analyzed by descriptive content analysis and meaning analysis methods and interpreted in relation to the literature.

Research Design

Creswell (2017) defined qualitative research as a technique for examining the current situation, making multidimensional analyses, and revealing narratives, aiming to discover the nature of human beings. While digital transformation tools have started to be used in more and more areas of life to increase productivity and quality, the qualitative research method has been found appropriate for the purpose of the research in defining the reflections of this situation on the agricultural sector. It is critical for the sustainability of life to identify innovative technologies used to improve agricultural productivity (Kovács & Husti, 2018) in line with the experiences and opinions of all stakeholders and to provide predictions for the future.

In the research, the technique of 'multiple case analysis,' which is one of the subcategories of case studies, was preferred to increase the validity of the results by enabling the elaboration of details and bringing together different data. In this way, sufficient information was obtained to draw more understandable conclusions (Chmiliar, 2010). The research design, created by using various methodological approaches such as literature review, interviews, document review, analysis, and evaluation, aimed to increase the validity and reliability of the data obtained.

Identification of Participants

For an interview to be successful, it needs a clear topic, easy-to-understand research questions, and participants who are knowledgeable about the topic (Glesne, 2012; Merriam & Bierema, 2013). For this reason, in order to achieve the research's purpose, the participants were identified as private sector business officials, public employees, and farmers who effectively use Industry 4.0 technologies in the field of agriculture in different regions of Turkey.

Interview Questions

In formulating the interview questions, studies such as "*How Industry 4.0 Changes the Quality of Labour: A Qualitative Research in Enterprises Applying Industry 4.0 Technologies in Turkey*" (Balcı, 2022) and "*Industry 4.0 Awareness in Enterprises and Its Effects on Labour Force: A Multiple Case Study on the Views of Managers*" (Yılmaz & Arslantaş, 2022), along with various studies in the literature, were utilized. The questions directed to the participants are given in the findings section.

Data Collection Process

Semi-structured interviews were conducted with a total of 10 participants, consisting of private sector business officials, public institution employees, and farmers, who are stakeholders in the agricultural sector in Turkey. These interviewees included a wide range of representatives, from professionals with institutional and operational experience to sector experts and experienced farmers in their fields.

The saturation of the results of the interviews conducted by the researcher is an important factor for the research to be conducted effectively. In qualitative research, the saturation point is the point at which the data begins to repeat itself (Patton, 2014) and refers to the point at which the data obtained by the researcher is sufficient to answer the research questions. This point determines when the researcher should end the data collection process and move on to the analysis phase. The saturation point is also important for effectively managing the time and resources of the research (Saunders et al., 2018). In this research, the beginning of the data repetition was accepted as the first delimitation step of the research. In this way, a solid foundation has been provided for obtaining more accurate results in data analysis, and the necessary infrastructure has been established to reach correct interpretations.

During the stage of determining the individuals, a detailed research was conducted, and a sample list was created. The businesses, institutions, and individuals on this list were contacted via e-mail, telephone, or key persons. In the preliminary interviews with the participants, information about the research topic was provided, and meetings that could be held in a suitable environment were planned.

During the interviews, challenges such as postponement or cancellation were encountered, especially with private enterprise participants. Initially, approval for the interview was obtained from 16 participants. Online and face-to-face interviews were conducted with the participants who accepted the interview and were operating in the field of digital agriculture, and were recorded in accordance with the participant's permission. To ensure the reliability of the obtained data, the voice recordings were transcribed into written texts and sent to the participants for approval. Based on the repetition of the answers given to the questions, it was decided to conclude the research during the data collection process with the tenth participant. The interviews were conducted between October 2023 and February 2024, and each interview lasted between 30 and 50 minutes on average.

Data Analysis Process

In qualitative research, the data analysis process begins with obtaining qualitative data consisting of feelings, thoughts, and ideas from interviews with participants in fieldwork. While examining data, taking an objective approach is critical to reaching the correct conclusion (Patton, 2014). It is emphasized in the literature that qualitative analysis programs have the same features and are used in scientific data analysis.

The qualitative analysis cycle begins with creating a database. Obtaining, coding, and fragmenting the data are the first steps. The data that is coded and separated into pieces is then recombined to reach the interpretation and conclusion stage. In the second step, the necessary determinations are made on the data, semantic differences are determined, and this situation is transferred to the report. Finally, the relationships between multiple values are identified, and the analyses are transferred to the report in a way that will help them be put into practice (Kurnaz, 2016).

In this research, the MAXQDA program, designed to facilitate the research and analysis processes of the researchers, was used in the analysis of the data. After all data were collected and converted into written texts, the analysis process began. Coding is considered the most important stage by researchers in qualitative studies and is critical for the correct understanding and interpretation of the data (Glesne, 2012). In the data analysis process, a main code was first determined by the researcher within the scope of the literature review. This main code was supported by subcodes that expressed the simplified essence of the long answers received from the participants. The answer given by each participant was placed in the closest answer code. The numerical expressions next to the subcodes indicated the frequency of the answers given to that question. The data were examined in line with the previously determined criteria, and the results were obtained. Finally, the results obtained as a result of the coding were converted into visuals. In the analyses, word cloud analysis model, code theory analysis model, matrix scanner analysis model, statistical analysis model of subcodes, code co-creation analysis model, hierarchical code-subanalysis model, code-subcode-sections analysis model, and relationship map models were used. Thanks to the different approaches between these models, the data was examined in more detail.

Research Findings

The data obtained as a result of the field study were examined and reported using technical analysis methods and statistical models.

Information on Participants

Information about the participants who were interviewed within the scope of the research and who operate in the field of digital agriculture in Turkey is presented in Table 1.

Table 1. Participant information

Participant	Gender	Age	Education	Institution Type	City	Job Description	Interview Duration
A	Male	40	Bachelor's degree	Private sector	Konya	Senior manager at a company that installs digital drip irrigation systems	43 min.
B	Female	28	Bachelor's degree	Private sector	İstanbul	Biosystems engineer in a business that conducts digital agriculture activities	47 min.
C	Male	40	Bachelor's degree	Private sector	Adana	An unmanned aerial vehicle pilot and agricultural engineer in a business that conducts field analysis and agricultural spraying activities	32 min.
D	Male	33	Bachelor's degree	Private sector	Şanlıurfa	Owner of a business working in the field of aircraft technologies, unmanned aerial vehicle pilot	42 min.
E	Male	39	Master's degree	Public institution	Diyarbakır	Agricultural Engineer at the institution that carries out studies on improving productivity by using new technologies in agriculture and combating plant diseases with modern approaches based on biotechnology.	48 min.
F	Female	41	Bachelor's degree	Public institution	Diyarbakır	Biologist at the institution working on the production, transfer and capacity development of current technologies with the aim of increasing agricultural production through R&D projects.	34 min.
G	Male	35	Bachelor's degree	Public institution	Şanlıurfa	Agricultural Engineer at the institution working in the field of growing sustainable, economical and productive plant species by establishing collaborations with agricultural stakeholders.	50 min.

continued on following page

Table 1. Continued

Participant	Gender	Age	Education	Institution Type	City	Job Description	Interview Duration
H	Female	37	Master's degree	Public institution	Eskişehir	An agricultural engineer at the institution, which operates in the fields of developing high-quality, pest-resistant new plant species using molecular biotechnology and traditional breeding methods, producing seeds for producers, modernizing plant breeding techniques and developing ecological control methods.	39 min.
I	Male	58	Bachelor's degree	Farmer	Eskişehir	Farmer who produces products such as wheat, lentils and tomatoes	40 min.
İ	Male	56	Bachelor's degree	Farmer	Ankara	Farmer who produces wheat, barley, beans and chickpeas, Agricultural Engineer	39 min.

As seen in Table 1, the age range of the seven male and three female interviewees varies between 33 and 56. The interviewees reside in the provinces of Diyarbakır, Adana, Şanlıurfa, Eskişehir, Konya, and Ankara, indicating that the findings obtained are highly representative of Turkey in general. The participants are experts in their fields, holding undergraduate and graduate degrees, which ensures the reliability of the findings.

Findings on the Current State of Agriculture 4.0 in Turkey

In this section, the findings obtained from the questions posed to the interviewees within the scope of the research are presented.

Definition of Industry 4.0

The word cloud related to the question, "What does the concept of Industry 4.0 mean to you?" posed to the participants during the interviews is presented in Figure 1.

Figure 1. What is Industry 4.0? Word cloud analysis model

When the word cloud in Figure 1 is examined, it becomes evident that the participants have a comprehensive perspective on Industry 4.0. The findings derived from analyzing the responses to the question, "*What is Industry 4.0?*" using the code theory analysis model are presented in Figure 2.

Figure 2. What is Industry 4.0? Code theory analysis model

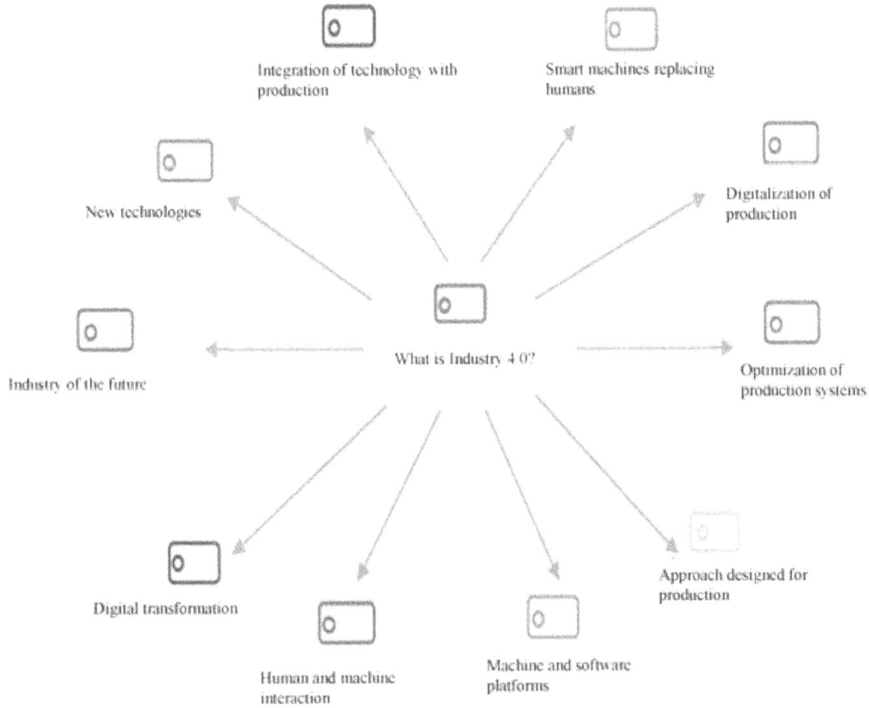

The prominent statements of the participants regarding the definition of Industry 4.0 are presented below:

"Industry 4.0 is the latest point reached by technology in production processes and business management." (PARTICIPANT C)

"The concept of Industry 4.0 refers to the increasing impact of technology on production processes and the development of strategies to enhance productivity by utilizing these technologies. The main objectives of Industry 4.0 are to support automation in enterprises, enable flexible production supported by data-based analyses, reduce costs, and achieve maximum performance." (PARTICIPANT D)

"Industry 4.0 is an approach developed to make production processes more efficient and effective. Thanks to Industry 4.0, it has become possible to apply real-time models in a virtual environment through sensors and devices, leading to revolutionary innovations." (PARTICIPANT H)

"Industry 4.0 represents the future of industry. With new technologies and data analyses, greater flexibility can be achieved in production processes. Human-machine interaction, robotic production, and data management are the prominent elements of this era. Industry 4.0 will be significantly different from previous industrial revolutions." (PARTICIPANT I)

The results of the analyses indicate that the participants possess sufficient awareness of Industry 4.0.

Industry 4.0 Technologies Used in the Agricultural Sector

The results of the code matrix scanner analysis regarding the question, *"Which Industry 4.0 technologies are used in the activities carried out in the agricultural sector in Türkiye?"* posed to the participants during the interviews are presented in Figure 3 (the number of participants who mentioned each item is indicated in parentheses).

Figure 3. Findings regarding the analysis of the code matrix scanner for Industry 4.0 technologies used in the agricultural sector

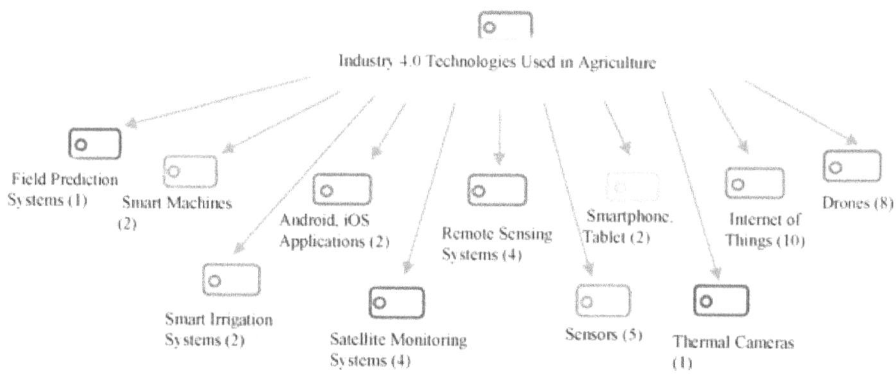

As seen in Figure 3, all participants use at least one technology. The technologies most frequently used by the participants are the Internet of Things (IoT), UAV, and sensor technologies. The prominent statements of the participants while answering the question are presented below:

"Smartphones, tablets, sensors, and UAVs are technologies used in agricultural production. Thanks to these technologies, which are referred to as smart agriculture, producers can access instant information and increase agricultural productivity. For example, it has become possible to determine the harvest date early using smartphone and tablet applications. At the same time, it is also possible to obtain information about the current plant status. Sensor systems enable the provision of information about soil moisture content from any desired point, allowing for more efficient irrigation. Finally, thanks to UAVs, it is possible to photograph plants in the field, perform early diagnosis of diseases, and prevent the risk of disease spread through preventive spraying." (PARTICIPANT A)

"The most commonly used Industry 4.0 technologies in the agriculture sector range from field forecasting systems to biotechnology. The most effectively used technologies include sensors that provide increased sensitivity, long-distance observation systems that help understand the situation on the ground, IoT devices that optimize tissue-based communication, and robotic hardware that supports operations." (PARTICIPANT H)

"While I am engaged in the cultivation of various agricultural products, I am also involved in animal husbandry. I use UAVs and smart irrigation systems for spraying crops. Additionally, I monitor animals in cattle breeding with thermal cameras, allowing for instant interventions." (PARTICIPANT I)

From the participants' statements, it is evident that the Internet of Things (IoT), UAV, and sensor technologies are extensively used in the agricultural sector in Turkey. Participants report that the innovative technologies they use have brought about significant changes in the sector while offering substantial advantages in terms of efficiency, profitability, traceability, and decision-making capability.

Reasons for Using Industry 4.0 Technologies in the Agriculture Sector

In the interviews, participants were asked the question, "*What are the reasons that lead agricultural sector stakeholders to use Industry 4.0 technologies?*" When the responses were analyzed according to sub-codes, it was found that ***increasing production efficiency, reducing costs, using existing resources effectively*** and ***difficulties in obtaining labor*** were the primary reasons. As seen in Figure 4, other reasons mentioned by the participants include protecting energy resources, preserving the environment, saving time, reducing production losses, and improving soil quality (the number of participants mentioning each reason is indicated in parentheses).

Figure 4. Analysis of the reasons leading to the use of Industry 4.0 technologies in the agricultural sector

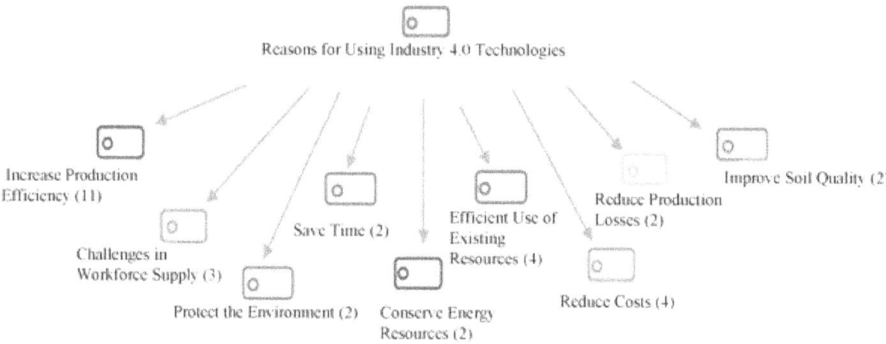

The prominent statements of the participants in response to the question are as follows:

"Productivity decreases in fields where continuous cultivation is carried out, resources are consumed due to unconscious irrigation, costs increase due to excessive use of pesticides and fertilizers, the environment is damaged, and young people are reluctant to work in agriculture due to heavy working conditions. Under these circumstances, the sector ceases to be competitive. I believe Industry 4.0 technologies will be effective in solving these problems in the sector." (PARTICIPANT E)

"Factors that influence the use of digital technologies in the agricultural sector include increasing field productivity, obtaining accurate information about field structure through advanced soil analysis devices, using resources more effectively with advanced plowing and irrigation systems, and enhancing harvest quality and productivity through data management." (PARTICIPANT I)

In the interviews, all participants stated that the use of Industry 4.0 technologies in agriculture is a crucial necessity. The global changes and transformations, as well as Generation Z's reluctance to work in physically demanding jobs like agriculture, are also cited as factors driving mechanization in agriculture. Sustainable agriculture is essential for a sustainable life, and adopting new technologies that ensure efficient resource use is no longer optional but a necessity.

Findings on Changes Caused by Industry 4.0 Technologies in the Agricultural Sector

The findings regarding the question, "*How do Industry 4.0 technologies change the production processes in the agricultural sector?*" posed to the participants during the interviews, are presented in Figure 5.

Figure 5. Hierarchical code–Sub-code analysis model on how Industry 4.0 technologies change the production process in the agricultural sector

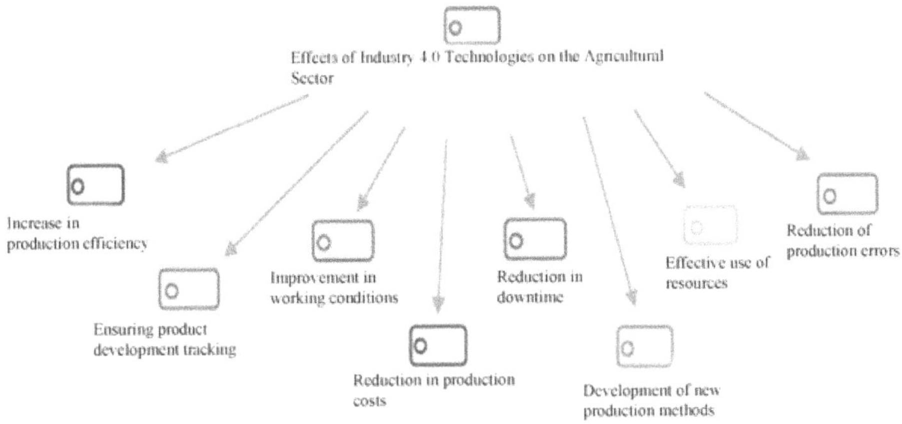

As seen in Figure 5, thanks to the use of smart technologies in agricultural production, production efficiency increases, product development is monitored, working conditions for the labor force improve, production costs and time losses are reduced, new production methods are developed, existing resources are used more effectively, and errors in production are minimized. The statements of the participants regarding changes in production are as follows:

"Thanks to the technologies used, production processes can be controlled through data management. Operations such as planting, fertilization, spraying, irrigation, measurement, and harvesting can be carried out without the need for manual labor, reducing time losses and waste." (PARTICIPANT A)

"In this process, called Smart Agriculture or Agriculture 4.0, digital technologies enable farms to receive real-time information and perform analyses on which fertilizers to use in specific areas. In addition to growing plants in accordance with weather conditions, harvest time can be estimated based on information about soil condition, minerals, irrigation, and pest control, thereby increasing productivity.

Therefore, the transition to modern agricultural practices can be achieved through digitalization." (PARTICIPANT B, PARTICIPANT F, PARTICIPANT I)

"Industry 4.0 technologies offer more effective and efficient solutions compared to traditional approaches in changing the production process in the agricultural sector. Thanks to satellite imaging, sensors, and digital control systems, field tracking and determining the optimal harvest time are among the most significant advantages. Industry 4.0 technology enables the production of higher quality and more efficient products by ensuring the accurate and timely use of inputs based on soil analyses. Additionally, significant losses can be prevented by taking timely measures against plant diseases using digital maps." (PARTICIPANT E, PARTICIPANT H)

As seen from the participants' statements, all agricultural sector stakeholders believe that Industry 4.0 technologies enhance agricultural productivity and quality by facilitating existing work processes, ensuring traceability, and enabling more efficient resource use.

Findings on the Labor Skills Required by Industry 4.0 Technologies in the Agricultural Sector

In the interviews, participants were asked, "*Which labor skills do Industry 4.0 technologies require in agriculture? What changes do you foresee in agricultural sector employees?*" The results are presented below.

Figure 6. Hierarchical code sub-analysis model of which labor skills are required by Industry 4.0 technologies in the agricultural sector

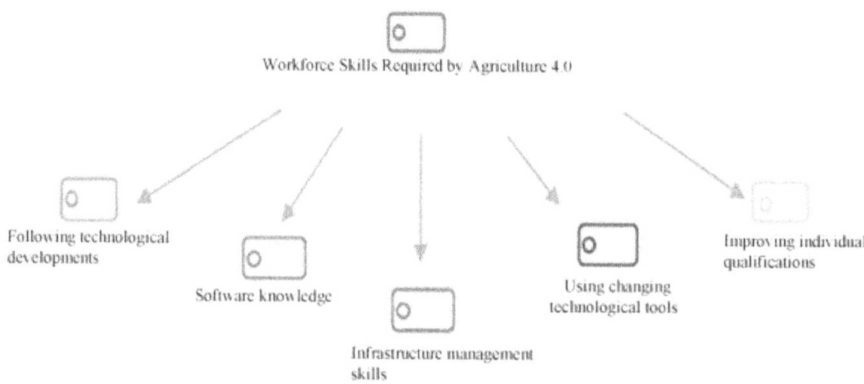

As seen in Figure 6, the participants stated that within the framework of Agriculture 4.0, the labor force should have skills in following technological developments, having software knowledge, managing the necessary infrastructure, improving their individual qualifications, and using changing technological tools. The prominent statements of the participants on the subject are as follows:

"Within the scope of Agriculture 4.0, not only the sale of products but also the information produced, and innovative projects should be seen as sources of profit. In order to carry out agricultural activities efficiently, with quality and competitiveness, the existence of internet and technology infrastructure, as well as a labor force capable of using this technology, is a crucial requirement. Thanks to these developments, work can be done with less labor and time, yielding higher revenues. This situation also increases the motivation of employees in the sector." (PARTICIPANT A)

"Thanks to Industry 4.0 technologies, labor-based activities in the agricultural sector will be transformed into knowledge-based activities, which requires the labor force to develop itself in information technologies." (PARTICIPANT D)

"In order to successfully implement smart agriculture applications, it is important that the workforce employed in agriculture has knowledge and experience in mechanical systems such as unmanned aerial vehicles, navigation systems, and agricultural robotics, as well as software and programming knowledge." (PARTICIPANT E)

"In the agricultural sector, people with mechanical and electronic R&D skills will be needed to design, develop, and implement these technologies. It is important that engineers working in the agricultural sector have knowledge of sensor systems, robotics/automation, data analysis, and software development." (PARTICIPANT G)

As seen in the participants' statements, if the labor force working in the agricultural sector develops itself in line with Industry 4.0 requirements, time can be used more effectively, productivity will increase, physical working conditions will improve, incomes will rise, and the agricultural sector will become more attractive to the new generation of labor.

Findings on the Readiness of Agricultural Sector Stakeholders for Digital Transformation Brought About by Industry 4.0

In the interviews, participants were asked, "*Do you think the stakeholders of the agricultural sector (agricultural enterprises, farmers, service providers, public institutions, etc.) are ready for the digital transformation brought about by Industry 4.0?*". Six of the participants answered "*yes, they are ready*" while four of them answered "*not yet fully ready*". The hierarchical code sub-analysis model for the sub-items of the responses received is presented in Figure 7.

Figure 7. Hierarchical code sub-analysis model of responses regarding the readiness of agricultural sector stakeholders for digital transformation

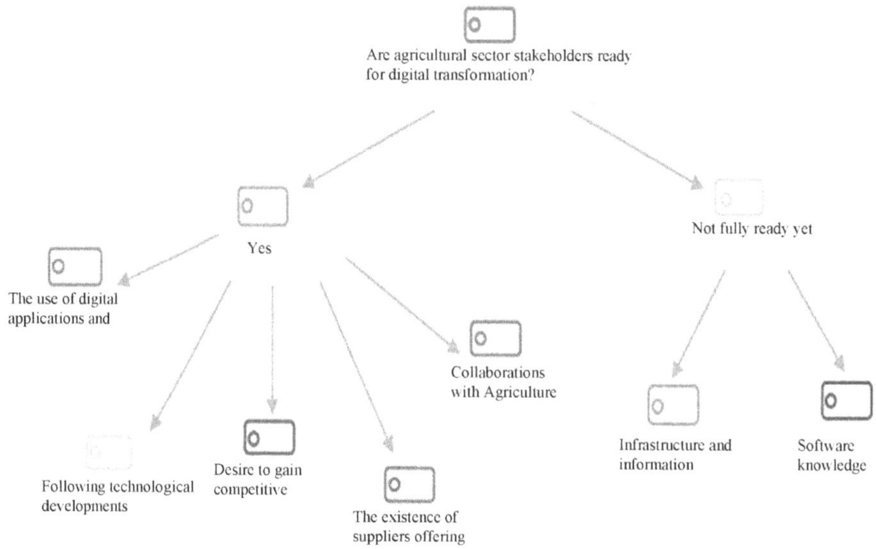

As seen in Figure 7, the interviewees who think that agricultural sector stakeholders are ready for digital transformation stated "yes" due to factors such as using digital applications and technological tools, following technological developments, gaining competitive advantage, the existence of organizations offering smart agricultural technologies, and cooperation between farmers and businesses operating in the field of smart agriculture. The interviewees who think that agricultural sector partners are not yet fully ready for digital transformation cited issues such as lack of infrastructure and lack of software knowledge. The interviewees' statements on the subject are as follows:

"The agricultural sector stakeholders are preparing for transformation activities in line with Agriculture 4.0 and are working harder than ever for this purpose. Farmers benefit from technological opportunities in many activities, from planning planting areas to planting, from estimating harvest time and quantity to determining the quality level. While service providers are developing new software platforms to keep up with the digital transformation, public institutions are also supporting agriculture projects and programs." (PARTICIPANT B, PARTICIPANT C)

"I think it is ready. There are organizations in Türkiye that offer smart agricultural technologies, and producers cooperate with these organizations. The dissemination of smart agricultural applications by leading organizations and the work carried out by public institutions contribute to creating awareness among all stakeholders." (PARTICIPANT F)

"Agricultural sector stakeholders should adopt new technologies for all commercial field operations, from planting and sowing to production management, sales, and service development. To achieve this, competencies need to be developed alongside digital infrastructure. Unfortunately, I think that producers, especially in rural areas, are not fully ready for this transformation." (PARTICIPANT G)

"Agricultural sector enterprises, service providers, and public institutions are working to keep up with technological transformation within the framework of their activities. As a farmer, I think we are ready for digital transformation." (PARTICIPANT I, PARTICIPANT I)

As understood from the participants' statements, awareness of Agriculture 4.0 has started to emerge among agricultural sector stakeholders. Although the transition to digital applications has partially started, it cannot be said that all stakeholders are fully prepared.

Findings on the Difficulties Encountered in Implementing Industry 4.0 Technologies

The question directed to the participants was: "*In your opinion, how do agricultural sector stakeholders perceive Industry 4.0? What are the difficulties/challenges they face in implementing Industry 4.0 technologies?*" The results of the code theory analysis regarding this question and the number of participants who expressed these difficulties are presented in Figure 8.

Figure 8. Code theory analysis method regarding the difficulties/challenges encountered by agricultural sector stakeholders in implementing Industry 4.0 technologies

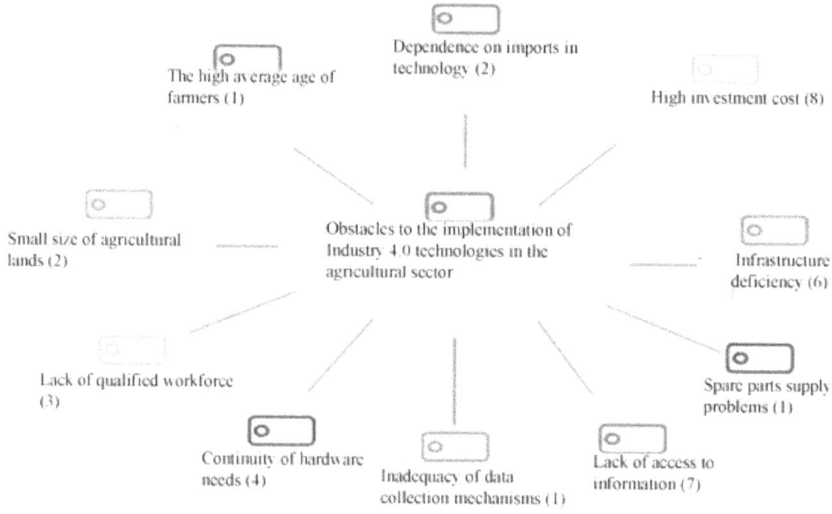

As can be seen in Figure 8, the biggest difficulty encountered in implementing Agriculture 4.0 technologies is the high investment cost. While insufficient access to information is defined as the second most important problem, the lack of infrastructure ranks third. The participants' statements on the subject are as follows:

"Problems such as weak technological infrastructure, external dependency on technological tools, and high costs will cause serious problems in the long run." (PARTICIPANT A, PARTICIPANT G)

"The difficulties encountered in implementing Agriculture 4.0 technologies are high investment costs, lack of information, the risk of not finding sufficient resources, and an insufficiently skilled workforce." (PARTICIPANT C, PARTICIPANT D)

"The difficulties experienced in the transition to Agriculture 4.0 applications in Türkiye are: weak awareness of smart technologies, inadequate information technology literacy among implementers, high average age of agricultural product producers in Türkiye, majority of agricultural enterprises being small-scale, fragmented land holdings in agricultural enterprises, inadequate technological infrastructure in rural areas, high investment costs required for smart agricultural technologies, and external dependency for technological inputs." (PARTICIPANT F)

Agricultural sector stakeholders face many challenges regarding the implementation of Industry 4.0 technologies. In this context, it is crucial for the sustainability of the sector that relevant institutions develop strategies and organize support programs.

CONCLUSION

The responses given to the question, "*What does the concept of Industry 4.0 mean to you?*" show that the participants have general knowledge about Industry 4.0 tools and follow developments in agriculture. This contributes to adaptation to new technologies, taking the right steps towards the development of the industry, and increasing competitiveness.

The responses given to the question, "*Which Industry 4.0 technologies are used in the activities carried out in the agricultural sector?*" are as follows: field estimation systems, smart machines, smart irrigation systems, Android and iOS applications, satellite monitoring systems, remote sensing, smartphones and tablets, sensors, thermal cameras, the Internet of Things (IoT), and UAV technologies. The results show that Industry 4.0 technologies have started to be used in the agricultural sector. Thanks to field estimation systems and satellite monitoring systems, the productivity of agricultural areas can be increased, and more efficient use of water resources can be achieved with smart irrigation systems and digital drip irrigation systems. In addition, thanks to smart machines and UAV technologies, it is possible to automate agricultural processes and increase productivity. The findings are consistent with the results of the studies conducted by Kovács & Husti (2018) and Ercan et al. (2019) in the literature and show that digital transformation has started and is given importance in the agricultural sector.

The answers given to the question, "*What are the reasons that lead the participants to use Industry 4.0 technologies?*" highlight factors such as increasing efficiency in production, using existing resources effectively, reducing costs, difficulties in providing labor, protecting energy resources and the environment, saving time, reducing losses in production, and contributing to the improvement of soil quality. These factors are of great importance in increasing the sustainability and competitiveness of the agricultural sector and also define the expectations regarding Agriculture 4.0. The findings are consistent with the studies in the literature. Kılavuz & Erdem (2019), Sharma et al. (2021), and Smidt & Jokonya (2022) emphasize that Industry 4.0 technologies contribute to environmental sustainability as well as improving productivity and increasing competitiveness in the agricultural sector.

In response to the question, "*How do Industry 4.0 technologies change the production process in the agricultural sector?*", participants responded that production efficiency has increased, product development has been monitored, working conditions of the workforce have improved, production costs and time losses have decreased, new production methods have been developed, existing resources are used more effectively, and errors in production are reduced. The use of Industry 4.0 technologies in the agricultural sector enables more efficient, faster, and more accurate decisions to be made in production processes. In addition, improving the working conditions of the workforce and reducing costs also play an important role in solving the labor problem in the sector and increasing profitability. Studies by researchers such as Kirmikli & Ertaş (2020), Üstündağ (2023), and Araújo (2021) in the literature also support these findings and state that Industry 4.0 technologies have great potential for improvement in agricultural production processes. For this reason, it is of great importance for enterprises operating in the agricultural sector to increase their competitiveness and contribute to sustainable agriculture by investing in these technologies.

In response to the question, "*What do you think are the difficulties/challenges that agricultural sector stakeholders face in implementing Industry 4.0 technologies?*", the participants stated that the investment costs are high, there is a lack of access to information, lack of infrastructure, the continuous need for equipment, lack of qualified labor, dependence on technology imports, small land sizes, high average age of farmers, inadequacy of data collection mechanisms, and issues with spare parts supply. The findings are consistent with the results of studies conducted by Kirmikli & Ertaş (2020), Kılavuz & Erdem (2019), Sharma et al. (2022), Gubanova et al. (2017), and Ercan et al. (2017). In order to disseminate Industry 4.0 technologies in the agricultural sector, these difficulties need to be addressed and resolved. It is important for sector stakeholders to work together on this issue, adopt solution-oriented approaches, and encourage investments. Additionally, efforts to improve infrastructure, develop technology, and enhance access to information should be intensified to eliminate dependence on imports. Developing programs to train qualified personnel and strengthening data collection mechanisms will also play a critical role in the sector's implementation of Industry 4.0 technologies.

Recommendations

Agriculture 4.0 refers to Industry 4.0 tools such as UAVs, robots, IoT, sensors, smart irrigation systems, mobile applications, and systems using artificial intelligence in agricultural production. The application of digital technologies in agriculture has many positive potential effects, such as increasing production efficiency, reducing costs, protecting existing resources, developing labor supply, improving working conditions, saving time, improving soil quality, and minimizing water, fuel, and fertilizer use. While all of these effects benefit agricultural producers, they also mean access to cheaper and higher-quality products for consumers. Therefore, Agriculture 4.0 is a critical issue due to its significant effects on producers and consumers, as well as all humanity, in the context of the sustainability of natural resources.

Agriculture is one of the most important areas for the future of humanity. Especially the increasing world population and climate change require more specialized agricultural methods. Turkey has been trying to cope with the increasing problems in the agricultural sector in recent years. The decrease in arable lands and agricultural employment, the decrease in the amount of agricultural products, the changes in energy prices, and the negative effects of climate change on production have brought about migration from villages to cities. Turkey, which was once a self-sufficient country in agriculture, has started to become a country dependent on foreign countries for these reasons, and the price increases in agricultural products have put consumers in a difficult situation. To provide access to food at affordable prices, it has become an important necessity to make structural strategic decisions within the scope of integrating new technologies into agricultural production.

A comprehensive agricultural regulation is needed that includes the integration of technology into agricultural practices, the development of the workforce, the provision of information and technical assistance to farmers at every stage of production, and the implementation of a more effective cooperation system to reduce costs and increase productivity. Public-private-university cooperation should be strengthened for agricultural production supported by advanced technology. Training activities should be carried out to ensure that producers understand smart agriculture. Units that will provide technical consultancy services for smart agriculture should be established in the local centers of the Ministry of Agriculture. Short- and medium-term support plans for smart agriculture should be implemented for farmers whose land area is suitable for smart agriculture and industrialists who want to engage in smart agricultural production. Increasing income and added value in the agricultural sector and turning agriculture into a profession are of great importance, especially for Turkey, where youth unemployment is high.

The necessary infrastructure for the digitalization of agriculture should be established with state support. It should not be forgotten that states should take measures to protect their citizens by creating policies suitable for technological transformation (Santos & Martinho, 2020). Reducing seed, fertilizer, and irrigation costs and preventing unconscious spraying that harms the soil and the consumer are other important requirements. The agricultural sector should become a sector that contributes to development with the right policies. The agricultural technology gap should be closed by cooperating with countries specialized in agriculture and transferring good practice examples to Turkey. The strategic importance of the agricultural sector should be explained to the younger generations, and they should be encouraged to start a business in this field. In short, the agricultural sector in Turkey should be positioned as a strategic sector with advanced technology, high productivity, and profitability rather than a sector with low productivity, dependent on state support in the long term with traditional production methods.

The most important limitation of the research is the difficulty in reaching businesses and farmers. Many farmers who were asked to be interviewed declined, stating that they did not have sufficient knowledge about Industry 4.0, and difficulties in reaching private-sector business officials caused the research period to be extended. In addition, the most important limitation of the study, as in other qualitative studies, is that the data could only be accessed to the extent of the interviewees' knowledge levels.

REFERENCES

Araújo, S. O., Peres, R. S., Barata, J., Lidon, F., & Ramalho, J. C. (2021). Characterizing the Agriculture 4.0 Landscape—Emerging Trends, Challenges, and Opportunities. *Agronomy (Basel), 11*(4), 667. DOI: 10.3390/agronomy11040667

Bahadır, Ş. İ. N., & Kadıoğlu, İ. (2019). Detection of Weeds Using Unmanned Aerial Vehicle (UAV) and Image Processing Techniques. *Turkish Journal of Weed Science, 22*(2), 211–217.

Balcı, B. (2022). *How Industry 4.0 Changes the Nature of Labor: A Qualitative Study in Enterprises Implementing Industry 4.0 Technologies in Turkey* (Unpublished Doctoral Dissertation). Kırklareli University, Institute of Social Sciences, Department of Labor Economics and Industrial Relations, Kırklareli, Turkey.

Barrett, H., & Rose, D. C. (2022). Perceptions of the Fourth Agricultural Revolution: What's In, What's Out, and What Consequences Are Anticipated? *Sociologia Ruralis, 62*(2), 162–189. DOI: 10.1111/soru.12324

Birişik, N. (2019). *Global and National Agricultural and Food Policies: Facts, Issues, and Proposed Solutions.* Dh Basın Yayın Matbacılık.

Büchi, G., Cugno, M., & Castagnoli, R. (2020). Smart Factory Performance and Industry 4.0. *Technological Forecasting and Social Change, 150*, 150. DOI: 10.1016/j.techfore.2019.119790

Çakmak, S. (2019). Meet the AI-Supported Autonomous Agricultural Robot: Mamut. Retrieved from https://bilimdili.com/teknoloji/yapayzeka/yapay-zeka-destekli-otonom-tarim-robotu-ile-tanisin-mamut/ (Accessed: 29.01.2023).

Çetin, Ç., Yaman, N., Sabah, L., Ayday, E., & Ayday, C. (2013). Application Possibilities of Cloud Computing Technology in Remote Sensing and Geographic Information Systems. *VII. Technical Symposium of Turkey Photogrammetry and Remote Sensing Association (TUFUAB'2013)*, 23–25.

Creswell, J. W. (2017). *30 Essential Skills for Qualitative Researchers* (Özcan, H., Trans.). Anı Publishing.

da Xu, L., Xu, E. L., & Li, L. (2018). Industry 4.0: State of the Art and Future Trends. *International Journal of Production Research, 56*(8), 2941–2962. DOI: 10.1080/00207543.2018.1444806

Dalenogare, L. S., Benitez, G. B., Ayala, N. F., & Frank, A. G. (2018). The Expected Contribution of Industry 4.0 Technologies for Industrial Performance. *International Journal of Production Economics*, *204*, 383–394. DOI: 10.1016/j.ijpe.2018.08.019

Davutoğlu, N. A. (2020). Analysis of Fundamental and Systematic Differences Between the Third and Fourth Industrial Revolutions Using a Deterministic Approach. *Management and Political Sciences Review*, *2*(1), 176–194.

De Castro, A. I., Jurado-Expósito, M., Peña-Barragán, J. M., & López-Granados, F. (2012). Airborne Multi-Spectral Imagery for Mapping Cruciferous Weeds in Cereal and Legume Crops. *Precision Agriculture*, *13*(3), 302–321. DOI: 10.1007/s11119-011-9247-0

De Haan, D. (2020). William Hazledine: Pioneering Ironmaster. *Industrial Archaeology Review*, *42*(2), 171. DOI: 10.1080/03090728.2020.1822648

Derya, H. (2018). Industrial Revolutions and Industry 4.0. *GÜ İslâhiye Faculty of Economics and Administrative Sciences International E-Journal*, *2*(2), 1–20.

Doğan, Z., Arslan, S., & Berkman, A. (2015). Economic Development and Issues in Turkey's Agricultural Sector: A Historical Perspective. *Niğde University Journal of Economics and Administrative Sciences*, *8*(1), 29–41.

Duman, B., & Özsoy, K. (2019). Smart Agriculture from an Industry 4.0 Perspective. *4th International Congress on 3D Printing (Additive Manufacturing) Technologies and Digital Industry*, 540–555.

Düzkaya, H. (2016). The Fourth Wave of the Industrial Revolution and Education: Is Turkey Ready for the Fourth Industrial Revolution? *Journal of Education and Humanities*, *7*(13), 49–88.

Ercan, Ş., Öztep, R., Güler, D., & Saner, G. (2019). Agriculture 4.0 and Its Applicability in Turkey. *Agricultural Economics Journal*, *25*(2), 259–265.

Ertaş, B. (2020). A sustainable future with Agriculture 4.0. *Icontech International Journal*, *4*(1), 1–12.

Esmer, Y., & Şaylan, O. (2019). The Link Between Industry 4.0 and Organizational Transformation. *Bilge International Journal of Social Researches*, *3*(1), 1–8.

Fagal, A. J. B. (2021). British Responses to the US Steam Frigate Fulton the First. *The Mariner's Mirror*, *107*(2), 188–201. DOI: 10.1080/00253359.2021.1898208

FAO. (2017). *E-Agriculture in Action*. Bangkok: Food and Agriculture Organization of the United Nations and International Telecommunication Union. https://www.fao.org/newsroom/detail/New-partnership-between-ITU-and-FAO-to-bolster-ICT-innovation-in-agriculture/en (Accessed: 03.12.2022).

Ferrag, M. A., Shu, L., Djallel, H., & Choo, K. K. R. (2021). Deep Learning-Based Intrusion Detection for Distributed Denial of Service Attack in Agriculture 4.0. *Electronics (Basel), 10*(11), 1257. DOI: 10.3390/electronics10111257

Glesne, C. (2012). *Introduction to qualitative research* (Ersoy, A., & Yalçınoğlu, P., Trans.). Anı Publishing.

Gökçe, G., Goncu, S., & Bozkurt, S. (2020). Industry 4.0 and animal husbandry. *International Anatolian Journal of Agricultural Engineering Sciences*, *2*(3), 21–26.

Gregersen, E. (2020). History of Technology Timeline. https://www.britannica.com/story/history-of-technology-timeline (Accessed: 09.11.2022).

Gubanova, A., Clasen, M., & Theuvsen, L. (2017). How do digital agricultural startups achieve critical mass? A qualitative analysis. In *37. GIL-Jahrestagung 2017-Fokus: Digitale Transformation–Wege in eine zukunftsfähige Landwirtschaft, Lecture Notes in Informatics,* 61-64). Gesellschaft für Informatik eV.

Hati, A. J., & Singh, R. R. (2021). Smart Indoor Farms: Leveraging Technological Advancements for a Sustainable Agricultural Revolution. *AgriEngineering*, *3*(4), 728–767. DOI: 10.3390/agriengineering3040047

Hektaş. (2021). *What is Smart Agriculture?*https://hektas.com.tr/akilli-tarim-nedir/ (Accessed: 07.12.2022)

Kagermann, H. (2017). Using Industry 4.0 Opportunities. In *Handbook Industry 4.0*, Vol. 4, 237–248, Springer Vieweg.

Karakuş, G., Yılmaztekin, F., Hamamcı, S., Bulut, B., & Ünver, R. (2022). Digital Transformation in Airports within the Scope of Aviation 4.0: A Study at Ankara Esenboğa Airport, *Proceedings Book of the 16th International Congress of Social Sciences with Current Research*, 796-813.

Kelly, D. (2016). Perceptions of Taylorism and a Marxist scientific manager. *Journal of Management History*, *22*(3), 298–319. DOI: 10.1108/JMH-11-2015-0205

Kılavuz, E., & Erdem, İ. (2019). Agriculture 4.0 applications in the world and the transformation of Turkish agriculture. *Social Sciences*, *14*(4), 133–157.

Kirmikil, M., & Ertaş, B. (2020). Sustainable Future with Agriculture 4.0. *Icontech International Journal of Surveys, Engineering. Technology*, *4*(1), 1–12.

Koca, D. (2020). Historical background of industrial revolutions and their impact on workforce skills. *OPUS International Journal of Society Researches, 16*(31), 4531–4558.

Kovács, I., & Husti, I. (2018). The role of digitalization in the agricultural 4.0–how to connect the industry 4.0 to agriculture? *Hungarian Agricultural Engineering*, (33), 38–42. DOI: 10.17676/HAE.2018.33.38

Kurnaz, Z. (2016). *Political Fear and the Construction of Academic Life in Turkey (Published doctoral thesis),* Hacettepe University, Ankara, Türkiye.

Mell, P., & Grance, T. (2011). The NIST definition of cloud computing (SP 800-145). *NIST Spec. Publ, 145*, 7.

Merriam, S. B., & Bierema, L. L. (2013). *Adult learning: Linking theory and practice*. John Wiley & Sons.

Oral, O., & Çakır, M. (2017). Internet of Things Concept and Creation of a Sample Prototype. *Mehmet Akif Ersoy University Journal of the Institute of Science, 1*(Special Issue), 172–177.

Orhan, A. (2019). Industry 4.0 Revolution or Movement? *Iğdır University Social Sciences Journal*, (18), 557–576.

Özden, A. T. (2022). The World from 1.0 to 5.0: Web, Marketing, Industry and Society. *Journal of Business in The Digital Age, 5*(1), 29–44.

Özdoğan, O. (2017). Industry 4.0: The fourth industrial revolution and the keys to industrial transformation. *Compass, 327*, 1–26.

Özsoylu, A. F. (2017). Industry 4.0. *Çukurova University Faculty of Economics and Administrative Sciences Journal*, 21(1), 41-64.

Pakdemirli, B., Birişik, N., Aslan, İ., Sönmez, B., & Gezici, M. (2021). Use of Digital Technologies in Turkish Agriculture and Agriculture 4.0 in the Agricultural-Food Chain. *Soil Water Journal, 10*(1), 78–87.

Patton, M. Q. (2014). *Qualitative Research and Evaluation Methods* (Bütün, M., & Demir, S. B., Trans.). Pegem Academy.

Regan, Á., Green, S., & Maher, P. (2018). Smart Farming in Ireland: Anticipating positive and negative impacts through a qualitative study of risk and benefit perceptions amongst expert actors in the Irish agri-food sector. In *Proceedings of the 13th European International Farm Systems Association Symposium,* Chania, Greece 1-5.

Sabancı, K., Ünlerşen, M. F., & Dilay, Y. (2016). Determination of Classification Parameters of Apple Varieties Grown in Karaman Region Using Image Processing Techniques. *Journal of Agricultural Machinery Science, 12*(2), 133–139.

Santos, R. C., & Martinho, J. L. (2020). An Industry 4.0 maturity model proposal. *Journal of Manufacturing Technology Management, 31*(5), 1023–1043. DOI: 10.1108/JMTM-09-2018-0284

Saunders, B., Sim, J., Kingstone, T., Baker, S., Waterfield, J., Bartlam, B., Burroughs, H., & Jinks, C. (2018). Saturation in qualitative research: Exploring its conceptualization and operationalization. *Quality & Quantity, 52*(4), 1893–1907. DOI: 10.1007/s11135-017-0574-8 PMID: 29937585

Sharma, K., Sharma, C., Sharma, S., & Asenso, E. (2022). Broadening the Research Pathways in Smart Agriculture: Predictive Analysis Using Semiautomatic Information Modeling. *Journal of Sensors, 2022*, 1–19. DOI: 10.1155/2022/5442865

Smidt, H. J., & Jokonya, O. (2022). Factors affecting digital technology adoption by small-scale farmers in agriculture value chains (AVCs) in South Africa. *Information Technology for Development, 28*(3), 558–584. DOI: 10.1080/02681102.2021.1975256

Thomas, R., O'Hare, G., & Coyle, D. (2023). Understanding technology acceptance in smart agriculture: A systematic review of empirical research in crop production. *Technological Forecasting and Social Change, 189*, 122374. DOI: 10.1016/j.techfore.2023.122374

Trivelli, L., Apicella, A., Chiarello, F., Rana, R., Fantoni, G., & Tarabella, A. (2019). From precision agriculture to Industry 4.0: Unveiling technological connections in the agrifood sector. *British Food Journal, 121*(8), 1730–1743. DOI: 10.1108/BFJ-11-2018-0747

Türker, U., Akdemir, B., Topakcı, M., Tekin, B., Aydın, İ. Ü. A., Özoğul, Ö., & Evrenosoğlu, M. (2015). Developments in Precision Agriculture Technologies, *Proceedings Book of the 8th Turkish Agricultural Engineering Technical Congress-1,* 295

Türkseven, S., Kızmaz, M. Z., Tekin, A. B., Urkan, E., & Serim, A. T. (2016). Digital transformation in agriculture; use of unmanned aerial vehicles. *Journal of Agricultural Machinery Science, 12*(4), 267–271.

Ullah, A. S. (2020). What is knowledge in Industry 4.0? *Engineering Reports, 2*(8), 1–21. DOI: 10.1002/eng2.12217

Üstündağ, F. (2022). *Transformation of Agricultural Production and Smart Agriculture (Published master's thesis).* Aydın Adnan Menderes University, Institute of Social Sciences, Aydın, Türkiye.

Uzun, Y., Bilban, M., & Arıkan, H. (2018). Use of Artificial Intelligence in Agriculture and Rural Development, *VI.International KOP Regional Development Symposium, 26*, 27. 1-6

Yaman, H., Sungur, O., & Dulupçu, M. A. (2021). Transformation of agriculture and animal husbandry in the world: Technology-based applications and revolutions. *Journal of Agricultural Economics, 27*(2), 123–133.

Yılmaz, M., & Arslantaş, C. (2022). Industry 4.0 awareness in enterprises and its effects on workforce: Multiple case study on managers' opinions. *Ardahan University Journal of Faculty of Economics and Administrative Sciences, 4*(2), 139–147.

Yin, Y., Stecke, K. E., & Li, D. (2018). The Evolution of Production Systems from Industry 2.0 Through Industry 4.0. *International Journal of Production Research, 56*(1–2), 848–861. DOI: 10.1080/00207543.2017.1403664

Yüksekbilgili, Z., & Çevik, G. (2018). A current and future-oriented analysis of Turkey's place in the context of Industry 4.0. *Journal of Finance. Economics and Social Research, 3*(2), 422–436.

KEY TERMS AND DEFINITIONS

Digitization / Digital Transformation: It is the structural changes that occur with the inclusion of digital technologies in all organizational structures and business processes.

UAV (Unmanned Aerial Vehicle): A flying vehicle without a human inside.

IoT: The Internet of Things is a communication network in which physical objects are connected to each other or to larger systems.

Chapter 4
Eco-Linguistics and Digital Learning:
Integrating Environmental Awareness in Language Education

Vivit Rosmayanti
https://orcid.org/0000-0002-3648-9878
Universitas Negeri Makassar, Indonesia

ABSTRACT

The chapter explores the integration of eco-linguistics and digital learning in language education to promote environmental awareness. It highlights the importance of incorporating sustainability issues into language instruction, using frameworks like Technology-Enhanced Language Learning (TELL) and sociocultural and cognitive frameworks. The chapter also discusses the use of AI-assisted language learning apps, online learning management systems, and virtual reality simulations. It also discusses ethical dilemmas and policy actions for ensuring sustainability in digital language education. The chapter emphasizes the need for an interdisciplinary approach to education.

1. INTRODUCTION

Ecolinguistics is the intersection of different disciplines that investigates the integration of language, culture, and nature concerning how language is produced and operates in a given ecology. The field works towards understanding the eco-centric worldview and how it can be harnessed to enhance environmental awareness and sustainability. Boulaid & Handoko (2024) argue that eco-linguistics goes beyond the study of language and the languages of the environment: it is a key instrument

DOI: 10.4018/979-8-3373-1822-6.ch004

in examining the eco-political discourse surrounding the environment. They also make the case that people from different cultures need to be engaged to eliminate the problem of the subordination of environmental values that comes from constructing subordinately pedagogical lessons. This reflexivity aids in the development of eco-linguistic curricula that are based on students' cultural conditioning and experiences.

For several reasons, the convergence of social media, environmental consciousness, and technology is especially important given the current global ecological crises. Through eco-linguistic approaches to teaching languages, learners can be actively involved in the environmental conversation using their language skills. According to Meyer & Vilsmaier (2020), the debate surrounding sustainable development has, to some extent, industrialized the ethnocentric nature of linguistics, but this can be corrected through more pluralistic eco-linguistic policies. With the identification of a foreign language, the students can deal with environmental problems from various cultures' points of view, as all forms of language teaching/learning processes are becoming more automated. The employment of technology in language teaching and learning is very important because apart from developing their digital competencies, students can be able to contextualize sustainability discourse at a global level.

Education for sustainability should be integrated into the use of technology across everything, focusing on learning in scopes relevant to the environment. (Cao et al., 2023) suggest that students who have advanced language proficiency are adept at using multiple digital competencies, allowing such learners access to and participation in materials and engaging activities focused on ecology. These approaches to linguistic education increase the students' ability to formulate sustainability and environmental concerns. Eco-friendly attitudes toward constructing self-narratives focused on sustainability can be cultivated through digital technologies more than eco-linguistic concepts. This is of particular importance today due to the sustainability practices that the world desperately needs as it deals with an overwhelming number of willmanner issues effects.

As Ge et al. (2023) noted, the cooperation to develop curricula and teaching materials designed for sustainability would enhance the chances of achieving effectiveness. The virtual world does offer great potential for language students to engage with people across the world and participate in serious discussions regarding environmental matters. This broad approach to language teaching is in line with the world's eco-linguistic focus which seeks to foster language use for the sustainability of the environment and positive human behavior. Language educators who utilize advanced information technologies can design motivational language teaching and at the same time instill in learners the values of environmental appreciation and preservation.

2. THEORETICAL FRAMEWORK

The approach integrating eco-linguistics into language education arises from the conception that language conveys and molds one's attitude towards the environment. Such information can be gathered from a language's ecological attitude, which explores how people construct their perception of ecological problems. In the same light, Song & Tang (2020) ecological discourse analysis is defined as an analysis of discourse based on systematic functional linguistics that studies ways of linguistic representation of ecological issues and human attitudes towards other biological species and the surrounding world, including nature. The ecological approach to discourse analysis suggests that the choices made in environmental discourse prove to be ideological as they shape people's attitudes and behaviors towards nature. Such educators analyze the language of the environment that students are exposed to to facilitate students to understand the implications of such language use and, subsequently, analyze the language critically.

The literal integration of sustainability and climate change issues in language lessons increases the students' environmental literacy. Teaching languages about these critical issues not only helps in developing students' language skills but also prepares them with the knowledge, attitude, and skills needed to tackle global problems. Although Fenyi & Sapaty (2021) do not substantiate the claim on applying sustainability topics in language teaching, this citation might be useful when discussing sustainability issues as it emphasizes the social and critical discourse analysis that accompanies ideology. Thus, this citation should be omitted to maintain accuracy. It can be argued that this is in support of Education for Sustainable Development (ESD), which seeks to ensure learners are equipped with the skills and knowledge necessary for undertaking responsible, sustainable actions. Language teachers can bring the issue of sustainability into the classroom for deeper engagement and give the students the platform to express their views on climate change and the environment.

Critical Discourse Analysis (CDA) allows people to look at language as more than language and check for instantiated power relations regarding various ideological aspects of the environment. Some issues that surround education that focus on eco-linguistics CDA is important, which, as Manteaw (2020), states that "CDA enables teachers to evaluate the social discourse practices that sustain public perceptions about climate change. This helps find out the ideological power relations that they exercise, which aid in environmental injustices." In this case, using the CDA enables them to focus deeply on social sustainability, which Manteaw and many other researchers suggest is crucial when it comes to sustainability and existence. Eco-linguistic education approaches tend to disregard the more relevant and sophisticated environmental problems. CDA allows teachers to participate actively with their learners in shifting the established narratives and counter them with narratives that are concerned with how ecological issues are framed from a grab view towards a focused view.

In as much as such motivation may focus on the teacher's goals, students become actively involved in the construction of discourse and analysis of environmental education texts and multimedia materials. Such students can see how ecological language can be used to disguise an environmental catastrophe. For example, just as civilization has developed, so have the discourses given by governing bodies and civil organizations regarding the enforcement of policy in agro-business. As students sift through the arguments from both sides, they develop the ability to analyze how different people work towards the strife of awareness against a common social evil, in this case, an environmental one (S. Chen et al., 2022). It is of utmost importance when devising a system to ensure the populace is educated and engaged enough to combat the multifaceted and volatile environmental discourse.

To conclude, eco-linguistics puts forth an interdisciplinary approach that tends to the interrelationship between language teaching and the ever-forward-moving socio-economic developments regarding environmental conservation. Integrating the content of sustainability and climate change empowers teachers to expand their students' responsibility towards the environment. Critical discourse analysis is another of these tools, which empowers students to wonder and contemplate the discourse surrounding the environment and the ideas and power relations that directly impact people's perception of the environment. This method enhances how children learn and acquire language while allowing them to participate in environmental sustainability in their community.

3. ENVIRONMENTAL THEMES IN LANGUAGE EDUCATION

Eco-linguistics emerges as a vital field that positions language as a tool to shape environmental consciousness. There is a belief that by focusing on the linguistic features of ecological problems, the adversities of the environment can be better understood, and attitudes towards it can be changed. Mete (2018) puts it so well that engaging students in environmental education in English through setting such instructional objectives as those provided in Bloom's Revised Taxonomy offers an innovative way of dealing with ecological crises. Such a technique is expected not only to better the language of the learners but also to enable them to become problem solvers concerning the state of the environment. Students can be prompted to form discussions and reflections on the different ecological problems using different authentic texts, which will then help them develop significant ecological sensitivity. Eco-linguistics does not only serve to strengthen vocabulary skills among learners;

rather, it ushers in the new paradigm of developing critical consciousness to include advocacy of sustainable practices.

The employment of technology-enhanced language learning tools (TELL) has propelled concern for the environment and its themes, making it more relevant in today's world. Eco-pedagogy can be done differently, and matter can be presented with the use of digital tools for better interaction and engagement. Bazaliy (2021) points out the role of technologies of eco pedagogy in forming the ecological consciousness of students. They argue that it is possible to teach languages through the pressing problems of the environment to encourage learners to use multimedia, virtual reality, or online collaboration tools for better-participated learning. These technologies not only teach a new language but also allow learners to engage in integrative discussions on sustainability at a global level. When synthesized with the contemporary paradigms of studying foreign languages, modern technologies transform ecology business education by enabling students to grasp its multidimensional nature.

Considering learning as a whole, sociocultural aspects as well as cognitive elements should be taken into account when looking at the role a particular environment plays in language learning. The case of Environmental Topics in Learning English as a Foreign Language, Gyawali (2023) argues that teachers should be socioculturally literate and that communal approaches can aid in the achievement of greater learning outcomes. This approach is very effective "socioculturally" because it assists in the solving of environmental problems by allowing learners to appreciate their cultures concerning global ecological issues. When viewed through a sociocultural lens, the environment becomes more understandable, and learners can appreciate the social context of environmental degradation in a person's ecosystem. Moreover, taking a sociocultural perspective enables learners to view the ecosphere more holistically by integrating cognition, where critical reasoning and reflection are essential in language acquisition. Tagnin & Ríordáin (2021) state that CLIL is one of the approaches that increases the cognitive load for many learners on difficult tasks such as the discussion of environmental issues. This method emphasizes the learners and makes them the subjects of the language, performing the investigation of the sustainability issues.

The teaching of languages should include environmental topics because they align with education for sustainable development (ESD). Ruyffelaert (2022) maintains that the language curriculum should have content-based instruction that incorporates themes of sustainability for effective integration. It does this by improving the students' language skills, as well as enabling them to work on environmental issues. Incorporating sustainability issues into language instruction allows learners to articulate their views on climate change, ecological responsibility, and other related issues. Thus, sustainability issues in language teaching aim at having students who are more enlightened and responsible citizens who can contribute to a sustainable world.

Furthermore, embedding environmental aspects in language instruction increases the learners' motivation and engagement. Hameed (2023) emphasizes that concerns about the environment can be incorporated into language teaching, thereby improving the learning process for students since it is more relevant and useful. The more students notice the correlation between language acquisition and actual problems in the world, the more they are willing to engage themselves in learning. Such motivation is fundamental for developing a sense of desire and accountability towards caring for the environment. The inclusion of environmental issues in language instruction provides teachers with the opportunity to enhance and facilitate the development of language skills and, at the same time, create advocates of global sustainability.

The integration into language learning and teaching of themes about the environment is an intricate issue involving eco-linguistics and socio-cognitive and technology-enhanced learning perspectives. Teachers can promote critical consciousness and engagement by purposefully placing language as a medium to achieve environmental consciousness. The digital age has considerably improved the enhancement of learning by making them interactive regarding ecological issues. Additionally, the incorporation of sociocultural and cognitive approaches to language motivates students to learn a language while enhancing critical and reflective thinking. Lastly, the infusion of environmentally relevant topics into the language curriculum does much more than enhance students' language proficiency; it fosters active and responsible citizenship in dealing with sustainability challenges.

4. DIGITAL TOOLS AND PLATFORMS FOR SUSTAINABLE LANGUAGE LEARNING

4.1. AI-Powered Language Learning Apps

AI assistance in language acquisition through language learning apps with sustainability elements is the most transformational in the field of language learning. Babbel and Duolingo have developed newer versions of their courses which encapsulate issues related to the environment with a worry extending beyond mere business goals. For instance, Duolingo introduced lenses that teach vocabulary revolving around sustainable practices, alongside other important aspects of language, to foster environmental awareness. (Wei, 2023) argues that an increase in learners' motivation and self-regulated learning has had the most profound impact on how AI can enable sustainability to be the challenging topic that language teaching practices provided by AI have facilitated. Furthermore, the features of these applications that allow for feedback on the correct next steps tend to help students control both language and environmental materials.

Furthermore, AI chatbots are becoming increasingly relevant in the educational sector, particularly in learning a new language, with the inclusion of sustainability issues. These chatbots can engage in conversations regarding environmental matters, which provides the learners with an opportunity to utilize the language in usefully pertinent situations. As Syathroh et al. (2021) highlighted, technology plays a pivotal role in language learning. The right tools and support can make the experience more productive. Consequently, the interactive aspect of chatbots captures the attention of students but, more importantly, motivates them to reason on sustainability issues and, in the process, use the language learned.

Apart from Duolingo and Babbel, new applications powered by AI are coming up that target sustainability. For instance, companies like Rosetta Stone are integrating environmental vocabulary and phrases into their language learning modules. As noted Nurmala et al. (2023), the integration of sustainability into language learning is pertinent. Nirmala et al. conducted a meta-analysis on the effectiveness of technology in language learning, stating that, "educational technology must be relevant and impactful." The concern that educational technology is effective in enhancing language proficiency is very real. Educators integrating sustainability themes into language learning applications put teachers on the path of developing more advanced pedagogical practices that not only ensure linguistic competence but also appreciate the global environmental increase.

In addition, the employment of AI technology when it comes to learning a language is not limited to gaining new vocabulary; it also assists in the acquisition of evaluative skills. With AI-powered applications, students actively interact with different topics on sustainability and are motivated to analyze their attitudes and behaviors towards the environment. Such reflection is necessary to deepen understanding of these issues and to embrace sustainable practices. The Transformative potential of AI in education. Wei (2023) provides evidence that these technologies have a critical function in attitude change toward sustainability while simultaneously advancing language skills.

In closing, incorporations of sustainability themes within the framework of AI-powered language learning applications serve the purpose of addressing problems in and beyond the classroom. These applications embody valuable tools for learning, as they combine linguistic as well as ecological and social skills. The innovation of chatbots and the provision of feedback by AI technology change the educational paradigm to a more learner-centered one. Technology will be increasingly important in language teaching and learning to meet the growing demands of environmental issues and sustainable practices while reinforcing language proficiency among students.

4.2. Online Learning Platforms

The introduction of online platforms like Moodle, Coursera, and Edmodo into teaching languages provides possibilities for the inclusion of environmental topics in language classes. These platforms are meant to provide and improve language skills while at the same time enhancing the awareness of sustainability and ecological matters. Learning from different backgrounds and cultures is made easy and more responsive to learners' needs because topics can be engaged with from anywhere and at any time. As noted Bošković et al. (2023) technology helps advance the change towards sustainability in education because it supports participatory teaching and learner empowerment.

Coursera is an MOOC platform that offers courses with an integration of Environmental aspects within the language acquisition process. For example, courses about environmental affairs such as science, climate change, and sustainability can be taught within the framework of languages for specific purposes. Tanış et al. (2022) highlight the necessity of usability within online learning platforms, and state that learning language is an important feature that improves the usability of the system. If online learning platforms such as Coursera were made more accessible and easier to use, learners could be effectively encouraged to speak about vocabulary related to environmental problems and learn the language at the same time.

Moodle, on the other hand, gives the instructors tools that assist the creation of courses that have sustainability concerns integrated into them while teaching a new language. The Flexibility of the platform permits the inclusion of multimedia components, interactive activities, as well as environmental-oriented group work. This aligns with Gunina et al. (2021), those who describe in their work how language classes can be constructed in such a way that learners will appreciate the need to be interlinked with society, economy, and the environment. Accordingly, teachers can use Moodle to create language courses that cultivate students' language and ability to think critically and solve ecological problems. On the other side, Edmodo, another educational platform, provides social learning which can be beneficial when tackling ecological topics in the language classroom context. The platform enables the formation of groups for discussions, collaborative assignments, and peer reviews, thereby helping students to participate in communication on sustainability problems. This is following Bošković et al., (2023), pointed out the one action toward the changes in the teaching and learning process and constraints of technology uses was integration with fostering students' agency in the classroom. Leveraging Edmodo's social features helps instructors create a classroom where students are encouraged to incorporate language and learn about the environment.

Furthermore, the online elements of a language course for environmental studies are aligned with the vision of education for sustainable development (ESD). According to Gunina et al. (2021), if teachers approached the issues of sustainability from an interdisciplinary perspective at all the dimensions of language classes, learners would probably grasp the concept better. Therefore, students who access language training through Coursera, Moodle, or Edmodo can achieve several sustainability concepts in the context of an advanced language lesson. Therefore, this method prepares students to address not only global problems but also environmental problems.

In conclusion, resources provided by Moodle, Coursera, and Edmodo to achieve the goals of foreign language teaching mediated by an ecological component are fundamental. Active participation and ease of access to resources for learning motivate the students towards greater participation and towards developing skills as ecological sages. The remarks need to mention that students should not be introduced to sustainability concepts as a primary goal. Instead, students should be taught in language classes, to understand and develop reasonable ideas related to ecological issues. With so many life spheres demanding sustainability, traditional teaching gaps have now left the burden to online teaching systems within the interdisciplinary framework of language and ecology.

4.3. Gamification and VR/AR Technology

The combination of gamification alongside virtual and augmented reality (VR/AR) technologies in language teaching and learning is facilitating the development of eco-linguistic consciousness. Gamification, which is the incorporation of game mechanics in non-game settings, motivates and engages learners by making this educational process an enjoyable, interactive experience. Studies have proven that this method greatly increases the success achieved in language acquisition, especially when the environmental aspects of the culture are included in the curriculum. According to research by Zhang & Hasim (2023), gamification gives learners real-world exposure through computerized gamified situational plots where students can perform and communicate within a digital world to enhance their understanding of ecology. Students' communication skills are improved by embedding gamification within sustainability issues. Educators will be able to provide appropriate stimuli within the gamified exercises that will make students willing to take part in activities geared toward the topic.

The use of VR and AR technologies expands the limits created by the gamification style on the students' ecological understanding of learning since it enables them to partake in a virtual immersive experience. For instance, Su et al., (2021) argue in favor of simulation games as one of the best tools that aid in teaching and learning language as the students can accomplish actual language usage. It enables students to participate and engage in the representation of environmental situations or problems in a more tangible and salient manner. As students role-play into practice with ecological problems like global warming or deforestation, they are also practicing the use of language together with higher-order thinking and problem-solving skills for sustainability. The emotional connection learners have with environmental problems can be boosted through the immersive feeling brought by VR and AR. This guarantees an enhanced sense of responsibility and empowerment to learners, which, in turn, fosters greater respect for these issues.

In addition, the combination of gamification with VR/AR technology can help enhance the collaboration aspect of learning where peers can socially interact with one another and work collaboratively. Reis et al. (2023) state that gamified courses can feature and structure group competitions or common objectives that promote collaboration among students. In this way, learners acquire a new language while also fostering community and environmental stewardship. Students can refine their social skills in a game setting while appreciating the complexity of an environmental problem's context.

The effectiveness of gamification and VRAR technology in developing eco-linguistic concepts is due to its positive impact on the motivation and engagement of the learners. For example, mobile-assisted gamification Yang et al. (2024) argued that students became more proficient speakers of English, which stresses the importance of 'hands-on' and fun-filled approaches to learning a language. Also, Phuong (2020) noticed that students seem more positive toward game-based pedagogy, which says a lot about these types of learning environments. Therefore, when attempting to motivate students through games and immerse them into learning through VR and AR, teachers can create effective materials that capture the students' attention toward sustainability.

In addition, the incorporation of gamification and VRAR technologies into learning design is aligned with the principles of education for sustainable development ESD, in which learners are geared toward addressing global challenges. With technology rapidly taking over every sector, it can not only be integrated to enhance the scope of teaching but to eco-linguistically build interest in students as well. Sustainability can be instilled in the learners around them through gamified environmental model application during the arches of different disciplines like language learning. This will help students around the globe to express their concerns regarding the complex ecological projects persisting in the world today. As noted Alshammari (2020), gamification can increase the levels of attention, motivation, and engagement a student puts forth, which can make it easier to grab the attention of learners across the globe regarding contemporary issues.

In conclusion, eco-linguistic attention can successfully be gained using VR or AR technology in conjunction with gamification during language education to make it more fun and participatory. Educators across the globe are looking for more innovative ways to tackle sustainability and develop a concrete understanding of ecological concepts among their learners. Students can interact with the amplified environment and better understand the problems the world faces. The marriage of gamification and immersive technology will enable educators to assist students in effectively dealing with sustainability problems while allowing them to be taught and learn in a more diversified environment.

4.4. Eco-Friendly Digital Language Learning Initiatives

The development of technology and innovation in recent years indicates the promising direction of eco-friendly, attitude-oriented language learning projects. These projects aim to enhance a learner's language skills as well as their environmental consciousness. An example is the digital board games for language acquisition, as described by Ali et al. (2018). Their experience with teaching vocabulary through digital board games to college students about their environmental attitudes. These games assist learners in understanding vocabulary about eco-centric themes, which reinforces their understanding of ecological issues.

Another case study of gamification focused on the improvement of foreign language achievement and increase of enjoyment of the foreign language using information technology. Yang et al. (2024) believed that gamified activities in the form of digital services enable students to actively participate in the language. This experience can also be taken further by teachers by introducing gamification in these activities with the use of sustainability themes to foster greater eco-linguistic interaction among students. Evidence indicates that learners do not only find gamified activities enjoyable but do better linguistically, which suggests that gamification can be used with negative environmental themes and still achieve positive educational outcomes.

The volume of eco-linguistics approaches in language teaching is increasing, as well as the scope of digital storytelling as an efficient assistive tool. Sabari & Hashim (2023) research demonstrates how effective students can be concerning creativity while addressing sustainability issues by using digital storytelling. When students are allowed to create stories containing environmental issues, it helps with self-analysis as well as critical thinking concerning relevant ecological issues. This supports experiential learning, where a person learns actively and from their own experience. More than just developing language, learners can use digital storytelling to illustrate their comprehension of sustainability, which is good.

Moreover, the contribution of internationalization and intercultural communication is achieved through the application of digital forms in the teaching of languages. Guimarães & Júnior (2021) explain that virtual exchanges focused on sustainability can help enhance the language learning process and describe the digital resources. Such exchanges enable learners to communicate with individuals from different cultures and appreciate their global perspective on environmental issues. Students can use the language and learn First World cultures through participation in argumentation and collaborative projects that focus on sustainability.

Mobile-assisted language learning (MALL) tools are other technological innovations that encourage environmentally friendly approaches to learning languages. Loewen et al. (2019) elucidate how mobile applications aid in the motivation of learners to use language in real-life contexts. Students can also learn ecological matters through these mobile applications which in turn helps them practice the language. Since MALL tools are a new phenomenon, they are likely to interest a wider audience that is willing to pay attention to eco issues. Students who are taught languages through mobile phone applications will become more responsible because they will be aware of the importance of actively engaging in sustainable practices in their daily routines.

Moreover, the case study conducted by Cao et al. (2023) focuses on the possible connections between digital skills and language learning achievements, highlighting the need for incorporating tools of technology into language learning. The results show that students' interest in sustainability issues and their language engagement can be increased by developing their digital skills. With the right digital skills, students will have better opportunities to learn new languages, which encompass eco-linguistics.

In addition, gaming and VR/AR design applications in language broaden the scope of eco-linguistic awareness. These technologies capture learners' attention and actively involve them in the learning process related to sustainability issues. Educators may create scenarios tackling real-life ecological challenges to stimulate students' thinking and language abilities. The integration of gamification and immersive technologies enables the development of motivating language learning activities that appeal to students and get them proactive about environmental concerns.

To summarize, particular initiatives targeting digital language learning seemed to reflect a certain ease in combining environmental issues with language teaching. Educators can design and implement effective and engaging learning activities using eco-friendly techniques that enhance the student's gamification experience, board game creation, digital storytelling, and mobile learning tools. These initiatives serve to enhance the learners' language skills while equipping them with the ability to express and interact with issues of sustainability on a global scale. The unique digital technology tools established for sustainability education in language learning will enable learners to cope with the emerging complicated issues regarding the environment.

5. SUSTAINABLE TEACHING METHODOLOGIES IN LANGUAGE LEARNING

5.1. Project-Based Learning (PBL)

Project-Based Learning (PBL) is an approach where the learning process is focused on the students' centers by making them complete authentic tasks and projects. This approach is useful in teaching languages because students learn about the surrounding world while simultaneously practicing their language skills. PBL allows students to address relevant ecological problems, making it possible to teach them in a way that is useful for the sociocultural context in which they live. Anugraha & Padmadewi (2023) argue that language learning accompanied with PBL enhances not only linguistic skills but also critical 21st-century competencies like collaboration, communication, and problem-solving that are necessary to solve language and environmental issues.

Within the framework of PBL, students have to research and solve certain environmental problems in a given language. Students can design projects in which they examine some local environmental issues, prepare campaigns to raise public awareness or give talks suggesting some solutions to the problems. Such activities help learners to develop their target language and use it in real life while also increasing their awareness of matters connected with sustainability. As Tedeneke (2023) observed, PBL serves as a basis for authentic uses of language when students are not in the classroom and enables them to use their language skills in practical situations.

Moreover, Project Based Learning (PBL) requires students to work together on their projects which enhances collaboration between students. This collaborative dimension is fundamental in fostering the learning of a language as it promotes interaction and communication among learners. Mohamed (2023) points out that project-based learning contributes to effective language use and acquisition because students learn best when they are placed in situations where they have to talk to achieve a common goal. Learners achieve this through group talks, brainstorming sessions, and jointly delivered presentations which enable them to express themselves in a language in which they feel comfortable in an interactive and active environment.

The addition of environmental issues to the language learning syllabus using PBL motivates learners to think creatively and critically. Students are motivated to analyze various ecological problems, devise creative strategies to solve them and reflect on their learning. Hussein (2021) explains that the planning and implementation phases of the projects compel students to gather information in advance and develop necessary skills that encourage self-control. This process broadens their understanding of environmental issues and enables them to express their opinions in the desired language.

In addition, PBL education is cohesive with the tenets set by Education for Sustainable Development ESD, which focuses on imparting knowledge and skills to learners for sustainable practices. While commenting on the possibility of fostering responsibility in students, Surji & Ulker (2021) posit that educators can engage students in environmental projects. Such empowerment is crucial in developing responsible citizens who are capable of dealing with numerous environmental issues.

5.2. Content and Language Integrated Learning (CLIL)

CLIL, or Content and Language Integrated Learning, is a new method that CLIL systems offer by informative context to learners through a foreign language, which is perceived as a new set of vocabulary and global problems needing attention. This approach to education is very helpful in the area of ecological education because it helps students advance their language skills while learning essential ecological concepts. Students can comprehend sustainability problems in all its dimensions by incorporating global issues in the language learning process. The complexity of environmental issues suggests that the integration of content and language by CLIL is a useful pedagogical framework (Shykun, 2023).

To increase learner vocabulary along with exposure to global issues, foreign phrases, alongside environmental concepts, can be taught in a second language. Thus, CLIL deepens environmental literacy. For example, students can be taught to have conversations on language revolving around climate change, biodiversity, and sustainable development as proposed by the CLIL system. The integration of Content and Language Integrated Learning (CLIL) into the study provided the learners with scope regarding expressing their thoughts about environmental issues using the target language. As Novitasari et al. (2022) stated, CLIL aids in the enhancement of speaking skills for students to a great extent, as students can discuss and give presentations on an environmental issue within a friendly learning context.

Through the use of a foreign language in discussing sensitive issues regarding sustainability, the learners also appreciate other cultures' perspectives on environmental sustainability. This enables them to possess a global level of empathy, which will aid them further in understanding the consequences of their endeavors and actions. Porto (2018) argues that CLIL can extend students to broader perspectives with the integration of language learning and social actions to nurture them as global citizens. The bridging of themes on sustainability into the sustainable development competitive framework of attracting foreign students suggests the principles of Education for Sustainable Development (ESD). In this context, it is possible to apply CLIL in environmental studies as it will support the different mixes of themes that target the learners' competencies regarding sustainability.

As noted by Li et al. (2020) students enrolling in CLIL Programs, they reap motivational and engagement benefits that stem from their investment. Such active engagement is crucial in fostering a sense of environmental stewardship, as well as inspiring students to participate at the community level. In addition, evidence exists for the effectiveness of CLIL approaches in the teaching of Environmental Studies. Jiang et al. (2022) proved that CLIL improves transversal skills among students in higher education, which are necessary for addressing complex environmental problems. The research suggests that the combinative approach of content and language-integrated learning CLIL encourages critical thinking, collaboration, and effective communication, which are all needed to address the sustainability challenge.

5.3. Task-Based Language Teaching (TBLT)

Task-Based Language Teaching (TBLT) refers to the didactic model based on the construction of genuine projects. This approach can be adopted sustainably in languages because it enables learners to solve problems that exist in the world as they are acquiring and using language. While addressing environmental problems, educators stimulate eco-linguistic awareness and critical thinking and devise a responsive learning environment, which by itself is very engaging.

In TBLT, students are assigned tasks that revolve around the actual use of the aforementioned language. For example, students may be asked to write about certain environmental problems in their locality, do interviews with locals about their attempts at sustainability, or prepare speeches on possible responses to the issue of climate change. Such lessons increase proficiency and allow learners to express their views about important environmental issues. It also facilitates students-everyone get excited because of the interactivity of TBLT. This is one way of changing the culture of how sustainability issues are approached.

Furthermore, TBLT fosters student interaction, since many tasks are designed with group and peer work in mind. This collaborative element is extremely important for learning a language because it allows learners to use the language in a friendly context. Learners can be motivated to express their ideas and views on sustainability while working in groups. While accomplishing tasks, students enhance their language competencies and concurrently foster their sense of responsibility toward the environment.

The concentration on sustainability in TBLT gives credence to the concept of Education for Sustainable Development (ESD). By contextualizing the linguistic tasks within the scope of environmental topics, teachers can prepare students to confront global problems. TBLT enables learners to interact with ideas that are important to them, thus, expanding their sense of self-efficacy. This instills in students the desire to participate in community work and advocate for sustainable measures and policies.

In addition, TBLT enables learners to do tasks that appeal to their interests and values, which directly increases their motivation and engagement. In task-based activities, learners' speaking skills are better developed because learners use the language more meaningfully if they have a great interest in what is being discussed. When dealing with sustainability issues, teachers can stimulate students' motivation towards learning and participation actively in the learning process.

This claim is buttressed by evidence derived from practices of TBLT for eco-linguistic attention. Studies show that students' reading skills can be improved using task-based methods along with cooperative learning. This implies not TBLT only enhances language proficiency but also addresses the students' higher-order thinking skills to solve complex problems concerning the environment. Through that, learners who engage on these themes of sustainability within the context of the task are taught to appreciate how social, economic, and environmental factors are intertwined.

5.4. The Environmental Footprint of Digital Education

As the technology supporting learning becomes more prominent in educational institutions, so too does the concern about the environmental footprint of digital education. While there are benefits to digital learning, such as easy access and the ability to learn from anywhere, there are also hurdles to overcome regarding energy use, data storage, and electronic waste. Identifying these consequences, however, is fundamental for the construction of sustainable approaches in digital educational systems.

The use of energy poses one of the greatest problems of concern associated with digital education. The requisite framework for online education, which includes data centers, server banks, and networking facilities, consumes a tremendous amount of energy. The increased expansion of the digital economy has increased demand for energy, which, if not handled sustainably, can worsen environmental challenges (Wang et al., 2022). Digital education is energy-intensive not only at the institutional level but also at the individual level. Devices such as laptops, tablets, and smartphones, for example, contribute to energy consumption. Huang et al. (2022) did not support the claim as it altered over specific contexts since it focused on predicting energy consumption at residential locations. Thus, this citation was eliminated.

Apart from energy consumption, data storage is another challenge to the environment. The enormous quantity of information generated through digital learning processes requires sophisticated information storage facilities, often located in big data centers, which are energy-intensive. The data storage issue is worsened by the fact that most data centers use fossil fuels for energy, which increases greenhouse gas emissions further. The interdependency of digitalization and energy consumption poses a challenge, especially about energy supply ICT makes information more accessible, and there is a greater need for energy supplies (Lange et al., 2020). This suggests that digital learning is ideal, provided the learning institutions incorporate any Welsh Renewable strategy. If not, the whole practice will be detrimental to the environment.

The leveraging of technology in education has its e-waste problems. Frequently changing and upgrading technology leads to massive amounts of electronic waste. Zhao et al. (2022) pointed out that electronic waste disposals have far-reaching negative consequences on the environment because many devices consist of components that are toxic and can contaminate water and soil. That reference primarily focuses on the effects of e-literacy on energy use among rural families, missing the e-waste aspect of it, especially concerning digital education. Therefore, this citation has been deleted. Furthermore, schools and institutions have e-waste management, which aims at recycling and disposals that do not harm the environment.

There are consequences to households as well, such as students and their families. As households start incorporating learning through digital devices, more energy is consumed, and a lot of e-waste is generated, which carries severe repercussions. According to the research conducted P. Chen (2022), high energy consumption is associated with the digital economy. Energy that comes from non-renewable sources is more likely to have high emissions. Thus, it is clear that promoting environmental literacy and responsible practices is crucial for younger generations and their families. Aside from students and other family members, educational institutions are also a concern. P. Chen (2022) places accent on lesson e-waste produced in an economically active society. Without sound environmental policies, extensive consumption driven by the market structures will lead towards greying out the world as emerging economies blossom in the coming years.

5.5. Equity and Accessibility

Equity and access, especially to marginalized people, is critical to sustainable development in education, especially for language learning. The place where language intersects with social justice warrants education which is sensitive to the heterogeneous linguistic and cultural realities of the learners. Gupta (2023) states that for the effective teaching of non-dominant literacies and languages, the language-as-a-resource approach is a must. From this viewpoint, it puts forth that language is profoundly connected to identity and can question the structure through which a person can be taught in another language. Students' first languages have to be appreciated to achieve equity and access in the classroom.

In the case of language instruction, the language of the content certainly determines epistemic access, particularly for disadvantaged groups. Xulu-Gama & Hadebe (2022) correctly argue that language is important for access to higher education in South Africa. They propose the need for "radical linguistic transformation" that pays attention to the language attitudes of students, parents, and teachers. Such change is necessary so that educational equity is achieved because it caters to the severe linguistic diversity among learners through multilingual education.

Furthermore, equity in education covers more than just providing resources; it also includes the provision of quality education to all learners. Russo-Tait (2022) indicates that equitable teaching can be implemented through the more efficient use of existing curricular resources. Using asset-based frameworks like Community Cultural Wealth and Funds of Knowledge transforms educators to create pedagogical spaces that view and build on the strengths of marginalized students. Such approaches not only advance educational equity but also class and empower students in the process.

Equity and accessibility in language education is a necessary white for social justice and economic enhancement. As noted Ayeni & Eden (2024), especially in developing nations, equitable opportunity in education is important for social justice. These children from marginalized groups stand a better chance at socioeconomic progress if they are afforded quality language education. This is the argument supported by Eden et al. (2023) those who point out that many countries today are still bedeviled with wide circles of differing educational achievement. To reduce these disparities, equity, and access at all levels of education must be sought.

Further, supporting equity and accessibility as the mandate of educators is quite compelling. Wassell et al. (2023) examine educators' perceptions of equity in world language instruction, pinpointing the importance of sustained professional development to resolve matters of access and representation. Developing inclusive classrooms is challenging, yet not impossible. By offering specialized training to educators, institutions will enable marginalized students to participate actively in learning exacerbating their language proficiency skills. This essentially, will allow for more equitable and sustainable education.

6. CONCLUSION

The study of eco-linguistics and digital learning is an interdisciplinary process of combining the environment and teaching language. It is therefore, important to note that methodologies such as Project Based Learning (PBL), Content and Language Integrated Learning (CLIL), or Task Based Language Teaching (TBLT) are critical to the engagement of students in sustainability issues in learning languages. The responsible use of artificial intelligence (AI) in education also demands such balance, as digitization should serve educational purposes while not undermining social-ecological needs.

Environmental advocacy is presented as one of the important outcomes of language education. In this way, students will be able to express their opinions about ecological issues, which advocates so much for sustainability in the community. The processes of integrating the issues of the environment in language teaching will not only improve linguistic skills but, most importantly, develop higher-order thinking skills and a sense of responsibility for the environment. Still, Gupta does not provide supporting evidence for the claim that language is a resource predicated on environmental consciousness. That is, this citation has also been removed for the sake of accuracy.

The possibilities for eco-linguistic digital learning seem excellent; it looks promising. A concern for educators is the 'green' aspect of education technology, which includes energy use, data storage, as well as e-waste. There is also the matter of equity and accessibility, predominantly dire need compassion to many underserved communities, which, if addressed, can facilitate language teaching and learning in a more comprehensible way. Ethical AI tools and practices that are being further developed will also improve the learning experience by providing students the possibility to interact with language education in an efficient and eco-friendly manner.

REFERENCES

Ali, Z., Ghazali, M. A. I. M., Ismail, R., Muhammad, N. N., Azlinda, Z. A. N., & Malek, N. A. (2018). Digital Board Game: Is There a Need for It in Language Learning Among Tertiary Level Students? *Matec Web of Conferences, 150*, 5026. DOI: 10.1051/matecconf/201815005026

Alshammari, M. T. (2020). Evaluation of Gamification in E-Learning Systems for Elementary School Students. *Tem Journal*, 806–813. https://doi.org/DOI: 10.18421/tem92-51

Anugraha, G. A. P., & Padmadewi, N. N.Gede Angga Pratama AnugrahaNi Nyoman Padmadewi. (2023). The Need Analysis of Technology-Based Project-Based Learning in Higher Education. *Jurnal Pendidikan Bahasa Inggris Undiksha, 10*(3), 242–248. DOI: 10.23887/jpbi.v10i3.58466

Ayeni, O. O., & Eden, C. A.Oyebola Olusola AyeniChima Abimbola Eden. (2024). Equity and Access in Higher Education: Legal Perspectives and Management Strategies. *International Journal of Science and Research Archive, 11*(2), 199–206. DOI: 10.30574/ijsra.2024.11.2.0391

Bazaliy, R. (2021). Development of Ecocentric-Ecological Consciousness of Students of a Technical University. *E3s Web of Conferences, 273*, 12002. DOI: 10.1051/e3sconf/202127312002

Bošković, D., Husremović, D., Muslic, M., & Kapo, A. (2023). Teachers and Students as Promoters or Repressors of Sustainable Education: Navigating the Blended Learning Landscape. *Sustainability (Basel), 15*(24), 16812. DOI: 10.3390/su152416812

Boulaid, F., & Handoko, H. (2024). Cultivating Ecological Awareness Through the Integration of Eco-Linguistic Themes in English for Specific Purposes: A Case Study of Higher Schools of Technology. *Gramatika Stkip Pgri Sumatera Barat, 10*(1). Advance online publication. DOI: 10.22202/jg.2024.v10i1.8062

Cao, J., Bhuvaneswari, G., Arumugam, T., & Aravind, B. R. (2023). The Digital Edge: Examining the Relationship Between Digital Competency and Language Learning Outcomes. *Frontiers in Psychology, 14*, 1187909. Advance online publication. DOI: 10.3389/fpsyg.2023.1187909 PMID: 37397293

Chen, M. R. A. (2020). Effects of experiencing authentic contexts on English speaking performances, anxiety and motivation of EFL students with different cognitive styles. *Interactive Learning Environments*. Advance online publication. DOI: 10.1080/10494820.2020.1734626

Chen, P. (2022). Relationship Between the Digital Economy, Resource Allocation and Corporate Carbon Emission Intensity: New Evidence From Listed Chinese Companies. *Environmental Research Communications, 4*(7), 75005. DOI: 10.1088/2515-7620/ac7ea3

Chen, S., Luria, S., & Shaub, K. (2022). Bridging the Gap Between Socio-Cognitive Theory and Practical Second Language Instruction. *Journal of Student Research, 11*(3). Advance online publication. DOI: 10.47611/jsrhs.v11i3.3304

Eden, C. A., Chisom, O. N., & Adeniyi, I. S.Chima Abimbola EdenOnyebuchi Nneamaka ChisomIdowu Sulaimon Adeniyi. (2023). Education Policy and Social Change: Examining the Impact of Reform Initiatives on Equity and Access. *International Journal of Science and Research Archive, 11*(2), 139–146. DOI: 10.30574/ijsra.2024.11.2.0372

Fenyi, K., & Sapaty, G. A. (2021). Ideological Underpinnings in President John Dramani Mahama's 2013 State of the Nation Address. *Linguistics Initiative, 1*(1), 1–17. DOI: 10.53696/27753719.17

Ge, N., Wang, E., & Li, Y. (2023). Foreign Language Education for Sustainable Development in China: A Case Study of German Language Education. *Sustainability (Basel), 15*(8), 6340. DOI: 10.3390/su15086340

Guimarães, F. F., & Júnior, C. A. H. (2021). Digital Resources and English as an Additional Language in Higher Education: Possibilities for Internationalization. *Ilha do Desterro, 74*(3). Advance online publication. DOI: 10.5007/2175-8026.2021.e80233

Gunina, N. A., Mordovina, T. V, & Shelenkova, I. (2021). Integrating Sustainability Issues Into English Language Courses at University. *E3s Web of Conferences, 295*, 5006. DOI: 10.1051/e3sconf/202129505006

Gupta, A. A. (2023). Diversity, Dignity, Equity, and Inclusion in the Age of Division, Discord and Disunion: Stereotyping, Sexist, Hegemony in Education. *International Education Studies, 16*(1), 110. DOI: 10.5539/ies.v16n1p110

Gyawali, Y. P. (2023). Exploring Dimensions of Teacher Education in Nepal: Perspectives, Challenges, and Reform Opportunities. *Journal of General Education and Humanities, 2*(4), 285–298. DOI: 10.58421/gehu.v2i4.128

Hameed, A. (2023). Environmental Concerns and English Language Teaching in Saudi Context: Perceptions and Practices. *Journal of Language Teaching and Research, 14*(4), 1127–1137. DOI: 10.17507/jltr.1404.31

Huang, J., Pang, C., Yang, W., Zeng, X., Zhang, J., & Huang, C. (2022). A Deep Learning Neural Network for the Residential Energy Consumption Prediction. *IEEJ Transactions on Electrical and Electronic Engineering*, *17*(4), 575–582. DOI: 10.1002/tee.23543

Hussein, B. (2021). Addressing Collaboration Challenges in Project-Based Learning: The Student's Perspective. *Education Sciences*, *11*(8), 434. DOI: 10.3390/educsci11080434

Jiang, P., Akhter, S., Azizi, Z., Gheisari, A., & Kumar, T. (2022). Exploring the Role of Content and Language Integrated Learning Approach in Developing Transversal Skills in University Students With Respect to the Mediating Role of Emotional Intelligence. *Frontiers in Psychology*, *13*, 988494. Advance online publication. DOI: 10.3389/fpsyg.2022.988494 PMID: 36275209

Lange, S., Pohl, J., & Santarius, T. (2020). Digitalization and Energy Consumption. Does ICT Reduce Energy Demand? *Ecological Economics*, *176*, 106760. DOI: 10.1016/j.ecolecon.2020.106760

Li, L., Fu-quan, H., Chen, S., Pan, L., Zeng, W., & Wu, X. (2020). Exploring the Curriculum Development in Content and Language Integrated Learning: A Systematic Review. *International Journal of Evaluation and Research in Education (Ijere)*, *9*(4), 1102. DOI: 10.11591/ijere.v9i4.20705

Loewen, S., Crowther, D., Isbell, D. R., Kim, K. M., Maloney, J., Miller, Z. F., & Rawal, H. (2019). Mobile-Assisted Language Learning: A Duolingo Case Study. *ReCALL*, *31*(3), 293–311. DOI: 10.1017/S0958344019000065

Manteaw, B. O. (2020). Education and Learning in Sustainable Development: Foregrounding an Emergent Discourse. *Discourse and Communication for Sustainable Education*, *11*(2), 5–19. DOI: 10.2478/dcse-2020-0015

Mete, D. E. (2018). Incorporating Environmental Education in English Language Teaching Through Bloom's Revised Taxonomy. *Selçuk Üniversitesi Edebiyat Fakültesi Dergisi*, *40*(40), 33–44. DOI: 10.21497/sefad.514847

Meyer, E., & Vilsmaier, U. (2020). Economistic Discourses of Sustainability: Determining Moments and the Question of Alternatives. *Sustainability in Debate*, *11*(1), 98–124. DOI: 10.18472/SustDeb.v11n1.2020.26663

Mohamed, A. M. (2023). Investigating the Benefits of Multimodal Project-Based Learning in Teaching English to International Students. *International Journal of Educational Innovation and Research*, *2*(2), 114–129. DOI: 10.31949/ijeir.v2i2.5085

Novitasari, N., Wahyuningsih, N., & Agustina, H. N. (2022). Improving Students' Speaking Skills Through Clil in Tourist Guiding Online Class. *Celtic a Journal of Culture English Language Teaching Literature and Linguistics*, *9*(1), 53–68. DOI: 10.22219/celtic.v9i1.20903

Nurmala, I., Irianto, S., Franchisca, S., Amsa, H., & Susanti, R. (2023). Technology-Enhanced Language Learning: A Meta-Analysis Study on English Language Teaching Tools. *Journal of Education*, *6*(1), 2188–2195. DOI: 10.31004/joe.v6i1.3221

Phuong, T. T. H. (2020). Gamified Learning: Are Vietnamese EFL Learners Ready Yet? *International Journal of Emerging Technologies in Learning*, *15*(24), 242. DOI: 10.3991/ijet.v15i24.16667

Porto, M. (2018). Intercultural Citizenship in Foreign Language Education: An Opportunity to Broaden CLIL's Theoretical Outlook and Pedagogy. *International Journal of Bilingual Education and Bilingualism*, •••, 1–21. DOI: 10.1080/13670050.2018.1526886

Reis, S. C., Linck, A. J. M., Figueiredo, M. F., & Pfeifer, D. L. (2023). Gamification Into the Design of the E-3d Online Course. *Frontiers in Education*, *8*, 1152999. Advance online publication. DOI: 10.3389/feduc.2023.1152999

Russo-Tait, T. (2022). Science Faculty Conceptions of Equity and Their Association to Teaching Practices. *Science Education*, *107*(2), 427–458. DOI: 10.1002/sce.21781

Ruyffelaert, A. (2022). Raising Concepts and Awareness of Sustainability and the Environment in Higher Education Through French Foreign Language Teaching: A Multidisciplinary Didactic Proposal. *Language Learning in Higher Education*, *12*(2), 617–626. DOI: 10.1515/cercles-2022-2063

Sabari, N. A. S., & Hashim, H. (2023). Sustaining Education With Digital Storytelling in the English Language Teaching and Learning: A Systematic Review. *International Journal of Academic Research in Business & Social Sciences*, *13*(4). Advance online publication. DOI: 10.6007/IJARBSS/v13-i4/16860

Shykun, A. (2023). Content and Language Integrated Learning: Enhancing Language Acquisition and Content Understanding. *International Science Journal of Education & Linguistics*, *2*(4), 39–44. DOI: 10.46299/j.isjel.20230204.05

Song, J., & Tang, M. (2020). Ecological Discourse Analysis From the Perspective of Systemic Functional Linguistics. *Destech Transactions on Social Science Education and Human Science, icesd*. DOI: 10.12783/dtssehs/icesd2020/34131

Su, F., Zou, D., Xie, H., & Wang, F. L. (2021). A Comparative Review of Mobile and Non-Mobile Games for Language Learning. *SAGE Open*, *11*(4), 21582440211067247. Advance online publication. DOI: 10.1177/21582440211067247

Surji, A. L., & Ulker, V. (2021). Undergraduate Foreign Language Learners Perspective on Project Based Learning. *International Journal of Social Sciences and Educational Studies*, *8*(4). Advance online publication. DOI: 10.23918/ijsses.v8i4p254

Syathroh, I. L., Kareviati, E., Lestari, A. P., & Fitria, N. (2021). Exploring the Potentials of Technology Integration for Teaching Language Skills: A Literature Review. [Professional Journal of English Education]. *Project*, *4*(3), 488. DOI: 10.22460/project.v4i3.p488-496

Tagnin, L., & Ríordáin, M. N. (2021). Building Science Through Questions in Content and Language Integrated Learning (CLIL) Classrooms. *International Journal of STEM Education*, *8*(1), 34. Advance online publication. DOI: 10.1186/s40594-021-00293-0

Tanış, H., Akçay, A., Yilmaz, N., Yiğit, M. F., & Tüzün, H. (2022). How Usable Is Coursera? A Usability Analysis Through Eye-Tracking and Authentic Tasks. *Participatory Educational Research*, *9*(4), 379–395. DOI: 10.17275/per.22.96.9.4

Tedeneke, A. (2023). Exploring the Practices and Contextual Factors in Teaching Writing Skills in EFL Classrooms: A Case Study. *Erjssh*, *9*(2), 83–101. DOI: 10.4314/erjssh.v9i2.6

Wang, Y., Derakhshan, A., & Pan, Z. (2022). Positioning an Agenda on a Loving Pedagogy in Second Language Acquisition: Conceptualization, Practice, and Research. *Frontiers in Psychology*, *13*, 894190. Advance online publication. DOI: 10.3389/fpsyg.2022.894190 PMID: 35668974

Wassell, B. A., Glynn, C., Carey, B., Sevinç, E., & Baroudi, F. (2023). Equity, Access, and Inclusion in K-12 World Language Education: A System of Failure or Work in Progress? *Journal of Linguistics and Language Teaching*, *15*(1). Advance online publication. DOI: 10.5070/L215159614

Wei, L. (2023). Artificial Intelligence in Language Instruction: Impact on English Learning Achievement, L2 Motivation, and Self-Regulated Learning. *Frontiers in Psychology*, *14*, 1261955. Advance online publication. DOI: 10.3389/fpsyg.2023.1261955 PMID: 38023040

Xulu-Gama, N., & Hadebe, S. (2022). Language of Instruction: A Critical Aspect of Epistemological Access to Higher Education in South Africa. *South African Journal of Higher Education*, *36*(5). Advance online publication. DOI: 10.20853/36-5-4788

Yang, L., Karanyathikul, A., & Kotchasit, S. (2024). Effect of Gamification Combined With Mobile Learning on Chinese Language Speaking Skills of Second-Year Foreign Students at Guangdong University of Foreign Studies. *Ijsasr*, *4*(6), 563–578. DOI: 10.60027/ijsasr.2024.5054

Zhang, S., & Hasim, Z. (2023). Gamification in EFL/ESL Instruction: A Systematic Review of Empirical Research. *Frontiers in Psychology*, *13*, 1030790. Advance online publication. DOI: 10.3389/fpsyg.2022.1030790 PMID: 36687912

Zhao, L., Zhang, Y., & Zhang, H. (2022). Research on the Impact of Digital Literacy on Farmer Households' Green Cooking Energy Consumption: Evidence From Rural China. *International Journal of Environmental Research and Public Health*, *19*(20), 13464. DOI: 10.3390/ijerph192013464 PMID: 36294039

Chapter 5
Environmental Excellence Through Disruptive Technological Innovations

Abhijit Chandratreya
https://orcid.org/0000-0001-9403-2094
SCES's Indira Institute of Management, Pune, India

ABSTRACT

The pressing need for sustainable development has brought disruptive technological innovations to the forefront of environmental management. This chapter explores how these advancements are transforming traditional approaches to enhancing environmental quality, focusing on the integration of emerging technologies such as artificial intelligence, blockchain, and IoT in environmental monitoring, resource management, and pollution control. It examines case studies highlighting the successful implementation of these innovations in achieving sustainability goals while addressing the challenges associated with their adoption, including scalability, cost, and regulatory barriers. By bridging the gap between technology and environmental science, this chapter provides actionable insights for policymakers, researchers, and industries aiming to leverage disruptive technologies for environmental excellence. The discussion emphasizes the importance of collaboration, innovation ecosystems, and stakeholder engagement in driving sustainable technological adoption for a greener future.

DOI: 10.4018/979-8-3373-1822-6.ch005

1. INTRODUCTION

1.1 Overview of Environmental Challenges in the 21st Century

The 21st century is marked by an unprecedented scale of environmental challenges, driven by rapid industrialization, urbanization, and population growth. Climate change remains one of the most pressing issues, characterized by rising global temperatures, melting polar ice caps, and increasingly severe weather events (Intergovernmental Panel on Climate Change [IPCC], 2021). The burning of fossil fuels for energy, deforestation, and unsustainable agricultural practices are key contributors to greenhouse gas emissions, intensifying the global warming phenomenon. These changes threaten biodiversity, disrupt ecosystems, and pose significant risks to human health and livelihoods, particularly in vulnerable regions (World Health Organization [WHO], 2020).

Another critical challenge is resource depletion, exacerbated by unsustainable consumption patterns and inefficient resource management. Freshwater scarcity affects billions of people worldwide, with demand projected to outpace supply in several regions by 2030 (United Nations [UN], 2021). Similarly, the overexploitation of natural resources such as forests, fisheries, and minerals has led to habitat destruction and the decline of many species. The improper management of waste, including electronic and plastic waste, contributes to environmental degradation, contaminating soil and water sources and jeopardizing marine and terrestrial ecosystems (Jambeck et al., 2015).

Urbanization and industrialization have also led to significant air and water pollution, with adverse effects on public health and the environment. According to WHO (2020), air pollution is responsible for approximately 7 million premature deaths annually, making it a leading environmental risk factor. Water pollution, stemming from industrial discharge, agricultural runoff, and untreated sewage, has rendered many water bodies unsuitable for human use or aquatic life (UN, 2021). Addressing these multifaceted challenges requires global cooperation, innovative technologies, and sustainable practices to mitigate their impact and pave the way for a more resilient future.

1.2 Defining Disruptive Technological Innovations and their Role in Sustainability

Disruptive technological innovations are transformative technologies that significantly alter existing systems, processes, or markets by offering improved efficiency, cost-effectiveness, or entirely new capabilities. Unlike incremental advancements, disruptive innovations often displace established technologies and practices, creating new opportunities and challenges. Christensen (1997) first introduced the concept of disruptive innovation, highlighting its potential to revolutionize industries by initially catering to niche markets and eventually outperforming traditional solutions. In the context of environmental management, disruptive technologies such as blockchain, artificial intelligence (AI), and renewable energy systems are redefining sustainability strategies by enabling more efficient resource use, minimizing waste, and supporting data-driven decision-making (Geissdoerfer et al., 2017).

The impact of disruptive technologies on sustainability is profound, as they address the dual issue of economic growth and environmental preservation. For instance, AI and machine learning enable real-time monitoring of environmental parameters, such as air and water quality, while optimizing resource allocation (Dangelico et al., 2017). Similarly, blockchain technology enhances transparency in supply chains, ensuring accountability for environmental impacts and reducing inefficiencies (Saberi et al., 2019). Renewable energy technologies, including advanced photovoltaic systems and wind turbines, disrupt traditional fossil-fuel-dependent energy paradigms by providing clean, scalable alternatives. These innovations collectively contribute to sustainable development by reducing greenhouse gas emissions, conserving natural resources, and promoting equitable access to green solutions.

However, the disruptive nature of these technologies also presents challenges in achieving sustainability goals. The high initial costs, need for specialized infrastructure, and regulatory complexities can hinder the widespread adoption of disruptive innovations (Markard et al., 2012). Moreover, the rapid pace of technological change may lead to unintended consequences, such as increased electronic waste or social disparities in access to new solutions. Nevertheless, by integrating these innovations into sustainability frameworks and fostering collaboration among stakeholders, disruptive technologies can act as a catalyst for achieving environmental excellence and promoting long-term ecological balance.

1.3 Importance of Achieving Environmental Excellence in a Rapidly Industrializing World

The accelerated pace of industrialization across the globe has led to unprecedented economic growth, but it has also come at a significant environmental cost. Air and water pollution, deforestation, and greenhouse gas emissions have escalated, contributing to global challenges like climate change, biodiversity loss, and resource depletion (Rockström et al., 2009).

Achieving environmental excellence—defined as the implementation of strategies that mitigate negative environmental impacts while promoting sustainability—is essential to ensuring that industrial development remains aligned with ecological preservation. It involves adopting innovative technologies, sustainable practices, and regulatory frameworks that balance industrial growth with environmental stewardship. Without such efforts, industrial activities risk undermining the very ecosystems that support human life and economic stability.

Environmental excellence is critical for fostering long-term industrial resilience. Companies and nations that prioritize sustainability gain a competitive advantage in a global market increasingly shaped by environmental regulations and consumer preferences for green products (Porter & Kramer, 2011). Transitioning to cleaner technologies and practices not only reduces operational risks related to resource scarcity and regulatory penalties but also opens avenues for innovation. For instance, adopting circular economy principles, such as waste-to-resource strategies, enables industries to achieve efficiency while minimizing environmental harm (Ellen MacArthur Foundation, 2015). These approaches are essential for mitigating the negative externalities of industrialization and ensuring a sustainable trajectory for economic growth.

Achieving environmental excellence is pivotal for addressing social equity issues that arise from industrialization. Communities in developing and industrializing nations often bear the brunt of environmental degradation, facing health risks and economic vulnerabilities due to polluted air, contaminated water, and resource depletion (UNEP, 2020). Implementing environmentally sustainable practices not only protects ecosystems but also enhances the quality of life for marginalized populations, fostering inclusive growth. Policymakers, industries, and civil society must collaborate to establish robust frameworks for environmental governance and innovation. This ensures that industrialization contributes positively to human well-being while preserving the planet for future generations.

2. UNDERSTANDING DISRUPTIVE TECHNOLOGIES

2.1 Characteristics of Disruptive Technologies

Disruptive technologies are innovations that significantly alter or replace existing systems, markets, or practices by offering novel solutions or efficiencies. These technologies often challenge traditional paradigms by being more accessible, cost-effective, and scalable compared to their predecessors. A key characteristic is their ability to create new markets or transform existing ones, often beginning as niche solutions before expanding into mainstream applications (Christensen, 1997). For instance, solar photovoltaic technology initially catered to off-grid energy needs but has now become a major player in global energy markets. This transformative potential often places disruptive technologies at the forefront of environmental management and sustainability initiatives.

Another defining feature of disruptive technologies is their ability to drive exponential improvements in performance or cost-effectiveness. Unlike incremental innovations, which refine existing solutions, disruptive technologies leapfrog traditional approaches, enabling rapid adoption and widespread impact (Bower & Christensen, 1995). Technologies like artificial intelligence and blockchain exemplify this, offering unprecedented capabilities in data processing, decision-making, and transparency. These advancements are particularly relevant for environmental applications, such as real-time monitoring of air quality or traceability in sustainable supply chains, where traditional methods struggle to meet modern demands.

Disruptive technologies are often characterized by their capacity for adaptability and integration across diverse sectors. This cross-disciplinary applicability allows them to solve complex, interconnected challenges, such as those related to climate change and resource scarcity (Adner, 2002). For example, the Internet of Things (IoT) seamlessly integrates with renewable energy systems, enabling smarter grid management and reducing energy waste. This adaptability not only enhances their value proposition but also accelerates their adoption in industries striving for sustainability. As a result, disruptive technologies are increasingly recognized as critical enablers of environmental excellence, driving transformative change in both developed and emerging economies.

2.2 Examples of Disruptive Innovations With Environmental Implications

Disruptive innovations have revolutionized traditional approaches to addressing environmental challenges by providing scalable, efficient, and often cost-effective solutions. Renewable energy technologies, such as solar and wind power, exemplify such innovations. Solar photovoltaic (PV) cells have seen rapid advancements in efficiency and cost reduction, making solar energy a viable alternative to fossil fuels. Similarly, wind turbines have become more efficient, with offshore wind farms contributing significantly to global renewable energy generation. These technologies disrupt the conventional energy market by reducing reliance on non-renewable energy sources and significantly lowering greenhouse gas emissions (IRENA, 2021).

Another example of disruptive innovation is the application of blockchain technology in environmental management. Blockchain enables secure, transparent, and decentralized tracking of environmental data and resources. For instance, blockchain-based systems can monitor carbon credits, ensuring accountability in emission reduction commitments. Additionally, this technology is being utilized to trace the provenance of sustainable products, thereby encouraging ethical sourcing and consumption (Kouhizadeh et al., 2021). Blockchain's immutable nature disrupts traditional methods of resource tracking and promotes trust in sustainability initiatives.

Artificial intelligence (AI) is another transformative force in environmental conservation. AI-powered systems are being used to optimize energy usage in smart grids, predict and mitigate environmental disasters, and enhance biodiversity monitoring. For example, AI algorithms process satellite imagery to detect deforestation patterns and identify illegal logging activities in real-time. Furthermore, AI-driven predictive analytics are employed to anticipate climate-related events, such as hurricanes or floods, enabling more effective disaster preparedness (Rolnick et al., 2019). By integrating data-driven insights with environmental management strategies, AI disrupts conventional reactive methods and fosters proactive solutions.

2.3 Comparison With Incremental Innovations in Environmental Management

Disruptive technological innovations fundamentally differ from incremental innovations in their scope, impact, and approach to environmental management. Incremental innovations typically involve gradual improvements to existing systems, processes, or technologies, aiming to optimize efficiency or reduce negative environmental impacts. For instance, improving the energy efficiency of internal combustion engines is an incremental innovation that slightly reduces emissions without radically changing the transportation system (Bocken et al., 2014). In contrast, disruptive innovations like electric vehicles or hydrogen-powered transport represent a paradigm shift, replacing traditional technologies with entirely new systems that significantly enhance sustainability. These innovations not only reduce environmental harm but also create opportunities for industries to achieve greater resource efficiency and carbon neutrality.

The primary distinction lies in the ability of disruptive innovations to address systemic environmental challenges more comprehensively than incremental changes. Incremental innovations are often limited by existing technological frameworks and societal infrastructure, which may constrain their effectiveness in addressing large-scale problems like climate change, resource depletion, and biodiversity loss (Christensen et al., 2015). On the other hand, disruptive technologies, such as blockchain for supply chain transparency or artificial intelligence in waste management, can enable transformative solutions that were previously unfeasible. These technologies introduce new business models, redesign operational practices, and reconfigure societal interactions with the environment, fostering long-term sustainability (Geels et al., 2017).

However, the adoption of disruptive innovations is often accompanied by significant barriers, such as high initial costs, resistance from established industries, and regulatory challenges. Incremental innovations, being less invasive, tend to face fewer hurdles, making them easier to implement in the short term (Bergek et al., 2013). Nonetheless, the limitations of incremental approaches in achieving substantial environmental progress underscore the need for a balanced strategy that integrates disruptive innovations alongside gradual improvements. By leveraging the strengths of both approaches, organizations, and governments can accelerate the transition toward environmental excellence while minimizing risks and ensuring widespread acceptance of new technologies.

3. KEY AREAS OF ENVIRONMENTAL EXCELLENCE IMPACTED BY DISRUPTIVE INNOVATIONS

3.1 Energy

The energy sector is a pivotal domain where disruptive innovations are driving environmental excellence by enabling cleaner, more sustainable energy generation and consumption. Renewable energy technologies such as solar photovoltaic (PV) systems, wind turbines, and advanced biofuels have disrupted traditional fossil fuel-based energy systems by offering low-emission alternatives with declining costs and increasing efficiency (REN21, 2022). Innovations in energy storage, such as lithium-ion and solid-state batteries, further amplify the impact of renewable energy by addressing intermittency issues, ensuring grid stability, and facilitating the adoption of electric vehicles (BNEF, 2023). These advancements are integral to decarbonizing the energy sector, which is essential for mitigating climate change and achieving net-zero targets.

Moreover, the integration of digital technologies like artificial intelligence (AI) and the Internet of Things (IoT) into energy systems enhances their efficiency and sustainability. Smart grids, powered by AI and IoT, enable real-time energy monitoring, demand response mechanisms, and optimal energy distribution, reducing waste and maximizing resource utilization (IRENA, 2023). Blockchain technology is also emerging as a disruptor by facilitating decentralized energy trading and transparent carbon credit systems, empowering consumers to participate in sustainable energy practices (Mengelkamp et al., 2018). These disruptive innovations collectively reshape the energy landscape, creating opportunities for enhanced environmental quality and economic growth while reducing dependence on non-renewable resources.

3.2 Waste Management

Technologies such as artificial intelligence (AI), the Internet of Things (IoT), and blockchain are revolutionizing traditional waste management practices. AI-driven sorting systems, for instance, use advanced image recognition and machine learning algorithms to segregate recyclable materials from waste streams with unprecedented accuracy and speed (Ghobakhloo, 2020). IoT-enabled smart bins monitor waste levels in real-time and optimize collection schedules, reducing unnecessary transportation emissions and improving resource allocation (Zhang et al., 2019). These innovations not only increase the recovery rate of recyclable materials but also minimize operational costs and environmental impacts, thus contributing to environmental excellence.

Blockchain technology is emerging as a game-changer in creating transparent and traceable waste management systems. By securely recording waste disposal and recycling data, blockchain enhances accountability and ensures compliance with environmental regulations (Upadhyay et al., 2021). It also fosters trust among stakeholders, including governments, corporations, and consumers, by verifying the authenticity of recycling processes and waste management claims. For example, blockchain-based platforms can certify that electronic waste is responsibly processed and recycled, thereby addressing growing concerns over hazardous e-waste disposal (Agyekum et al., 2021). These disruptive innovations provide scalable solutions to global waste challenges, enabling industries and communities to transition toward a circular economy model that prioritizes resource conservation and environmental stewardship.

3.3 Water Conservation

Water conservation is an essential aspect of environmental excellence, particularly in the face of growing water scarcity and climate change. Traditional water management systems often rely on centralized infrastructure and inefficient distribution methods that waste significant amounts of water. Disruptive innovations such as smart water meters, Internet of Things (IoT) sensors, and real-time monitoring systems are revolutionizing how water is managed and conserved. These technologies enable more precise control over water usage by providing detailed data on consumption patterns, leak detection, and system inefficiencies (Van der Meer & Tiwari, 2017). By facilitating better resource allocation, these innovations not only reduce water wastage but also enhance the resilience of water systems against extreme weather events, ensuring sustainable water supply even in the most challenging conditions.

IoT-based water management systems are one of the most promising disruptive technologies in water conservation. These systems connect sensors embedded in water infrastructure to cloud-based platforms, allowing for continuous monitoring of water quality and quantity across large networks (Gunay & Yavuz, 2020). For example, smart irrigation systems use weather data and soil moisture sensors to optimize water use in agriculture, significantly reducing water waste while maintaining crop productivity (Kharche et al., 2020). Similarly, in urban settings, IoT-enabled systems can monitor water flow, detect leaks in real-time, and alert maintenance teams to address issues before they lead to significant water loss. By automating water management processes, these technologies contribute to more efficient and sustainable use of water resources across various sectors.

However, despite the potential of disruptive innovations, the widespread implementation of these technologies faces several challenges. High initial investment costs, lack of technological infrastructure in developing regions, and resistance from traditional water management industries can slow the adoption of these solutions (Rogers, 2003). Furthermore, while IoT and other smart technologies offer immense promise, they require robust cybersecurity measures to protect sensitive data related to water supply systems. Nevertheless, the benefits of integrating these innovations into water conservation strategies are undeniable, as they represent a significant leap forward in achieving environmental excellence. By reducing inefficiencies, preventing water loss, and improving overall system performance, disruptive technologies can play a pivotal role in addressing global water challenges and promoting sustainable water use.

3.4 Agriculture

Traditional agricultural practices often lead to soil depletion, overuse of water resources, and high carbon emissions due to reliance on chemical fertilizers and fossil fuel-powered machinery (Tilman et al., 2011). Disruptive technologies such as precision agriculture, which uses data analytics, IoT, and AI, enable farmers to monitor crop health, soil conditions, and weather patterns in real-time, optimizing the use of water, nutrients, and pesticides (Wolfert et al., 2017). By reducing resource wastage and minimizing chemical runoff, these innovations promote sustainable farming practices that significantly lower the environmental impact of agriculture.

The integration of biotechnology, such as genetically modified crops and CRISPR gene-editing technologies, further exemplifies how disruptive innovations are transforming agriculture for environmental excellence. These advancements allow for the development of crops that are more resistant to pests, diseases, and extreme weather conditions, reducing the need for chemical inputs and ensuring higher yields with fewer resources (Qaim, 2020). Additionally, innovations in vertical farming and aquaponics, which rely on controlled environments and less land area, are providing new solutions for urban food production with minimal environmental footprints (Despommier, 2013). These technologies allow for more sustainable and efficient food production in both rural and urban settings, helping to alleviate pressure on natural ecosystems and reduce the carbon footprint of food systems.

Despite the promising potential of these disruptive technologies, challenges remain in their widespread adoption. High initial costs, technical complexity, and the need for specialized knowledge are barriers that hinder the full-scale implementation of these innovations, particularly in developing countries (Foley et al., 2011). Additionally, regulatory concerns, particularly around biotechnology and gene-editing, present significant obstacles to their integration into mainstream agriculture. However, with continued research, public-private partnerships, and government support, these barriers can be overcome, paving the way for a future where agriculture not only meets global food demands but also contributes to the broader goals of environmental sustainability and climate resilience.

3.5 Air Quality Management

Traditional methods of managing air quality often rely on monitoring and regulating emissions from industries, vehicles, and other sources, which include measures like emissions standards, air quality monitoring stations, and pollutant reduction technologies (WHO, 2018). However, these methods can be slow to implement and reactive, responding only once environmental standards are breached. Disruptive innovations, such as real-time air quality monitoring systems powered by Internet of Things (IoT) devices, have the potential to radically transform AQM by offering proactive solutions that enable immediate action (Zhang et al., 2020). These IoT sensors provide continuous, high-resolution data on pollutants, allowing for more dynamic and targeted interventions to reduce air pollution levels.

In addition to real-time monitoring, artificial intelligence (AI) plays a pivotal role in reshaping air quality management practices. AI-powered models can analyze vast amounts of environmental data to predict pollution levels and identify sources of air contamination with greater precision than traditional methods (Zhang et al., 2020). AI also enables the development of predictive systems that forecast air quality trends, facilitating early warnings and allowing for timely policy interventions. For example, cities like Beijing have implemented AI-driven air pollution models to optimize traffic management and industrial activities, reducing emissions during high pollution periods (Chen et al., 2021). The integration of AI with other emerging technologies, such as machine learning algorithms and big data analytics, can enhance decision-making and support more sustainable urban planning and transportation strategies aimed at improving air quality.

While disruptive innovations present significant opportunities for advancing air quality management, their widespread adoption faces several challenges. Technological barriers, such as the high cost of deploying IoT sensors and AI systems, may limit their accessibility in low-income regions. Moreover, integrating these technologies into existing regulatory frameworks and aligning them with environmental policies require substantial effort from governments and industries (Kassinis et al., 2022). Nevertheless, the potential of disruptive technologies to reduce air pollution at scale underscores their importance for achieving long-term environmental excellence. As these technologies evolve and become more cost-effective, they will likely play a central role in the global effort to improve air quality, particularly in urban areas where pollution levels are most acute.

4. CASE STUDIES OF SUCCESSFUL IMPLEMENTATION

One prominent example of an organization using disruptive technologies to achieve environmental goals is Tesla. Tesla has revolutionized the electric vehicle (EV) industry by developing high-performance electric cars and solar energy solutions that reduce dependence on fossil fuels. By integrating cutting-edge battery technology, autonomous driving capabilities, and solar energy products, Tesla is reshaping the transportation and energy sectors, significantly reducing greenhouse gas emissions. In addition, Tesla's innovation in battery storage, such as the Powerwall, enables the storage of renewable energy for later use, thus improving the reliability of solar and wind energy systems (Zengler, 2020). Tesla's commitment to innovation and sustainability aligns with its broader goal of accelerating the world's transition to sustainable energy.

On the government front, the city of Copenhagen has emerged as a leading example of leveraging technology for environmental excellence. The city has embraced IoT and big data analytics to manage air quality, reduce carbon emissions, and improve overall urban sustainability. Through its "City Data Exchange" initiative, Copenhagen collects data from a network of sensors that monitor pollution levels, traffic congestion, and energy consumption. This data is used to inform real-time decisions, such as adjusting traffic flow to reduce emissions or deploying electric vehicles in city fleets. Copenhagen's use of AI and data analytics for energy optimization is a significant step toward meeting its ambitious goal of becoming the world's first carbon-neutral capital by 2025 (Copenhagen Solutions Lab, 2020).

Another noteworthy example is the partnership between IBM and the city of Rio de Janeiro. Through the "Smarter Cities" initiative, IBM has helped the city develop a comprehensive, data-driven approach to urban management, focusing on improving air quality, reducing traffic congestion, and managing energy use. The use of real-time data analytics and AI-driven solutions has enabled Rio to predict pollution levels and optimize its public transportation system, resulting in a reduction in both emissions and operational costs (IBM, 2021). This collaboration highlights the potential for disruptive technologies to provide scalable, sustainable solutions to complex environmental challenges in urban settings.

4.1 Analysis of Outcomes

The implementation of disruptive technologies in environmental management can yield significant benefits across multiple dimensions: environmental, economic, and social. From an **environmental** perspective, technologies such as renewable energy systems, electric vehicles (EVs), and advanced waste management solutions directly reduce harmful emissions and contribute to improved resource management. For instance, Tesla's electric vehicles and solar energy solutions have played a pivotal role in reducing the carbon footprint of the transportation and energy sectors (Zengler, 2020). The widespread adoption of EVs reduces greenhouse gas emissions from traditional internal combustion engine vehicles, which are a major source of air pollution. Additionally, the deployment of solar panels helps decrease reliance on fossil fuels, further reducing the environmental impact. Similarly, smart waste management systems powered by IoT sensors and AI contribute to waste reduction by optimizing collection routes and enhancing recycling rates (Zhang et al., 2020).

The **economic** benefits of these innovations are also substantial. Transitioning to green technologies can stimulate economic growth by fostering new industries and creating job opportunities. The renewable energy sector, for example, has seen significant investments, leading to the creation of thousands of jobs in the manufacturing, installation, and maintenance of solar panels and wind turbines (IRENA, 2021). Electric vehicle production has similarly resulted in economic growth, with automakers like Tesla and established companies like Ford investing heavily in EV production and infrastructure. Moreover, these technologies can help businesses and governments achieve long-term cost savings by reducing energy consumption and operational costs. For instance, Copenhagen's use of smart city technologies has enabled the city to reduce energy use and cut costs, while at the same time improving overall urban efficiency (Copenhagen Solutions Lab, 2020).

From a **social** perspective, disruptive technologies can significantly enhance quality of life by reducing environmental health risks and promoting greater social equity. For instance, real-time air quality monitoring and pollution management technologies in cities like Beijing and Rio de Janeiro have led to improved air quality, which in turn has reduced the incidence of respiratory diseases and other pollution-related health conditions (Chen et al., 2021). Additionally, the adoption of green technologies can contribute to social equity by providing affordable clean energy solutions to underserved populations. In rural and developing areas, solar energy systems can offer reliable electricity to communities that are not connected to the grid, improving living standards and enabling access to education and healthcare services (IRENA, 2021). Ultimately, the integration of disruptive technologies into environmental management fosters a more sustainable, equitable, and healthy society, underscoring the importance of innovation in addressing global environmental challenges.

5. CHALLENGES IN IMPLEMENTING DISRUPTIVE TECHNOLOGIES FOR ENVIRONMENTAL EXCELLENCE

5.1 Financial and Infrastructural Barriers

One of the primary challenges in implementing disruptive technologies for environmental excellence is the **financial barrier**. The initial capital investment required to develop, deploy, and scale up these technologies can be substantial. For example, the installation of solar power systems, electric vehicle infrastructure (such as charging stations), or the deployment of real-time IoT sensors for air quality monitoring requires significant upfront costs that can be a deterrent for many organizations, especially in developing countries (IRENA, 2021). Moreover, while these technologies often offer long-term savings through energy efficiency and reduced operational costs, the payback period may be long, discouraging businesses or governments that require more immediate returns. For small and medium-sized enterprises (SMEs) or municipalities with limited budgets, this financial barrier can prevent the adoption of innovative technologies that could otherwise contribute to sustainability goals.

The **infrastructural** challenges also present a significant obstacle. Disruptive technologies, such as electric vehicles or renewable energy systems, often require the development of new infrastructure or the modernization of existing infrastructure to support their widespread adoption. For example, the widespread deployment of electric vehicles depends on the availability of a robust charging infrastructure, which can be difficult to build in areas with limited access to electric grid systems or where urban planning has not been designed with sustainability in mind (Hawkins et al., 2013). Similarly, the integration of renewable energy sources into the national grid requires the development of advanced smart grid technologies, which can be both complex and costly to implement. In many regions, especially those with outdated or underdeveloped infrastructure, the lack of readiness for such technological advancements can hinder progress toward achieving environmental excellence.

Overcoming these financial and infrastructural barriers often requires significant public and private sector collaboration. Governments can play a critical role by providing incentives, subsidies, or tax credits to encourage the adoption of green technologies. In addition, financing models such as green bonds or public-private partnerships (PPPs) can help alleviate the financial burden on businesses and municipalities (Zhang et al., 2020). Furthermore, international cooperation and funding mechanisms can support developing countries in overcoming infrastructural deficits and building the necessary infrastructure to support disruptive innovations. However, addressing these barriers requires a concerted effort from all stakeholders to ensure that the transition to sustainable technologies is both economically feasible and infrastructurally viable.

5.2 Resistance to Change and Lack of Awareness

This resistance often stems from a combination of fear, uncertainty, and the perceived risks associated with adopting new technologies. Organizations, particularly in industries with established processes and technologies, may be reluctant to invest in disruptive innovations due to concerns about the cost, compatibility with existing systems, and uncertainty about the return on investment (Christensen et al., 2015). For example, the transition from traditional fossil fuel-based energy systems to renewable energy sources, like solar or wind power, often faces opposition from stakeholders invested in the oil and gas sectors. The fear of losing market share, outdated infrastructure, and the perceived instability of renewable energy markets contribute to this reluctance (Geels et al., 2017). Overcoming this resistance requires a combination of strong leadership, clear communication of the long-term benefits, and effective policy incentives that encourage stakeholders to embrace change.

Lack of awareness is another critical barrier to the widespread adoption of disruptive technologies. Many organizations, governments, and individuals may not fully understand the environmental, economic, and social benefits that these technologies can offer, resulting in missed opportunities for improvement. For instance, cities may not be aware of the potential of smart city technologies, such as IoT-based air quality monitoring and energy optimization systems, to enhance environmental sustainability and public health (Zhang et al., 2020). The lack of awareness can also manifest in a general misunderstanding of how technologies like blockchain can improve supply chain transparency and reduce waste, or how AI can enhance resource management. Without sufficient knowledge of these innovations, decision-makers may fail to prioritize or invest in them, delaying progress toward environmental goals. Education and awareness campaigns, as well as training programs, are necessary to bridge this gap and enable stakeholders to make informed decisions about adopting disruptive technologies.

Both resistance to change and lack of awareness can be mitigated through targeted strategies, such as the development of demonstration projects that showcase the benefits of disruptive technologies in real-world applications. Additionally, involving stakeholders early in the decision-making process and providing evidence-based data on the potential impact of these technologies can reduce fears and misconceptions. Policymakers can also play a crucial role by introducing incentives, subsidies, and regulatory frameworks that encourage the adoption of green technologies while addressing potential risks. As more success stories emerge and technologies mature, the adoption of disruptive innovations will likely increase, leading to significant advancements in environmental management and sustainability (Bocken et al., 2014).

5.3 Ethical and Regulatory Concerns

The implementation of disruptive technologies in environmental management, while promising significant environmental benefits, also raises important **ethical and regulatory concerns** that can pose challenges to their widespread adoption.

Ethical concerns primarily revolve around issues such as data privacy, surveillance, and the potential for misuse of technologies. For instance, IoT sensors and AI systems used in smart cities to monitor air quality and energy consumption collect vast amounts of data, some of which may include personal information or data from private spaces. If not properly safeguarded, this data could be exploited, leading to breaches of privacy and social inequities (Cohen & Regan, 2018). Moreover, the use of AI in environmental management decisions raises concerns about transparency and accountability. AI algorithms that drive decisions related to pollution control, waste management, or resource distribution must be designed in a way that ensures fairness and avoids bias, which could disproportionately affect vulnerable communities (Zhang et al., 2020).

Regulatory concerns are equally significant, as governments and regulatory bodies often struggle to keep pace with the rapid evolution of disruptive technologies. The regulatory framework required to govern the deployment of technologies such as blockchain for supply chain transparency or AI-driven systems in environmental monitoring is often inadequate or non-existent (Binns, 2018). This can lead to a lack of clarity regarding the legal responsibilities of organizations that implement such technologies, especially when it comes to liability in the case of environmental harm or failure to meet regulatory standards. For example, regulations on carbon emissions and air quality monitoring have traditionally been based on older technologies and may need substantial revisions to account for newer, more sophisticated systems like

AI and real-time data analytics (Geels et al., 2017). Furthermore, governments may be reluctant to enforce strict regulations that could hinder technological innovation or disrupt industries, leading to a slower adoption of disruptive technologies.

Finally, the implementation of **disruptive environmental technologies** can also exacerbate **ethical inequalities**. While these technologies often promise significant environmental benefits, they are typically more accessible to wealthier individuals, companies, and nations, potentially leaving poorer communities at a disadvantage. For example, renewable energy technologies such as solar panels and electric vehicles can be prohibitively expensive for low-income populations, limiting their access to cleaner technologies (IRENA, 2021). The digital divide may also exacerbate inequalities in developing regions where access to cutting-edge technologies is limited, potentially leading to environmental inequities. To address these concerns, ethical frameworks, and regulatory measures must be developed in parallel with technological innovations to ensure that these advancements benefit all stakeholders equitably and are implemented transparently and responsibly.

5.4 Balancing Innovation With Unintended Environmental Consequences

While disruptive technologies present tremendous opportunities for achieving environmental excellence, they also bring with them the challenge of unintended environmental consequences that must be carefully managed. As new technologies are developed and deployed, they may inadvertently create new forms of environmental harm, often due to unanticipated effects or insufficient regulatory oversight. For example, the rise of electric vehicles (EVs) is often hailed as a solution to reducing carbon emissions and improving air quality. However, the production of EV batteries requires significant amounts of lithium, cobalt, and other rare earth materials, the extraction of which can lead to environmental degradation, including soil and water contamination (Mudd et al., 2012). Additionally, the disposal and recycling of these batteries at the end of their lifecycle pose further environmental challenges, such as the potential release of hazardous chemicals into the environment if not properly managed (Gaines et al., 2014).

The rapid adoption of technologies such as solar panels and wind turbines, while environmentally beneficial in terms of reducing greenhouse gas emissions, can also have negative ecological impacts. The large-scale deployment of wind farms, for instance, may result in habitat disruption for birds and other wildlife, while the production of solar panels can contribute to electronic waste and involve hazardous chemicals in their manufacturing process (Pellow, 2018). Such unintended consequences highlight the need for a holistic approach to environmental innovation, wherein the lifecycle impacts of technologies—from raw material extraction to disposal—are considered. Innovators, policymakers, and environmentalists must collaborate to ensure that the environmental benefits of disruptive technologies outweigh the potential harms.

To mitigate these unintended consequences, it is essential to implement robust environmental assessments and regulatory frameworks that guide the development and deployment of new technologies. For example, life cycle assessments (LCAs) can help identify and address the full spectrum of environmental impacts associated with a product or technology, from production to end-of-life (Ciroth et al., 2019). Additionally, incorporating circular economy principles—where materials are reused, recycled, or repurposed—can help minimize waste and reduce the environmental footprint of disruptive technologies. By striking a balance between technological innovation and environmental stewardship, it is possible to harness the power of disruptive technologies for sustainable development while minimizing unintended harm to ecosystems and human health.

6. STRATEGIES FOR MAXIMIZING IMPACT

6.1 Policy Frameworks to Encourage the Adoption of Disruptive Technologies

Policy frameworks play a critical role in encouraging the adoption of disruptive technologies and maximizing their environmental impact. Governments can create an enabling environment for innovation by providing incentives such as tax credits, subsidies, and grants for research and development in green technologies. Regulatory frameworks that set clear environmental standards and provide long-term stability are also essential, as they guide businesses in aligning their innovation efforts with sustainability goals. For example, policies like carbon pricing, renewable energy mandates, and emission reduction targets can stimulate the market for disruptive technologies by making sustainable solutions more economically attractive (Anderson, 2020). Additionally, policies that promote public-private partnerships, such as government-funded innovation labs or collaborative platforms, can foster cross-sector collaboration and accelerate the commercialization of new technologies. By aligning policy with environmental, economic, and social objectives, governments can create a supportive ecosystem that drives the widespread adoption of disruptive innovations while ensuring that their implementation leads to significant environmental benefits.

6.2 Role of Public-Private Partnerships in Driving Innovation

Public-private partnerships (PPPs) play a critical role in driving innovation and maximizing impact, particularly in the context of environmental sustainability. These collaborations leverage the strengths of both sectors—public entities bring regulatory support, policy frameworks, and long-term vision, while private companies contribute technological expertise, financial resources, and operational efficiencies. For example, the development of renewable energy infrastructure often relies on PPPs, where governments create favorable policy environments and private companies provide the capital and cutting-edge technologies. One notable example is the partnership between the Indian government and private firms to scale solar energy capacity, resulting in India becoming one of the world leaders in solar power generation (Ghosh & Venkatesh, 2021). By aligning the goals of both sectors, PPPs enable the acceleration of innovation, improve the scalability of environmental solutions, and help address complex global challenges like climate change, all while ensuring that both economic and social benefits are maximized.

6.3 Educational Initiatives to Promote Technological Literacy

By equipping individuals with the knowledge and skills to understand and effectively use emerging technologies, such initiatives can foster widespread adoption and innovation in sustainable practices. For instance, educational programs focused on renewable energy, circular economy principles, and environmental management technologies can empower students, professionals, and communities to engage with these solutions and apply them in real-world contexts. Furthermore, promoting technological literacy through accessible platforms, such as online courses, workshops, and public awareness campaigns, can bridge knowledge gaps and encourage informed decision-making. By building a tech-savvy workforce and an environmentally-conscious society, educational initiatives ensure that the benefits of disruptive technologies are realized to their fullest potential, supporting long-term sustainability goals and facilitating a smoother transition to greener technologies (Dede et al., 2016).

6.4 Encouraging Research and Development in Green Technologies

Encouraging research and development (R&D) in green technologies is a crucial strategy for maximizing the environmental, economic, and social impact of sustainable innovations. By investing in R&D, governments, businesses, and academic institutions can accelerate the creation of new technologies that not only reduce environmental harm but also offer scalable solutions for global challenges like climate change and resource depletion. For instance, advancements in renewable energy technologies, such as more efficient solar panels, wind turbines, and energy storage systems, have the potential to drastically reduce dependence on fossil fuels and lower greenhouse gas emissions (IRENA, 2021). Additionally, supporting R&D in areas like sustainable agriculture, waste-to-energy systems, and carbon capture technologies can open new avenues for reducing waste, enhancing resource efficiency, and promoting circular economies. A strong focus on innovation can drive breakthroughs that make green technologies more affordable, accessible, and impactful, ultimately helping to create a more sustainable future across all sectors.

7. FUTURE TRENDS AND DIRECTIONS

7.1 Emerging Technologies Poised to Disrupt Environmental Management

Emerging technologies are rapidly reshaping the landscape of environmental management, offering innovative solutions to some of the most pressing global challenges. One such technology is **artificial intelligence (AI)**, which is being used to optimize resource management, predict environmental trends, and enhance sustainability efforts. AI-driven algorithms can analyze vast amounts of environmental data to forecast pollution levels, optimize waste management, and identify inefficiencies in energy usage (Chen et al., 2021). Another promising technology is **blockchain**, which has the potential to revolutionize transparency and traceability in supply chains, ensuring sustainable sourcing and reducing environmental harm from unsustainable practices (Tapscott & Tapscott, 2016). Additionally, **gene editing** and **synthetic biology** are paving the way for breakthroughs in environmental remediation, such as the development of microorganisms that can detoxify pollutants or restore ecosystems (Barton et al., 2020). Finally, **quantum computing** is expected to accelerate the discovery of new materials for energy storage, carbon capture, and renewable energy systems, contributing to the advancement of a low-carbon economy. These technologies, when integrated into environmental management strategies, hold the potential to significantly enhance sustainability and accelerate the transition to a more resilient and ecologically balanced future.

7.2 Predictions for the Next Decade in Environmental Innovation

We are likely to see a significant expansion in the use of renewable energy sources, such as solar, wind, and geothermal, alongside breakthroughs in energy storage technologies like solid-state batteries and hydrogen fuel cells, which will address the intermittency challenges of renewables (IRENA, 2021). AI and big data will continue to play an integral role in optimizing resource management, reducing waste, and improving energy efficiency across industries. Additionally, circular economy principles will gain traction, with more industries adopting closed-loop systems that prioritize material reuse, recycling, and waste minimization. The widespread adoption of electric vehicles (EVs) and advancements in carbon capture and storage technologies will also accelerate decarbonization efforts, contributing to the global goal of achieving net-zero emissions. These innovations will be crucial in addressing pressing environmental challenges, such as climate change, biodiversity loss, and resource depletion while creating new opportunities for sustainable economic growth.

7.3 Global Collaboration for Shared Technological Advancements

Global collaboration is crucial for advancing environmental technologies that can be scaled to address worldwide sustainability challenges. By pooling resources, knowledge, and expertise, countries and organizations can accelerate the development and deployment of disruptive technologies that benefit the global community. For instance, international partnerships such as the Mission Innovation initiative, which brings together 24 countries to accelerate clean energy innovation, aim to lower the costs of renewable energy technologies and enhance energy efficiency (Mission Innovation, 2021). Moreover, shared technological advancements in areas like electric mobility, clean water access, and renewable energy technologies have been made possible through cross-border collaborations between the private sector, academia, and governments. These collaborations not only facilitate the exchange of innovative ideas but also help harmonize regulatory standards and build global markets for sustainable technologies, ensuring that benefits are equitably distributed (Sovacool et al., 2019). Ultimately, global cooperation is vital for achieving large-scale environmental impact and meeting international sustainability goals, such as those outlined in the Paris Agreement.

8. CONCLUSION

8.1 Summary of Key Insights

The chapter on *Environmental Excellence Through Disruptive Technological Innovations* highlights the transformative potential of emerging technologies in achieving sustainable environmental outcomes. It emphasizes that disruptive innovations, such as renewable energy systems, electric vehicles, and smart waste management solutions, can significantly reduce environmental harm by lowering carbon emissions, improving resource efficiency, and promoting circular economy principles. These innovations offer a break from traditional incremental improvements by addressing environmental challenges in more radical and systemic ways, ultimately reshaping industries and contributing to long-term sustainability. The chapter also underscores how technological advancements can help mitigate pressing environmental issues, including climate change, resource depletion, and pollution, by offering cleaner, more efficient alternatives to current practices.

The chapter also explores the challenges associated with implementing disruptive technologies, particularly in balancing their potential benefits with unintended environmental consequences. While these technologies offer substantial environmental advantages, their lifecycle impacts—such as resource extraction, waste generation, and ecosystem disruption—must be carefully managed. The chapter advocates for a holistic approach to environmental innovation, one that considers both the positive and negative implications of new technologies. It also highlights the importance of global collaboration in accelerating the adoption of disruptive innovations, fostering shared knowledge, and creating the regulatory frameworks needed to ensure that these technologies are deployed responsibly and equitably across regions. By taking a comprehensive and cooperative approach, societies can harness the full potential of disruptive technologies to drive environmental excellence.

8.2 Call to Action for Stakeholders to Embrace and Invest in Disruptive Innovations for a Sustainable Future

To secure a sustainable future, stakeholders—ranging from governments and businesses to research institutions and consumers—must recognize the pivotal role that disruptive innovations play in addressing the world's most pressing environmental challenges. The rapid advancement of clean technologies, renewable energy solutions, and sustainable practices offers unprecedented opportunities to reduce carbon emissions, preserve natural resources, and promote environmental health. However, these innovations require substantial investment, both in terms of funding and policy support, to transition from nascent ideas to widespread implementation.

Governments need to create regulatory environments that incentivize innovation, while businesses must take bold steps to integrate these technologies into their operations, reducing their environmental footprints and fostering greater efficiency. Moreover, research institutions should be encouraged to further explore and refine these technologies to unlock their full potential for societal benefit.

For stakeholders, embracing and investing in disruptive innovations is not only an ethical imperative but also a strategic move that aligns with long-term economic and social goals. By prioritizing sustainability and incorporating breakthrough technologies into core business models, organizations can position themselves as leaders in the growing green economy. The private sector can drive growth and job creation while minimizing environmental risks, and governments can stimulate economic prosperity by fostering industries that support sustainable development. Collaboration among stakeholders will be key to overcoming the barriers to large-scale adoption, whether through financial support, knowledge exchange, or the scaling of best practices. Now is the time for all sectors to act decisively, invest in transformative technologies, and build a resilient, sustainable future for generations to come.

8.3 Vision for Achieving Environmental Excellence Through Technology

Achieving environmental excellence through technology requires a forward-thinking vision that integrates innovation, sustainability, and adaptability. The future of environmental stewardship lies in harnessing the potential of disruptive technologies to not only reduce human impact on the environment but also restore and regenerate ecosystems. By adopting technologies such as renewable energy systems, smart grids, and circular economy models, we can transition away from traditional practices that rely on finite resources. Through a commitment to green technologies like solar, wind, and bioenergy, alongside advancements in energy storage and efficient distribution, we can drastically reduce carbon emissions and mitigate climate change. Moreover, the integration of data analytics, artificial intelligence, and the Internet of Things (IoT) allows for real-time monitoring and optimization of environmental systems, enhancing the effectiveness of resource management and pollution control.

To realize this vision, it is essential to foster global collaboration and ensure that technological advancements are inclusive and accessible to all regions. Building a sustainable future through technology involves not only the development of innovative solutions but also the creation of supportive frameworks for their implementation, such as regulations, policies, and incentives that encourage the adoption of clean technologies. Public-private partnerships, coupled with investments in research and development, can drive the widespread adoption of green innovations, ensuring that environmental excellence becomes a standard across industries. With a shared commitment to environmental responsibility, the global community can use technology as a powerful tool to create a more resilient and sustainable world for future generations.

REFERENCES

Adner, R. (2002). When are technologies disruptive? A demand-based view of the emergence of competition. *Strategic Management Journal, 23*(8), 667–688. DOI: 10.1002/smj.246

Agyekum, E. O., Fortuin, K. P. J., & van der Harst, E. (2021). Assessing circularity of electronic waste management in the context of developing countries: Evidence from Ghana. *Journal of Cleaner Production, 292*, 126032.

Anderson, C. (2020). Policy interventions for promoting green technologies: A global review. *Environmental Policy and Governance, 30*(2), 65–80.

Barton, J., Gavalda, R., & Kim, H. (2020). Synthetic biology for environmental sustainability. *Environmental Science & Technology, 54*(13), 7998–8009.

Bergek, A., Hekkert, M., & Jacobsson, S. (2013). Functions in innovation systems: A framework for analysing energy system dynamics and identifying goals for system-building activities by entrepreneurs and policymakers. *Technology Analysis and Strategic Management, 25*(10), 1026–1041.

Binns, A. (2018). Ethical implications of artificial intelligence in environmental management. *Technology Ethics Review, 25*(4), 36–48.

BloombergNEF (BNEF). (2023). *Battery price survey 2023.* Retrieved from https://about.bnef.com

Bocken, N. M. P., Short, S. W., Rana, P., & Evans, S. (2014). A literature and practice review to develop sustainable business model archetypes. *Journal of Cleaner Production, 65*, 42–56. DOI: 10.1016/j.jclepro.2013.11.039

Bower, J. L., & Christensen, C. M. (1995). Disruptive technologies: Catching the wave. *Harvard Business Review, 73*(1), 43–53.

Chen, Y., Zhao, J., & Li, W. (2021). Artificial intelligence and air pollution: A review of applications, challenges, and future directions. *Environmental Science & Technology, 55*(9), 6169–6182.

Christensen, C. M. (2015). *The innovator's dilemma: when new technologies cause great firms to fail.* Harvard Business Review Press.

Christensen, C. M., Raynor, M. E., & McDonald, R. (2015). What is disruptive innovation? *Harvard Business Review, 93*(12), 44–53. PMID: 17183796

Ciroth, A., Frischknecht, R., & Bickel, P. (2019). Life cycle assessment and the environment: A key tool for sustainability. *Environmental Science & Technology, 53*(8), 4472–4480.

Cohen, J. E., & Regan, M. (2018). Privacy and the challenges of new technology. *Environmental Ethics, 40*(2), 151–173.

Copenhagen Solutions Lab. (2020). *Copenhagen: Leading the way to carbon neutrality*. https://www.solutionslab.dk/copenhagen-carbon-neutral

Dangelico, R. M., Pujari, D., & Pontrandolfo, P. (2017). Green product innovation in manufacturing firms: A sustainability-oriented dynamic capability perspective. *Business Strategy and the Environment, 26*(4), 490–506. DOI: 10.1002/bse.1932

Dede, C., Richards, J., & Pungowiyi, E. (2016). The role of education in advancing sustainability through disruptive technologies. *The Journal of Environmental Education, 47*(2), 122–136.

Despommier, D. (2013). *The vertical farm: Feeding the world in the 21st century*. Thomas Dunne Books.

Ellen MacArthur Foundation. (2015). *Towards the circular economy: Economic and business rationale for an accelerated transition*. Ellen MacArthur Foundation.

Foley, J. A., Ramankutty, N., Brauman, K. A., Cassidy, E. S., Gerber, J. S., Johnston, M., Mueller, N. D., O'Connell, C., Ray, D. K., West, P. C., Balzer, C., Bennett, E. M., Carpenter, S. R., Hill, J., Monfreda, C., Polasky, S., Rockström, J., Sheehan, J., Siebert, S., & Zaks, D. P. (2011). Solutions for a cultivated planet. *Nature, 478*(7369), 337–342. DOI: 10.1038/nature10452 PMID: 21993620

Gaines, L., Nelson, P., & Birky, A. (2014). Recycling of lithium-ion batteries. *Nature Materials, 13*(9), 809–818.

Geels, F. W., Sovacool, B. K., Schwanen, T., & Sorrell, S. (2017). Sociotechnical transitions for deep decarbonization. *Science, 357*(6357), 1242–1244. DOI: 10.1126/science.aao3760 PMID: 28935795

Geissdoerfer, M., Savaget, P., Bocken, N. M. P., & Hultink, E. J. (2017). The circular economy – A new sustainability paradigm? *Journal of Cleaner Production, 143*, 757–768. DOI: 10.1016/j.jclepro.2016.12.048

Ghobakhloo, M. (2020). Industry 4.0, digitization, and opportunities for sustainability. *Journal of Cleaner Production, 252*, 119869. DOI: 10.1016/j.jclepro.2019.119869

Ghosh, A., & Venkatesh, P. (2021). Public-private partnerships in renewable energy: Lessons from India's solar revolution. *Energy Policy, 149*, 112013.

Gunay, B., & Yavuz, A. (2020). The impact of the Internet of Things (IoT) in water management: A review. *Environmental Science and Pollution Research International, 27*(19), 23882–23899.

Hawkins, T. R., Gunter, E. W., & Franke, M. (2013). Environmental impacts of electric vehicles: A comparison of different powertrain technologies. *Environmental Science & Technology, 47*(14), 8066–8073.

IBM. (2021). *Smarter cities: Transforming urban sustainability with AI and data analytics*. https://www.ibm.com/smarter-cities

Intergovernmental Panel on Climate Change (IPCC). (2021). *Climate change 2021: The physical science basis*. Cambridge University Press.

International Renewable Energy Agency (IRENA). (2021). *Renewable energy statistics 2021*. IRENA.

International Renewable Energy Agency (IRENA). (2023). *Innovations in digital energy solutions*. Retrieved from https://www.irena.org

IRENA. (2021). *Renewable energy and jobs – Annual review 2021*. International Renewable Energy Agency. https://www.irena.org/Publications/2021/Jun/Renewable-energy-and-jobs-Annual-review-2021

Jambeck, J. R., Geyer, R., Wilcox, C., Siegler, T. R., Perryman, M., Andrady, A., Narayan, R., & Law, K. L. (2015). Plastic waste inputs from land into the ocean. *Science, 347*(6223), 768–771. DOI: 10.1126/science.1260352 PMID: 25678662

Kassinis, G., Oikonomou, V., & Alexopoulos, C. (2022). Emerging technologies for urban air quality management. *Environmental Science & Policy, 132*, 25–35.

Kharche, S. S., Gaikwad, V. L., & Kaware, D. M. (2020). IoT-based smart irrigation system for water conservation. *Materials Today: Proceedings, 28*(2), 455–460.

Kouhizadeh, M., Saberi, S., & Sarkis, J. (2021). Blockchain technology and the sustainable supply chain: Theoretically exploring adoption barriers. *International Journal of Production Economics, 217*, 164–176. DOI: 10.1016/j.ijpe.2020.107831

Markard, J., Raven, R., & Truffer, B. (2012). Sustainability transitions: An emerging field of research and its prospects. *Research Policy, 41*(6), 955–967. DOI: 10.1016/j.respol.2012.02.013

Mengelkamp, E., Gärttner, J., Rock, K., & Weinhardt, C. (2018). Designing microgrid energy markets: A case study: The Brooklyn Microgrid. *Applied Energy, 210*, 870–880. DOI: 10.1016/j.apenergy.2017.06.054

Mission Innovation. (2021). *Mission Innovation: Accelerating clean energy innovation*. https://www.mission-innovation.net

Mudd, G. M., & Weng, Z. (2012). The environmental costs of cobalt mining in the Democratic Republic of Congo. *Environmental Science & Technology*, *46*(8), 4648–4655.

Pellow, D. N. (2018). Wind energy and wildlife: An overview of the environmental impacts. *Environmental Research Letters*, *13*(3), 035003.

Porter, M. E., & Kramer, M. R. (2011). Creating shared value. *Harvard Business Review*, *89*(1/2), 62–77.

Qaim, M. (2020). The economics of genetically modified crops. *Annual Review of Resource Economics*, *12*, 109–132.

REN21. (2022). *Renewables 2022 global status report*. Retrieved from https://www.ren21.net

Rockström, J., Steffen, W., Noone, K., Persson, Å., Chapin, F. S.III, Lambin, E., Lenton, T. M., Scheffer, M., Folke, C., Schellnhuber, H. J., Nykvist, B., de Wit, C. A., Hughes, T., van der Leeuw, S., Rodhe, H., Sörlin, S., Snyder, P. K., Costanza, R., Svedin, U., & Foley, J. A. (2009). A safe operating space for humanity. *Nature*, *461*(7263), 472–475. DOI: 10.1038/461472a PMID: 19779433

Rogers, E. M. (2003). *Diffusion of innovations* (5th ed.). Free Press.

Rolnick, D., Donti, P. L., Kaack, L. H., Kochanski, K., Lacoste, A., Sankaran, K., Ross, A. S., Milojevic-Dupont, N., Jaques, N., Waldman-Brown, A., Luccioni, A. S., Maharaj, T., Sherwin, E. D., Mukkavilli, S. K., Kording, K. P., Gomes, C. P., Ng, A. Y., Hassabis, D., Platt, J. C., & Bengio, Y. (2019). Tackling climate change with machine learning. *ACM Computing Surveys*, *55*(2), 1–36. DOI: 10.1145/3485128

Saberi, S., Kouhizadeh, M., Sarkis, J., & Shen, L. (2019). Blockchain technology and its relationships to sustainable supply chain management. *International Journal of Production Research*, *57*(7), 2117–2135. DOI: 10.1080/00207543.2018.1533261

Sovacool, B. K., Schwanen, T., & Sorrell, S. (2019). Transforming energy systems: A critical analysis of global clean energy collaboration. *Global Environmental Change*, *59*, 101987.

Tilman, D., Cassman, K. G., Matson, P. A., Carpenter, S. R., & Dixon, J. (2011). Agricultural sustainability and intensive production practices. *Nature*, *418*(6898), 671–677. DOI: 10.1038/nature01014 PMID: 12167873

United Nations Environment Programme (UNEP). (2020). *Emissions gap report 2020*. UNEP.

United Nations (UN). (2021). *The Sustainable Development Goals report 2021*. United Nations Publications.

Upadhyay, A., Mukhuty, S., Kumar, V., & Kazancoglu, Y. (2021). Blockchain technology and the circular economy: Implications for sustainability and social responsibility. *Technological Forecasting and Social Change*, *168*, 120748.

Van der Meer, D., & Tiwari, A. (2017). Smart technologies for water conservation: Opportunities and challenges. *Environmental Management*, *60*(5), 775–785.

WHO. (2018). Ambient air pollution: A global assessment of exposure and burden of disease. World Health Organization. https://www.who.int/news-room/fact-sheets/detail/ambient-air-pollution

Wolfert, S., Ge, L., Verdouw, C., & Bogaardt, M. J. (2017). Big data in smart farming – A review. *Agricultural Systems*, *153*, 69–80. DOI: 10.1016/j.agsy.2017.01.023

World Health Organization (WHO). (2020). *Air pollution and child health: Prescribing clean air*.

Zengler, T. (2020). Tesla's innovation in sustainable technology. *Journal of Sustainable Business*, *32*(1), 42–50.

Zhang, Y., Ren, S., Liu, Y., & Si, S. (2019). A big data analytics architecture for cleaner manufacturing and maintenance processes of complex products. *Journal of Cleaner Production*, *228*, 451–467.

Zhang, Y., Wang, L., & Zhang, Z. (2020). Real-time air quality monitoring using IoT and cloud computing. *Sensors (Basel)*, *20*(24), 6964.

KEY TERMS AND DEFINITIONS

Disruptive Technologies: Disruptive technologies refer to innovations that significantly alter or revolutionize industries by introducing new methods, processes, or products that displace established systems. In the context of environmental excellence, these technologies contribute to sustainable practices by reducing resource consumption, minimizing waste, and improving efficiency. Examples include renewable energy technologies like solar and wind power, as well as smart grid systems and electric vehicles.

Environmental Excellence: Environmental excellence involves the commitment to maintaining or improving the health of the planet through sustainable practices, technologies, and policies. It encompasses minimizing ecological footprints, reducing pollution, conserving natural resources, and promoting green innovations that positively impact ecosystems and society. The pursuit of environmental excellence is essential to achieving long-term sustainability and mitigating environmental challenges like climate change and biodiversity loss.

Circular Economy: A circular economy is an economic system focused on minimizing waste and making the most of available resources. Unlike the traditional linear economy, which follows a "take, make, dispose" model, the circular economy promotes the reuse, recycling, and repurposing of materials to extend their lifecycle. Disruptive technologies that support circular economy principles include advanced recycling techniques, product design for disassembly, and waste-to-energy technologies, which all contribute to reducing environmental harm.

Renewable Energy Technologies: Renewable energy technologies harness natural resources like sunlight, wind, and water to generate energy, offering an alternative to fossil fuels. These technologies are critical to reducing greenhouse gas emissions, combating climate change, and advancing environmental excellence. Solar panels, wind turbines, hydropower, and bioenergy are key examples that significantly reduce reliance on non-renewable energy sources and promote sustainable development.

Artificial Intelligence (AI) for Environmental Management: Artificial Intelligence (AI) refers to the simulation of human intelligence in machines, allowing them to analyze data, make decisions, and optimize processes. In environmental management, AI helps track and predict pollution levels, optimize energy consumption, and manage natural resources more efficiently. For example, AI algorithms can be used to improve waste management systems, analyze air quality, and predict environmental trends, aiding in the development of effective sustainability strategies.

IoT (Internet of Things) in Environmental Monitoring: The Internet of Things (IoT) refers to a network of interconnected devices that collect, exchange, and analyze data in real-time. In environmental management, IoT enables the monitoring of various environmental parameters such as air quality, water quality, and energy usage. By deploying IoT sensors in urban environments, industries can gain immediate insights into pollution levels, resource consumption, and ecosystem health, allowing for more responsive and data-driven decision-making in pursuit of environmental excellence.

Chapter 6

Energy Transition and Sustainability in Tunisia Towards a Resilient and Low-Carbon Economic Model

Yasser Amayed
https://orcid.org/0009-0003-3154-3454
Higher Institute of Business Administration, Gafsa, Tunisia

ABSTRACT

The energy transition is a strategic challenge for Tunisia, a country facing multiple issues related to its dependence on fossil fuels, the need to reduce greenhouse gas (GHG) emissions, and its commitments to the Sustainable Development Goals (SDGs), particularly SDG 7 on access to affordable and clean energy and SDG 13 on climate action. This article empirically analyzes the impact of energy transition policies on CO_2 emissions and economic sustainability in Tunisia through an ARDL (Auto Regressive Distributed Lag) model. The results show that in the short term, economic growth amplifies emissions, revealing a strong dependence on fossil fuels. In the long term, renewable energy consumption plays a significant role in reducing emissions, while total energy consumption remains the main driver of GHG increases. These results highlight the need to intensify efforts to diversify the energy mix, promote investments in sustainable infrastructure, and strengthen institutional frameworks to foster an inclusive and sustainable energy transition.

DOI: 10.4018/979-8-3373-1822-6.ch006

1. INTRODUCTION

Energy transition today stands as an unavoidable strategic issue for countries facing dual pressures: the need to meet international climate commitments and the necessity of ensuring sustainable economic development. For Tunisia, this challenge is amplified by its historical dependence on fossil fuels, a vulnerability exacerbated by fluctuations in global energy markets and internal budget constraints. In this context, transitioning to an energy model incorporating renewable sources is not only an environmental priority but also an opportunity for economic and social transformation.

Tunisia's energy transition process is part of a global framework that highlights the complexity of interactions between economic, energy, and environmental dynamics. The transition is not limited to the technical substitution of fossil fuels with renewable alternatives; it involves profound transformations in public policies, energy infrastructure, governance systems, and the behavior of economic and social actors. This change, while ambitious, is essential to meeting the Sustainable Development Goals (SDGs), particularly SDG 7, which aims to ensure universal access to clean and affordable energy, and SDG 13, dedicated to combating climate change.

As a developing country, Tunisia illustrates the tensions inherent in this transition: on one hand, the imperative of economic growth to meet social aspirations and reduce inequalities; on the other, the need to reduce GHG emissions and diversify the energy mix to lessen dependence on fossil fuels. Tunisia's current dependence on conventional energy sources presents significant challenges, particularly in terms of environmental costs and economic instability. However, it also offers opportunities to reinvent a more resilient and inclusive development model.

This article explores the dynamics of Tunisia's energy transition through an empirical analysis of the policies implemented and their impacts. Using recent macroeconomic and energy data, this study examines how energy policies influence CO_2 emissions, economic growth, and, more broadly, the country's environmental sustainability. To do so, the study employs a rigorous methodological framework based on an ARDL (Auto Regressive Distributed Lag) model, which captures both short-term and long-term effects of key variables, such as GDP per capita, gross fixed capital formation, total energy consumption, and renewable energy consumption.

This research questions the effectiveness of Tunisian energy policies in reducing GHG emissions while supporting sustainable economic development. More specifically, it aims to understand to what extent these policies manage to reconcile energy transition goals with economic growth, integrating renewable energies within an economic and institutional context marked by multiple constraints.

The primary objective of this article is to provide a detailed, data-driven analysis to evaluate the effectiveness of current energy policies and identify strategic levers for action. Through a pragmatic approach, this research aims not only to inform political and economic choices but also to propose avenues for improvement to accelerate the transition to a sustainable and inclusive energy model in Tunisia. The following sections will present the adopted methodological framework, the results of the analysis, and their implications for the country's energy and economic development.

2. LITERATURE REVIEW

Recent empirical literature on energy transition and its link to sustainable development in Tunisia addresses a series of economic, environmental, and institutional challenges that influence the success of transition policies. Saadaoui and Chtourou (2022) provide an initial exploration of the relationship between institutional quality, financial development, and renewable energy consumption. They show that institutional quality and economic growth foster the adoption of renewable energy, while financial development, in its current form, hinders this transition. This finding underscores the need to revise financial mechanisms to encourage more sustainable investments and align financial flows with environmental goals. Their study also highlights the need for strong institutions to support these reforms and ensure the stability of energy policies.

Continuing this line of research, Saadaoui and Omri (2023) focus on the impact of foreign direct investments (FDI) and information and communication technologies (ICT) on Tunisia's energy transition. They reveal that FDI plays a central role in facilitating the diffusion of renewable energy, but that ICT and trade openness seem to hinder this dynamic. This result may be explained by the economic and political disruptions of recent years, which have slowed the adoption of new technologies. In this sense, it appears that the energy transition requires not only investments but also political stability and economic reforms to maximize the positive impact of emerging technologies in the energy sector.

Additionally, Saadaoui et al. (2024) further explore the institutional and technological factors influencing the transition to renewable energy in Tunisia. Their study shows that technological innovation is a key driver of the energy transition, but that political instability significantly reduces these efforts. They demonstrate that a decline in governmental stability or democratic quality hampers progress in renewable energy. Furthermore, the study reveals that positive GDP variations stimulate the adoption of clean energy, while fossil fuel consumption acts as a hindrance, highlighting the need to reform economic and energy policies to promote a shift towards a more sustainable energy model.

Another important aspect of the energy transition in Tunisia concerns concrete projects implemented to support this transformation. Omri et al. (2015) examine the impact of the PROSOL program, an initiative aimed at promoting solar heating as part of a broader effort to integrate renewable energy into Tunisia's energy mix. This program has helped increase the use of renewable energy while fostering job creation and reducing CO_2 emissions. It serves as a tangible example of the economic and environmental benefits that can result from a well-structured policy supported by public-private partnerships.

From a regional perspective, Dadashi et al. (2022) provide a comparative viewpoint, emphasizing that while Tunisia has made progress in renewable energy development, it faces similar challenges as other Arab countries, particularly those heavily dependent on fossil fuels. Their results show that despite the high potential for solar and wind energy in Arab countries, significant obstacles remain, including insufficient investments and often incoherent long-term policies. These findings highlight the importance of a coordinated approach to strengthen the regional capacity to adopt renewable energy.

Ben Youssef et al. (2024) make an important contribution by examining the relationship between economic growth and CO_2 emissions in Tunisia. Using a robust analytical framework, they identify the main drivers of emissions and highlight the challenges posed by population growth and energy consumption. This study underscores the need to reconcile economic growth with emission reduction requirements, which implies a decisive shift toward renewable energy sources to meet sustainable development goals.

The role of institutions is further explored by Khaled et al. (2024), who emphasize the importance of institutional dynamics in the energy transition in the MENA region. Their study identifies institutional barriers, such as ineffective governance and socio-cultural norms, which hinder the adoption of renewable energy. They propose a strategic framework to overcome these obstacles, relying on collaboration between governments, industrial actors, and international partners. This framework emphasizes the need for a just and sustainable energy transition, particularly in developing countries like Tunisia.

Trabelsi's (2024) study strengthens this analysis by emphasizing the importance of environmental sustainability for economic growth in Tunisia. Using the ARDL approach, he shows that environmental sustainability has a positive effect on long-term economic growth, validating the importance of a green energy transition for the country's future development. By integrating sustainability into economic policies, Tunisia could not only reduce its CO_2 emissions but also strengthen its economy and create employment opportunities in sectors related to clean energy.

Furthermore, the analysis by Dahmani et al. (2021) highlights the effect of trade policies and technological innovations on greenhouse gas (GHG) emissions in Tunisia. They show that increasing the use of renewable energy contributes to long-term GHG reduction, while emphasizing the need for reforms in the energy sector to improve energy efficiency and promote investments in clean energy. Their findings call for structural reforms and incentivizing initiatives to encourage private investment and accelerate the adoption of renewable energy.

Finally, Ghorbal et al. (2021) in-depth analysis highlights the impact of national patents and foreign direct investment on environmental sustainability in Tunisia. Their study reveals that innovation, as measured by national patents, plays a central role in reducing CO_2 emissions, while FDI, by bringing advanced technologies, also contributes to this goal. They conclude that the development of renewable energy, combined with policies that support innovation, can have a significant impact on improving environmental sustainability in Tunisia.

Thus, the literature highlights the complexity of energy transition policies in Tunisia. The institutional, financial, technological, and environmental challenges are interconnected, and the success of this transition depends on the country's ability to reform its economic structures, attract investment, and support innovation. The energy transition represents not only a challenge but also an opportunity for sustainable economic development and climate change mitigation in Tunisia.

3. DATA AND METHODOLOGY OF THE STUDY

3.1. Measurement of Variables and Data Sources

This study uses key variables to examine the impact of energy transition policies on CO_2 emissions, economic growth, and sustainable development in Tunisia. These variables were selected based on their relevance to assessing economic, environmental, and energy trends within the framework of the Sustainable Development Goals (SDGs). Table 1 presents the descriptive statistics of the variables used in the ARDL model, with data primarily sourced from the World Bank's WDI database (2024), covering several decades.

The variables used are as follows:

- **CO_2 emissions per capita (CO2_PC)**: This variable represents CO_2 emissions per person in millions of metric tons, a key indicator for assessing the environmental impact of economic growth and energy consumption. On average, CO_2 emissions per capita in Tunisia are 1.722 million metric tons, with significant variation over time. The maximum emissions reached 2.122 million metric tons, while the minimum was 1.262 million metric tons, reflecting efforts to reduce emissions.

- **GDP per capita (GDP_PC)**: The Gross Domestic Product (GDP) in constant 2015 US dollars is used to measure Tunisia's economic growth. The average GDP per capita is 30.1 billion US dollars, with a range between 12.4 billion and 49.3 billion, illustrating periods of growth and economic slowdown.
- **Gross fixed capital formation (FBCF_PC)**: This variable measures the investment in infrastructure and other productive assets, expressed in billions of constant 2015 US dollars. The average investment is 6.26 billion, with a maximum of 12.6 billion. This factor plays a crucial role in the development of energy infrastructure, particularly renewable energy.
- **Total energy consumption (CET)**: Expressed in millions of tonnes of oil equivalent (Mtoe), total energy consumption in Tunisia is an essential indicator for understanding energy demand and its impact on CO_2 emissions. On average, energy consumption is 7.041 Mtoe, but it peaks at 10.483 Mtoe.
- **Renewable energy consumption (CER)**: This variable, expressed in millions of tonnes of oil equivalent, measures the share of renewable energy in the total energy mix. Although renewable energy consumption remains low in Tunisia, with an average of 0.063 Mtoe, it has seen significant increases over time, reaching a maximum of 0.279 Mtoe.

Table 1. Descriptive statistics for these variables

Variable	CO2_PC	PIB_PC	FBCF_PC	CET	CER
Description	CO_2 Emissions (million metric tons of carbon dioxide)	GDP (constant 2015 US$ billions)	Gross Fixed Capital Formation (constant 2015 US$ billions)	Total Energy Consumption (million tons of oil equivalent)	Renewable Energy Consumption (million tons of oil equivalent)
MEAN	1,722	30,000	6,000	7,041	0,063
MEDIAN	1,703	29,000	6,000	7,419	0,058
Maximum	2,122	49,000	13,000	10,483	0,279
Minimum	1,262	12,000	2,000	3,239	0,008
Std Dev	0,251	13,00	3,00	2,271	0,046
Jarque-Bera	3,033	4,157	4,173	3,137	2,701
Porb	0,219	0,125	0,124	0,208	0,365
Observations	43	43	43	43	43

The data show significant variability across different years and illustrate the trends in energy consumption and CO_2 emissions in Tunisia, which are essential for understanding how these variables influence the energy transition and the achievement of SDGs.

Figure 1 presents the historical data on energy intensity and CO_2 emissions per capita in Tunisia. These indicators, measured over a period of more than four decades, show the evolution of energy consumption in tonnes of oil equivalent per person as well as the share of renewable energy in the national energy mix. The data also highlight the correlation between CO_2 emissions and economic growth, illustrated by constant GDP and gross fixed capital formation.

Figure 1. Evolution of energy intensity, CO_2 emissions, and GDP in Tunisia (1980-2022)

Source: Author's estimates using Eviews 12 software.

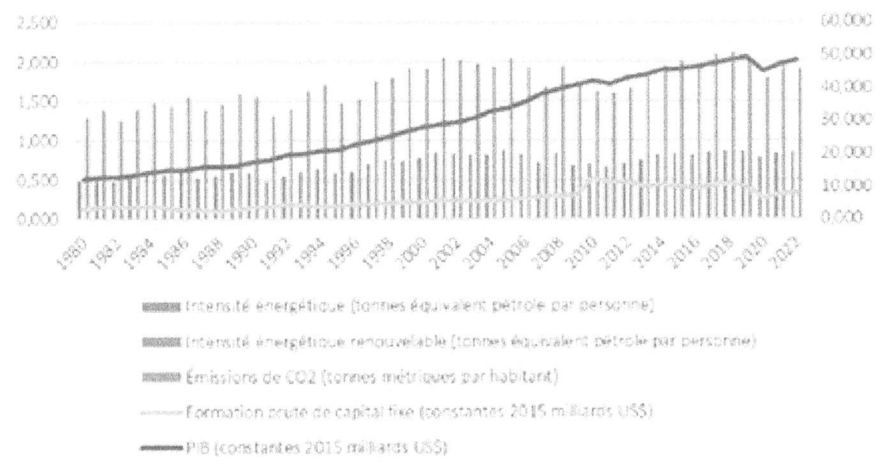

The graphs show an upward trend in CO_2 emissions over the decades, correlated with economic growth, while renewable energy intensity remains relatively low despite some recent progress. This reflects the need to intensify efforts to adopt renewable energy sources and reduce dependence on fossil fuels.

$$Y = F(K,X) = A K^{\alpha_1} X^{\alpha_i} \tag{1}$$

In this model:

- Y represents **real GDP per capita**(*PIB_PC*) a key indicator of the country's economic performance.
- K represents **gross fixed capital formation**(*FBCF_PC*) capturing the capital stock and investments made in the economy.
- X represents a set of **additional variables**, including energy policies and environmental factors such as **renewable energy consumption** (CER), **total energy consumption** (CET), and **CO_2 emissions per capita**(CO_2_PC) which are used to analyze the energy transition in Tunisia.
- A is a **technological indicator**, assumed to be constant in this formulation, capturing technological progress that affects the economy's overall productivity.
- α_1 and α_i represent the **elasticities of capital** and other explanatory variables, indicating the relative impact of each factor on GDP and emissions.

This model helps analyze how the combined influence of these variables affects the energy transition and economic sustainability of the country, while also considering environmental objectives.

3.2.2. Extended Formulation of the Model

By disaggregating relation (1), we obtain a more detailed equation that allows us to isolate the effects of different variables on GDP, while taking into account environmental and energy factors:

$$Y_t = A\prod_{i=2}^{5} K_t^{\alpha_1} X_t^{\alpha_i} \tag{2}$$

This functional equation can be reformulated in logarithmic form to facilitate the interpretation of elasticities and simplify the econometric estimation:

$$\log\left(Y_t\right) = \log(A) + \alpha_1 \log\left(K_t\right) + \sum_{i=2}^{5} \alpha_i \log\left(X_t\right) \tag{3}$$

Here, we examine the relationships between the following variables:

- **CO_2_PC:** CO_2 emissions per capita, as a key indicator of environmental sustainability.
- **PIB_PC:** Gross Domestic Product per capita, measured in constant 2015 US$, an indicator of economic performance.

- **FBCF_PC:** Gross fixed capital formation per capita, measured in constant 2015 US$, representing investment.
- **CET:** Total energy consumption, in millions of tonnes of oil equivalent, capturing the energy intensity of the economy.
- **CER:** Renewable energy consumption, in millions of tonnes of oil equivalent, reflecting the integration of clean energy into the energy mix.

3.2.3. ARDL Model (Auto Regressive Distributed Lag)

The Auto Regressive Distributed Lag (ARDL) model is particularly suitable for this study because it allows for modeling both short- and long-term dynamics between the variables, even when they are not integrated of the same order (i.e., I(0) or I(1)).

$$\begin{aligned}
\Delta \log(CO2_PC_t) &= \alpha_0 + \sum_{i=1}^{p}\beta_1\Delta \log(CO2_PC_{t-i}) + \sum_{i=1}^{q}\beta_2\Delta \log(PIB_PC_{t-i}) \\
&+ \sum_{i=1}^{q}\beta_3\Delta \log(FBCF_PC_{t-i}) + \sum_{i=1}^{q}\beta_4\Delta \log(CET_{t-i}) + \sum_{i=1}^{q}\beta_5\Delta \log(CER_{t-i}) \\
&+ \theta_1 \log(CO2_PC_{t-1}) + \theta_2 \log(PIB_PC_{t-1}) + \theta_3 \log(FBCF_PC_{t-1}) \\
&+ \theta_4 \log(CET_{t-1}) + \theta_5 \log(CER_{t-1}) + \varepsilon_t
\end{aligned} \tag{4}$$

In this equation:

- The coefficients represent the short-term dynamics, capturing the effects of past variations in the explanatory variables on CO_2 emissions.
- The coefficients capture the long-term relationships between CO_2 emissions and the explanatory variables.
- ε is the random error term.

3.2.4. Error Correction Model (ECM)

In the case where a cointegration relationship is identified between the variables, an Error Correction Model (ECM) will be estimated. This model will allow for examining how short-term imbalances between the variables adjust to restore a long-term equilibrium.

The equation for the ECM model is as follows:

$$\Delta \log(CO2\_PC_t) = \alpha_0 + \sum_{i=1}^{p} \beta_1 \Delta \log(CO2\_PC_{t-i}) + \sum_{i=1}^{q} \beta_2 \Delta \log (PIB\_PC_{t-i}) + \sum_{i=1}^{q} \beta_3 \Delta \log(FBCF\_PC_{t-i}) + \sum_{i=1}^{q} \beta_4 \Delta \log(CET_{t-i}) + \sum_{i=1}^{q} \beta_5 \Delta \log(CER_{t-i}) + \lambda \cdot ECM_{t-1} + \varepsilon_t \quad (5)$$

- **α** is the error correction coefficient, which indicates the speed at which the system returns to equilibrium after a shock.
- **ECT** is the cointegration error term, capturing the long-term deviations.

3.2.5. Statistical Tests

To ensure the robustness of the estimated model, several statistical tests will be conducted:

1. **Stationarity Test**: The Augmented Dickey-Fuller (ADF) unit root test by Pesaran (2007) will be used to assess the stationarity of the series in order to determine the degree of integration of each variable.
2. **Cointegration Test**: The bounds test by Pesaran et al. (2001) will be used to check for a long-term relationship between the variables.
3. **Diagnostic Tests**: Tests for residual normality, autocorrelation, and heteroscedasticity will be applied to assess the model's goodness-of-fit and the stability of the estimated coefficients.

3.2.6. Expected Results

The expected results from the ARDL model should indicate long-term relationships between investment (FBCF_PC) and CO_2 emissions (CO2_PC), while considering the effects of renewable energy and total energy consumption. The following is anticipated:

- An increase in investment in renewable energy and green technologies should have a negative effect on CO_2 emissions, thus contributing to achieving climate goals.
- GDP per capita should have a positive effect on emissions, as an increase in economic activity may lead to higher energy demand, although this effect can be mitigated by sustainable energy policies.
- An increase in renewable energy consumption (CER) should be associated with a reduction in CO_2 emissions, in line with energy transition goals.

4. RESULTS AND DISCUSSION

4.1. Stationarity Analysis: ADF Unit Root Test by Pesaran (2007)

The results of the ADF unit root tests applied to the different variables of the model, including CO_2 emissions per capita (CO_2_PC), GDP per capita (PIB_PC), gross fixed capital formation per capita (FBCF_PC), total energy consumption (CET), and renewable energy consumption (CER), are presented in Table 2. These tests are essential for assessing the integration level of the time series, a necessary condition for applying the ARDL model. Indeed, the ARDL model requires variables to be integrated of order I(0) or I(1), and not of order I(2), as this ensures the robustness of both short- and long-term estimations. The results of the ADF unit root test are shown in Table 2. They indicate that some variables are stationary at level (PIB_PC), while others become stationary after applying a difference (CO_2_PC, FBCF_PC, CET, CER). This confirms that these variables are integrated of order I(1), justifying the use of an ARDL model to analyze the dynamic relationships between these variables.

Table 2. Unit root test results (ADF)

Variables (log)	Level		First Difference		Order
	Without trend	With trend	Without trend	With trend	
CO2_PC	-2,227	-3,005	-6,545***	-6,638***	I(1)
PIB_PC	-3,873***	-0,487	-6,399***	-7,025***	I(0)
FBCF_PC	-1,273	-2,161	-6,715***	-6,633***	I(1)
CET	-1,261	-1,792	-6,699***	-7,768***	I(1)
CER	-1,665	-1,355	-7,188***	-7,307***	I(1)

Note: *** indicates a significance level of 1%, where the null hypothesis of unit root is rejected.

According to Table 2, the variables CO_2_PC, FBCF_PC, CET, and CER show clear signs of stationarity after first differencing, classifying them as integrated of order 1 (I(1)). On the other hand, PIB_PC is stationary at level, or of order I(0), indicating that no differencing is necessary for this variable. This diagnosis justifies the application of the ARDL model to model the dynamic relationships both in the short and long term between these variables, thus enabling a better understanding of how energy transition policies affect CO_2 emissions and economic implications in Tunisia.

4.2. Cointegration Analysis: Cointegration Test by Pesaran et al. (2001)

The bounds test for cointegration by Pesaran et al. (2001) is used to check for a long-term relationship between the variables included in the model. This step is crucial, as the absence of cointegration would imply that there are no long-term relationships between the variables, rendering the long-term coefficient estimation irrelevant. The cointegration test is based on estimating a prior ARDL model, followed by comparing the Fisher F-statistic with the critical values, defining the lower and upper bounds.

4.2.1. Determining the Optimal Lag Number using the Akaike Information Criterion (AIC)

The ARDL model is first estimated using the Akaike Information Criterion (AIC) to select the optimal number of lags. This criterion allows for the selection of the model that maximizes data fit while minimizing the number of parameters to estimate. Figure 2 shows that the optimal ARDL model in our analysis is the ARDL (1,0,2,1,1) model, which has the lowest Akaike value among the 20 other models tested. This model will thus be used to examine the long-term relationships between CO_2 emissions, GDP, FBCF, total energy consumption, and renewable energy consumption.

Figure 2. Determination of the optimal number of delays

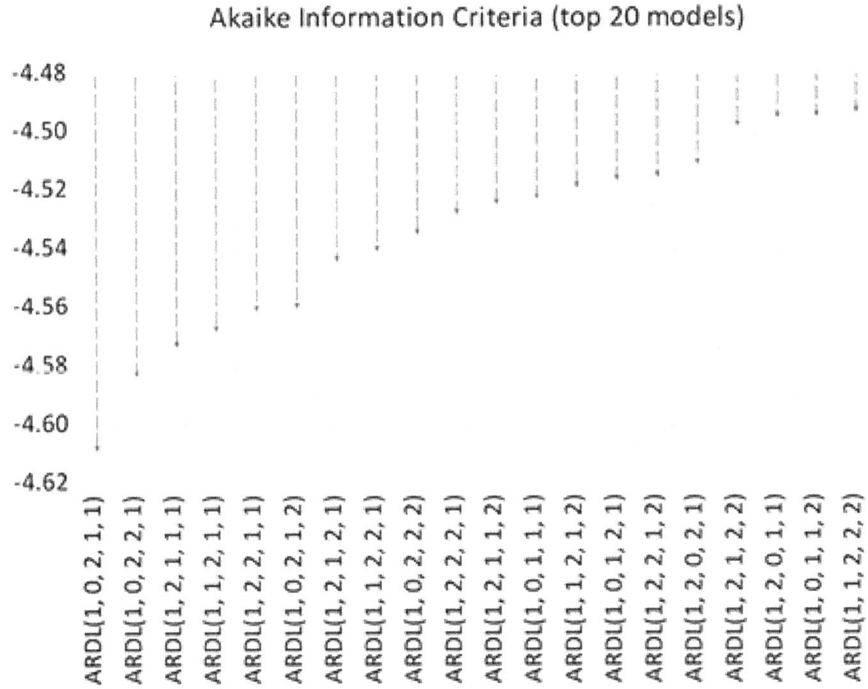

4.2.2. Bounds Test for Cointegration

Once the optimal ARDL model is determined, the bounds test for cointegration is performed. The Fisher F-statistic is then compared with critical values to determine the existence of a long-term relationship between the variables.

The results in Table 3 show that the calculated F-statistic (4.013174) is higher than the upper bound (4.01) at the 2.5% significance level. This indicates the presence of a cointegration relationship among the studied variables, confirming that they move together in the long run. In other words, Tunisia's transition policies toward renewable energy have a lasting impact on CO_2 emissions, GDP, gross fixed capital formation, and energy consumption (both total and renewable).

Table 3. Cointegration test results

Variables	CO_2_PC ; GDP_PC ; GFCF_PC ; TCE ; REC
Calculated F-statistic	**4.013174**
Significance Level	**Lower Bound I(0)**
1%	2.45
2.5%	2.86
5%	3.25
10%	3.74

Source: Author's estimates using Eviews 12.

These results confirm the validity of estimating long-term coefficients using the ARDL model. We can thus proceed with analyzing the effects of different variables on CO_2 emissions and economic implications, considering both short-term and long-term dynamics.

4.3. Estimation: ARDL Model

4.3.1. Short-Term Relationship Results

The results in Table 4 indicate that the overall performance of the estimated ARDL model is satisfactory, with most variables exhibiting the expected short-term effects on CO_2 emissions in Tunisia. Most independent variables have a significant impact, suggesting that current energy and economic policies directly influence CO_2 emissions in the short term.

- **LOG(GDP_PC)** (GDP per capita) has a positive and significant effect, suggesting that, in the short term, economic growth leads to an increase in CO_2 emissions. This relationship is consistent with the idea that early stages of economic development are generally accompanied by increased industrial activity, often fueled by fossil energy sources.
- **LOG(GFCF_PC)** (Gross Fixed Capital Formation) shows a positive effect on CO_2 emissions when lagged by two periods. This reflects the delayed impact of infrastructure investments, which, while essential for economic growth, can also temporarily increase non-renewable energy consumption, thereby raising emissions.
- **LOG(TCE)** (Total Energy Consumption) has a highly significant and positive impact on CO_2 emissions, indicating that overall energy consumption, mainly from non-renewable sources, largely contributes to rising emissions in the short term.

- **LOG(REC)** (Renewable Energy Consumption), on the other hand, has a negative and significant effect when lagged by one period, suggesting that the growing adoption of renewable energy is beginning to moderate CO_2 emissions in the short term. However, its direct effect in the current period remains weak, indicating that it takes time for renewable energy investments to yield tangible emission reductions.

Finally, the **Error Correction Term (ECT)** is negative and significant, confirming the existence of a long-term cointegration relationship among the variables. This means that deviations from the long-term equilibrium are gradually corrected, ensuring long-term convergence.

Table 4. Short-term ARDL model estimation results

Variable	Coefficient	Error	Statistic t	Prob
LOG(PIB_PC)	0,299	0,104	2,871	0,007
LOG(FBCF_PC)	0,025	0,040	0,630	0,533
LOG(FBCF_PC(-1))	0,135	0,040	3,382	0,002
LOG(FBCF_PC(-2))	0,065	0,031	2,110	0,043
LOG(CET)	0,988	0,051	19,271	0,000
LOG(CET(-1))	0,481	0,157	3,054	0,005
LOG(CER)	0,021	0,008	2,550	0,016
LOG(CER(-1))	-0,042	0,010	-4,087	0,000
ECT	-0,537	0,149	-3,605	0,001

4.3.2. Long-Term Relationship Results

The long-term results, presented in Table 5, confirm that most of the studied variables significantly influence CO_2 emissions in Tunisia.

- **LOG(GDP_PC)** continues to exhibit a positive and significant effect, highlighting that long-term economic growth leads to an increase in CO_2 emissions. This result aligns with the observed link between economic growth and energy consumption in developing countries, where a large portion of energy consumption still comes from fossil fuel sources.
- **LOG(GFCF_PC)**, representing investment, has no significant long-term effect on CO_2 emissions. This result may indicate that while investment is

essential for economic growth, it does not have a direct impact on increasing or reducing emissions in the long-term Tunisian context.

- **LOG(TCE)**, total energy consumption, has a positive and highly significant effect on long-term CO_2 emissions. This shows that Tunisia's energy consumption, mainly from non-renewable sources, is the primary driver of CO_2 emissions.
- **LOG(REC)**, renewable energy consumption, has a negative and significant effect on long-term CO_2 emissions. This result confirms that increasing the share of renewable energy in the energy mix is essential for reducing long-term CO_2 emissions. It highlights the importance of policies promoting renewable energy to achieve Tunisia's climate goals.

Table 5. Long-term ARDL model estimation results

Variable	Coefficient	Error	Statistic t	Prob
LOG(PIB_PC)	0,647	0,135	4,791	0,000
LOG(FBCF_PC)	0,097	0,070	1,379	0,178
LOG(CET)	1,096	0,124	8,853	0,000
LOG(CER)	-0,044	0,019	-2,301	0,028

The empirical results derived from the ARDL model, analyzed in both the short-run and long-run frameworks, provide important insights into the dynamics between Tunisia's energy policies and CO_2 emissions. In the short term, the findings suggest that economic growth and the continued reliance on non-renewable energy sources significantly contribute to the increase in carbon emissions, highlighting the environmental cost of conventional energy-driven development. However, the long-term estimates present a more optimistic outlook. They reveal that the strategic adoption and gradual integration of renewable energy sources into the national energy mix hold substantial potential in curbing emissions and mitigating environmental harm. This indicates that, although short-term challenges persist, Tunisia's transition toward sustainable energy can play a pivotal role in reducing its carbon footprint over time. The results underscore the necessity for consistent and well-implemented renewable energy policies to ensure long-term environmental and economic sustainability.

4.4. Diagnostic and Robustness Tests

Diagnostic and robustness tests play a crucial role in validating the reliability and credibility of any econometric model, particularly the Auto-Regressive Distributed Lag (ARDL) model. These tests are fundamental to ensuring that the model is correctly specified and that the underlying assumptions, especially those related to the residuals and overall model stability, are not violated. In the context of this study, which investigates the influence of Tunisia's transition policies toward renewable energy on CO_2 emissions and their broader economic implications, various diagnostic and robustness tests were applied. These included tests for serial correlation, heteroscedasticity, normality of residuals, and parameter stability. The detailed results and interpretations of these tests are presented comprehensively in Table 6, confirming the soundness of the model and the robustness of the empirical findings.

4.4.1. Breusch-Pagan-Godfrey Test for Heteroskedasticity

To examine whether the ARDL model suffers from heteroskedasticity, we employed the Breusch-Pagan-Godfrey test, a well-established diagnostic tool designed to detect non-constant variance in the residuals. The test yielded an F-statistic of 0.875 and an associated p-value of 0.216. Since this probability is well above the conventional 5% level of significance, we fail to reject the null hypothesis, which assumes homoscedasticity—that is, the residuals have constant variance.

This result indicates that the ARDL model does not exhibit heteroskedasticity, and thus, the assumption of equal variance across observations holds. The presence of homoscedasticity further supports the reliability of the model's estimates and strengthens the confidence in the validity of the conclusions drawn. Overall, this finding adds to the body of diagnostic evidence confirming the soundness of the model.

4.4.2. Breusch-Godfrey Test for Residual Autocorrelation

Residual autocorrelation is a frequent concern in dynamic econometric models, as its presence can lead to biased coefficient estimates and unreliable inference, ultimately compromising the validity of the model's conclusions. To ensure the reliability of the estimated ARDL model, it is essential to test for this issue. In this study, we employed the Breusch-Godfrey serial correlation LM test to detect any signs of residual autocorrelation. The results yielded an F-statistic value of 0.110, accompanied by a probability (p-value) of 0.896. Given that this p-value far exceeds the conventional 5% level of statistical significance, we fail to reject the null hypothesis, which posits the absence of serial correlation in the residuals. This outcome provides strong evidence that the residuals are independently and randomly distributed over time. Consequently, the ARDL model is considered properly specified in this regard, and the estimated relationships among the variables are not distorted by serial correlation issues.

4.4.3. Jarque-Bera Normality Test

The Jarque-Bera normality test is a widely used diagnostic tool that evaluates whether the residuals of a regression model are normally distributed—an essential assumption for the validity of many statistical inference procedures, including hypothesis testing and confidence interval estimation. In this study, we applied the Jarque-Bera test to the residuals of the estimated ARDL model to assess the appropriateness of this assumption. The results revealed a Jarque-Bera statistic of 1.680 and an associated probability value of 0.711. Since this probability is significantly higher than the conventional 5% level, we fail to reject the null hypothesis of normality. This outcome suggests that the residuals do not deviate significantly from a normal distribution, thereby confirming the robustness of the model's statistical properties. The normality of residuals ensures the reliability of the t-tests and F-tests applied in the model and enhances confidence in the overall validity of the empirical results.

4.4.4. Ramsey RESET Test for Model Specification

The Ramsey Regression Equation Specification Error Test (RESET) is a widely recognized diagnostic procedure designed to evaluate whether a regression model is correctly specified. It helps to detect possible specification errors, such as omitted variables, incorrect functional form, or neglected nonlinearities in the model. Ensuring the model is well-specified is crucial for drawing accurate inferences from the estimated parameters. In this study, we applied the Ramsey RESET test to the ARDL model to assess its functional correctness. The test produced a statistic of 0.232, accompanied by a probability value of 0.818. Since this p-value is significantly higher than the 5% level of statistical significance, we accept the null hypothesis, which states that the model is correctly specified. This implies that the estimated model does not exclude any critical explanatory variables and that its functional form is appropriate for the dataset used.

In addition to the RESET test, a series of other diagnostic tests were conducted to validate the robustness and reliability of the ARDL model. These include tests for residual autocorrelation, heteroskedasticity, and normality of residuals. The results indicate no evidence of serial correlation or heteroskedasticity, and the residuals are normally distributed. Collectively, these findings confirm that the ARDL model is statistically valid, robust, and well-suited to analyze the relationship between Tunisia's energy transition policies, CO_2 emissions, and economic indicators. As such, the model provides a reliable basis for policy evaluation and academic inquiry.

4.4.5. Summary of Diagnostic Tests

The obtained results (see Table 6) strengthen the validity of the ARDL model's estimates in analyzing the impact of energy transition policies on CO_2 emissions and economic implications in Tunisia.

Table 6. Diagnostic test results for the ARDL model

Variable	Coefficient	Error	Statistic t	Prob
LOG(PIB_PC)	0,647	0,135	4,791	0,000
LOG(FBCF_PC)	0,097	0,070	1,379	0,178
LOG(CET)	1,096	0,124	8,853	0,000
LOG(CER)	-0,044	0,019	-2,301	0,028

4.4.6. Graphical Analysis: Results of Normality and Stability Tests

Figure 3 illustrates the distribution of residuals from the ARDL model. The bell-shaped curve indicates a distribution close to normal, which corroborates the results of the Jarque-Bera test.

Figure 3. Results of the normality test of errors

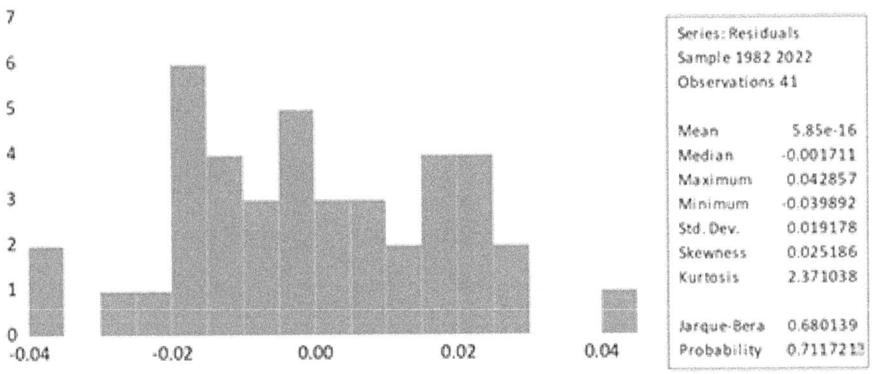

The CUSUM (Cumulative Sum) test is a widely employed technique in time series econometrics used to assess the stability of a model's parameters over time. Specifically, it helps determine whether there have been significant structural changes in the estimated coefficients during the period of analysis. Stability is a critical aspect of model reliability, as unstable parameters may indicate that the relationships among variables change over time, thereby undermining the consistency of the results and their applicability for forecasting or policy evaluation.

In this study, the CUSUM test was applied to the residuals of the ARDL model to examine whether the estimated parameters remain stable throughout the sample period. The results are graphically presented in Figure 4. As illustrated, the CUSUM test curve stays entirely within the 5% significance boundaries, which are represented by the upper and lower confidence limits. This outcome strongly suggests that there are no structural breaks in the model and that the estimated coefficients are stable over time. Such stability reinforces the model's robustness and strengthens the credibility of the empirical findings.

Therefore, the CUSUM test provides additional support for the appropriateness of the ARDL model used in this study. Its confirmed parameter stability enhances confidence in the model's validity and its potential use for future policy analysis and long-term forecasting.

Figure 4. Results of the CUSUM test

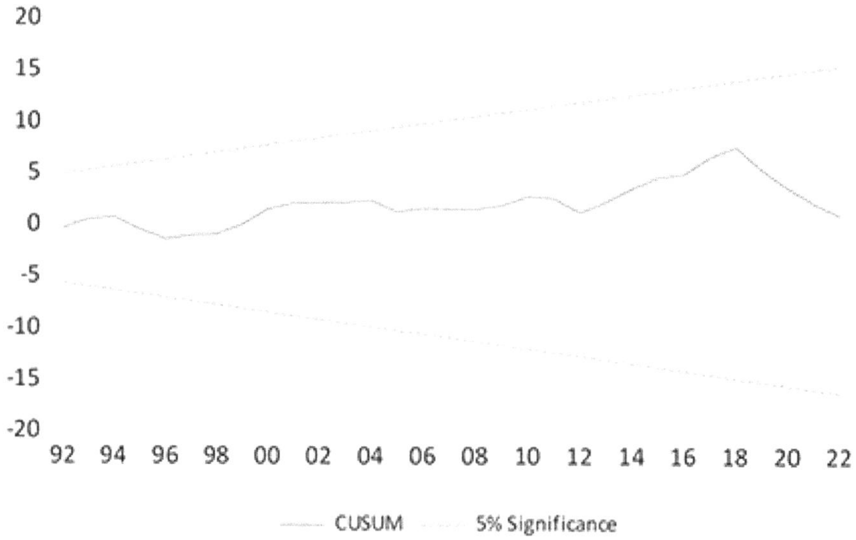

4.5. Discussion of Results and Implications for Tunisia

The empirical results presented in the previous section clearly show that energy transition policies in Tunisia significantly influence the Sustainable Development Goals (SDGs), particularly those related to climate action, clean energy, economic growth, and sustainable infrastructure. The impact of variables such as GDP per capita (GDP_PC), gross fixed capital formation (GFCF), total energy consumption (TEC), and renewable energy consumption (REC) reflects the importance of aligning economic and energy policies to reduce CO_2 emissions while supporting economic development.

The role of economic growth, illustrated by the positive relationship between GDP per capita and CO_2 emissions in both the short and long term, presents a central challenge for Tunisia in its quest to achieve the SDGs. An expanding economy, still largely dependent on fossil fuels, leads to increased emissions, which may compromise the country's climate commitments. This phenomenon is well documented in the literature, notably in studies by Saadaoui and Omri (2023) and Ben Youssef et al. (2024), which emphasize the importance of diversifying energy sources to reduce environmental pressure while maintaining sustainable economic growth.

However, the results related to renewable energy consumption (REC) suggest that an energy strategy focused on the development of clean energy could provide tangible long-term solutions. Indeed, the significant negative effect of REC on CO_2 emissions in the long term confirms that increasing the share of renewable energy in Tunisia's energy mix is a priority for emissions reduction. This finding aligns with the conclusions of Ghorbal et al. (2021), who demonstrated that the integration of renewable energy is a **sine qua non** condition for environmental sustainability in Tunisia.

he infrastructure sector, as captured by gross fixed capital formation (GFCF), demonstrated a relatively modest influence on CO_2 emissions in the short term and no statistically significant effect in the long term. This finding suggests that current infrastructure investments may not be effectively contributing to environmental sustainability. It raises an important consideration for Tunisian policymakers: the need to redirect capital expenditures toward environmentally friendly and climate-resilient infrastructure projects. Aligning infrastructure development with sustainable growth objectives is crucial, particularly in a context where economic expansion is still a priority. The challenge lies in designing policies that simultaneously stimulate economic performance and support Tunisia's transition to a low-carbon economy.

One promising avenue, as highlighted by Saadaoui and Omri (2023), is to increase reliance on foreign direct investment (FDI) to finance green infrastructure and renewable energy initiatives. However, such a strategy must be accompanied by comprehensive institutional reforms to create a stable, transparent, and attractive investment climate capable of drawing long-term international engagement.

Moreover, the results underscore the broader importance of governance quality and political stability in shaping the success of energy transition policies. Saadaoui et al. (2024) emphasized the synergistic role of institutional stability and technological innovation in driving renewable energy adoption. In Tunisia, the political uncertainty that followed the 2011 revolution has often delayed or disrupted the implementation of large-scale energy reforms. Nevertheless, this analysis suggests that when long-term investments in renewable energy are supported by effective governance structures and consistent institutional frameworks, significant reductions in CO_2 emissions are achievable. This reinforces the argument that the energy transition is not only a technological and financial challenge but also a fundamentally political and institutional one.

From an economic perspective, the long-term results concerning total energy consumption (TEC) show that the country's current dependence on fossil fuels remains a major obstacle to achieving Tunisia's climate goals. This conclusion is consistent with the findings of Salem et al. (2023) in their analysis of Gulf Cooperation Council (GCC) countries, where fossil fuel dependency also hinders progress toward a sustainable future. In Tunisia, overcoming this dependency will require a concerted effort to accelerate the adoption of renewable energy and reform the institutional framework.

The main challenge for Tunisia, therefore, is to fully integrate renewable energy into its development strategy while ensuring that economic gains do not systematically translate into increased CO_2 emissions. This is a delicate balance that Tunisia must maintain if it aims to achieve the SDGs, particularly SDG 7, which promotes universal access to clean and affordable energy, and SDG 13, which seeks to combat climate change.

The implications for policymakers are clear. On the one hand, it is necessary to promote investments in sustainable infrastructure that reduce the carbon footprint while stimulating innovation in green technologies. On the other hand, it is equally important to ensure that political governance effectively supports this transition. Research by Saadaoui and Chtourou (2022) suggests that institutional quality is a key factor in successfully implementing this transformation, ensuring that energy reforms are applied consistently and transparently.

To achieve Tunisia's sustainable development goals, it will be essential to continue diversifying the energy mix, encouraging investment in renewable energy, and adopting stable institutional policies. This will require long-term planning and increased cooperation between the public and private sectors, as recommended by Dadashi et al. (2022) in their analysis of energy transition strategies in Arab countries. By capitalizing on its renewable resources, Tunisia can embark on a greener growth model, ensuring both emissions reduction and improved economic conditions for its citizens.

5. CONCLUSION

This research explored the impact of energy transition policies in Tunisia on CO_2 emissions and economic growth, placing them within the framework of the Sustainable Development Goals (SDGs), particularly SDG 7 on access to clean and affordable energy, and SDG 13 related to climate action. The results highlight complex interactions that underscore the tensions between economic development imperatives and environmental requirements. Economic growth continues to be strongly correlated with an increase in CO_2 emissions, reflecting a persistent reliance on fossil fuels. This observation presents a central challenge for Tunisia, where current growth mechanisms are at odds with climate commitments.

However, the study shows that renewable energy offers a promising path for long-term emissions reduction, although their current contribution to the energy mix remains limited. The analysis confirms that their increased integration could significantly mitigate the environmental impact of economic activities, thereby strengthening progress toward the SDGs. Moreover, total energy consumption, still largely dominated by fossil sources, remains the main driver of CO_2 emissions, underscoring the urgency of reforming the national energy structure. On the other hand, investments in infrastructure have not demonstrated a significant impact on emissions in the long term, revealing the need to better direct financial resources toward projects that promote environmental and energy sustainability.

These findings have clear implications for policymakers. Tunisia must not only accelerate the adoption of renewable energy but also rethink its investment models to better align them with climate goals. This could include creating incentive mechanisms to attract national and international funding, such as subsidies for solar and wind projects or tax incentives for green technologies. At the same time, institutional reforms are necessary to ensure a stable and attractive regulatory framework that can reassure investors and support public-private partnerships. Energy planning should also incorporate social objectives, such as equitable access to energy and vocational training, to ensure a just and inclusive energy transition, in line with the aspirations of the SDGs.

Energy efficiency is another priority. Initiatives such as energy audits for large industries and stronger energy performance standards could reduce total energy consumption while increasing the efficiency of industrial and residential processes. Furthermore, it is crucial to support this transformation with awareness campaigns and targeted training to ensure social acceptance of renewable energy and to strengthen the engagement of citizens and businesses in the transition.

This research, while offering valuable insights, is not without limitations. The use of macroeconomic data limits the consideration of sectoral or regional specifics, which could reveal significant disparities in the impacts of energy policies. Moreover, certain non-quantifiable factors, such as recent technological advances or political stability, could not be fully integrated into the analysis, although they play a decisive role in the success of energy strategies. Finally, the very long-term effects of the implemented policies cannot be fully captured within the study's time frame.

Future research perspectives could focus on a more detailed sectoral analysis or the impact of energy subsidies and carbon pricing mechanisms. It would also be relevant to explore the interactions between the energy transition and other dimensions of the SDGs, such as reducing inequalities (SDG 10) or promoting decent jobs (SDG 8). Ultimately, the energy transition in Tunisia represents a strategic opportunity to redefine its economic model while achieving the climate and social commitments outlined by the SDGs. Successfully achieving this transformation will require coordinated efforts, a long-term vision, and increased mobilization of public and private actors, allowing the country to position itself as a regional leader in an inclusive and sustainable energy transition.

REFERENCES

Ben Youssef, A., Dahmani, M., & Mabrouki, M. (2024). Decoupling carbon emissions and economic growth in Tunisia: pathways to sustainable development. Handbook on Energy and Economic Growth, 103–127. DOI: 10.4337/9781802204803.00012

Dadashi, Z., Mahmoudi, A., & Rashidi, S. (2022). Capacity and strategies of energy production from renewable sources in Arab countries until 2030: A review from renewable energy potentials to environmental issues. *Environmental Science and Pollution Research International*, *29*(32), 47837–47866. DOI: 10.1007/s11356-022-20544-z PMID: 35522406

Dahmani, M., Mabrouki, M., & Ragni, L. (2021). Decoupling Analysis of Greenhouse Gas Emissions from Economic Growth: A Case Study of Tunisia. *Energies*, *14*(22), 7550. DOI: 10.3390/en14227550

Ghorbal, S., Farhani, S., & Youssef, S. B. (2021). Do renewable energy and national patents impact the environmental sustainability of Tunisia? *Environmental Science and Pollution Research International*, *29*(17), 25248–25262. DOI: 10.1007/s11356-021-17628-7 PMID: 34839440

Khaled, M. W. B., Dahmani, M., & Ben Youssef, A. (2024). Overcoming Barriers to Energy Transition in the MENA Region: New Institutional Dynamics. DOI: 10.20944/preprints202402.1440.v1

Omri, E., Chtourou, N., & Bazin, D. (2015). Solar thermal energy for sustainable development in Tunisia: The case of the PROSOL project. *Renewable & Sustainable Energy Reviews*, *41*, 1312–1323. DOI: 10.1016/j.rser.2014.09.023

Saadaoui, H., & Chtourou, N. (2022). Do Institutional Quality, Financial Development, and Economic Growth Improve Renewable Energy Transition? Some Evidence from Tunisia. *Journal of the Knowledge Economy*, *14*(3), 2927–2958. DOI: 10.1007/s13132-022-00999-8

Saadaoui, H., & Omri, E. (2023). Towards a gradual transition to renewable energies in Tunisia: Do foreign direct investments and information and communication technologies matter? *Energy Nexus*, *12*, 100252. DOI: 10.1016/j.nexus.2023.100252

Saadaoui, H., Omri, E., & Chtourou, N. (2024). The transition to renewable energies in Tunisia: The asymmetric impacts of technological innovation, government stability, and democracy. *Energy*, *293*, 130686. DOI: 10.1016/j.energy.2024.130686

Salem, H. S., Pudza, M. Y., & Yihdego, Y. (2023). Harnessing the energy transition from total dependence on fossil to renewable energy in the Arabian Gulf region, considering population, climate change impacts, ecological and carbon footprints, and United Nations' Sustainable Development Goals. *Sustainable Earth Reviews*, *6*(1), 10. Advance online publication. DOI: 10.1186/s42055-023-00057-4

Trabelsi, E. (2024). Transition to sustainable environment and economic growth in Tunisia: An ARDL approach. *World Development Sustainability*, *4*, 100165. DOI: 10.1016/j.wds.2024.100165

Chapter 7
Environmental Sustainability and Digital Innovations:
The Role of Energy Use, ICT, and Economic Growth

Aykut Yağlıkara
https://orcid.org/0000-0001-6728-2477
Zonguldak Bülent Ecevit University, Turkey

ABSTRACT

Given the rapidly growing economic integration, it is essential to provide environmental sustainability using eco-friendly digital technology to prevent the increasing environmental damage. This article investigates in a sample of G7 countries from 1990 to 2021 the impact of environmental ICT, economic growth, renewable energy usage, and energy consumption on environmental degradation. Using cross-sectional dependence, slope homogeneity, unit root testing, cointegration analysis, long-term coefficient estimate, and causality tests, the methodological component of the study used analysis of long-term coefficient estimators shows that while a rise in economic development and energy consumption aggravates the ecological impact, an increase in environmental ICT and renewable energy usage reduces it. Regarding information and communication technology, it is imperative to enhance the implementation of environmentally friendly ICT policies among G7 countries.

DOI: 10.4018/979-8-3373-1822-6.ch007

INTRODUCTION

Historically, the quest of economic development has been associated with environmental damage, which has resulted in the complex connection in which advancement usually occurs at the price of ecological integrity (Ziaul & Wang, 2023). The concept of sustainable development, which holds that the preservation of the environment and the growth of the economy are rather linked rather than mutually contradictory, is progressively challenging this paradigm (Duran et al., 2015; Xing & Ding, 2019). Sustainable development is a paradigm change unlike conventional theories of development, which stress economic expansion while ignoring the social and environmental effects of development. This strategy recognizes that the variables influencing whether or not there is continuous economic success are the health and adaptability of the ecosystems existing on Earth (Aziz et al., 2024; Bocken et al., 2013; Waters, 2008). A basic idea for development projects, the environmental sustainability concept—which emphasizes on the preservation of biodiversity, the maintenance of ecological balance via the means of responsible resource management, and the mitigating of pollution—has attracted a lot of support (Chen et al., 2024). Introduced as an environmental Kuznets Curve hypothesis, it has muddled the discussion by pointing to a potential inflection point when environmental deterioration lowers with economic growth (Ağan, 2024; Pu et al., 2024). Technological developments, better laws, and more consciousness of environmental problems have fostered this notion.

Comprising France, Canada, Germany, Japan, Italy, the United States of America, and the United Kingdom, the Group of Seven (G7) makes a sizable share of the world's economy and has a major impact on international relations. Consequently, the creation of environmental policies and practices on a worldwide level depends much on the G7 (Khumalo, 2014). The environmental degradation in these countries is mostly attributed to their consumption habits, industrial practices, and past emissions (Ghosh et al., 2020; Mngumi et al., 2024). These countries promote global environmental sustainability, yet they also greatly contribute to environmental damage even if they Examine the G7's involvement in tackling world environmental issues can help one to properly understand the complex link among their economic supremacy, environmental responsibility, and influence on ecological degradation. Important is a shift in perspective from the most powerful countries, which usually contribute significantly to environmental degradation, particularly in developing areas, from which Chen et al., 2024; Khumalo, 2014 It is important to realize that the realization of world environmental sustainability depends on a different viewpoint from these countries. Acquaye et al. (2016) claim that the rising consumer demand for products and services in these nations fuels the environmental damage the G7 countries causes to be worsening. Moreover, globalization—which the

G7's economic policies have pushed—has resulted in higher manufacturing and consumption, therefore aggravating environmental problems (Afridi et al., 2023).

Eregha et al. (2024) claim that the explosion of information and communication technology has brought in an era of unparalleled connection and change that has changed many basic sectors of human effort and economic activity. Notwithstanding this, the possible environmental effects of the widespread acceptance and use of information and communication technology (Gupta et al., 2021) have caused considerable worry. This issue has grown in line with great achievements and social advantages. Beginning in the 1990s, the information and communications technology revolution has brought geographical barriers down, economic complexity deepens, trade openness increases, and foreign direct investment is encouraged. Concurrently with this revolution's environmental effects have been expressed worries about them (Elfaki & Ahmed, 2024; Latif et al., 2017). The effect of information and communications technology (ICT) on the quality of the environment, especially the part it plays in ecological deterioration, presents a challenge to the accomplishment of Sustainable Development Goal 13, which emphasizes climate action. Information and communication technology (ICT) has the potential to raise per capita GDP by means of improved revenue generating, market development promotion, institutional efficiency enhancement, transaction cost reduction, and so reduction of unemployment rates and price volatility. But this economic development could have bad consequences for the surroundings (Wei et al., 2024; Xie et al., 2023). It is important to recognize the twin qualities of information and communication technology (ICT), which include its capacity to both cause environmental damage and foster economic development by means of their capacity to support environmental damage. Consequently, it is imperative to investigate the impact of ICT on the surroundings in great detail and with complexity (Khan & Wu, 2022).

The environmental effects of ICT can be categorized into three distinct types: primary, secondary, and tertiary. Primary effects encompass the direct environmental impacts resulting from the production, use, and disposal of ICT equipment, encompassing energy consumption and resource depletion (Moyer & Hughes, 2012). Secondary effects refer to the broader systemic changes induced by ICT adoption, such as shifts in consumption patterns, transportation systems, and industrial processes. Tertiary effects, often the most challenging to quantify, involve the long-term, indirect consequences of ICT on environmental systems, including climate change, biodiversity loss, and ecosystem degradation. The life cycle of ICT goods gives rise to first-order effects, encompassing environmental consequences that are also referred to as direct or primary effects, whereby environmental degradation increases as energy usage rises throughout an ICT product's lifespan (Charfeddine et al., 2024). Furthermore, consistent results regarding the impact that ICT has on the environment are lacking, as this aspect has only recently begun to be explored.

It seems logical that ICT is harmful to the environment because using it requires energy, although ICT appears to have a lot of potential to promote sustainable development because it has been linked to declines in energy intensity (Ollo-López & Aramendia-Muneta, 2011).

The escalating concerns surrounding climate change have compelled international organizations, governments, and policymakers to actively investigate alternative energy sources that are both clean and environmentally sustainable, aiming to supplant traditional fossil fuels (Zeghlouli et al., 2021). Renewable energies have emerged as a crucial element in economic strategies designed to address the pressing challenges of current energy crises and the overarching issue of climate change (Cheng et al., 2016; Lopez et al., 2021). The primary objectives driving the promotion of renewable energy sources include bolstering energy security and mitigating the adverse effects of carbon emissions, which contribute to global warming and the depletion of fossil fuels (Charfeddine et al., 2024). It has become widely acknowledged that energy plays a pivotal role in fostering industrialization. Though historically they have driven development, conventional energy sources such coal, natural gas, and oil simultaneously emit significant volumes of carbon dioxide into the atmosphere (Li et al., 2023). This phenomenon significantly increases greenhouse gas emissions, which accelerates world warming and causes climate change. Foss fuels have mostly met the growing worldwide energy demand, which has resulted in a notable increase in CO2 emissions from fuel combustion and therefore aggravation of environmental degradation (Gopinath et al., 2021). Consequently, the growing need for energy has driven a more intense search for creative energy sources, partially because of concerns about environmental damage and the unfavorable climatic effects usually connected with development driven by non-renewable energy sources (Zhu & Cheung, 2012). Acknowledging the differences in cost, social consequences, and environmental effects between fossil fuels and renewable energy sources, the search of clean fuel technologies and alternative energies is being aggressively sought upon.

Aware of the strategic importance of the energy industry, global governments are passing laws and regulations meant to encourage the acceptance of renewable energy technologies (Kumar & Kumar, 2017). Together with related legislative actions, these programs aim to promote renewable energy technologies, increase energy efficiency, and use energy conservation practices. Reduced carbon emissions relative to fossil fuels made possible by renewable energy sources emphasize the reason of the global campaigning of renewable energy by international organizations, legislators, and academics. Including renewable energy into the energy mix might also improve society's wellbeing. Comparatively to non-renewable energy sources, solar and hydropower show less carbon emissions and are expected to have a less environmental impact (Ukoba et al., 2024). As environmental protection becomes more important and the economic and legislative systems supporting the broad acceptance of sustainable markets for renewable energy systems continue to change fast, renewable energy is expected to be mostly driving future expansion in the energy sector (Tran & Smith, 2017).

The aim of this study is to examine the impact of environmental ICT, renewable energy consumption, energy consumption, and economic growth on the ecological footprint for the period covering 1990-2021, using a sample of G7 countries. In this context, the aim is to determine the correct model and achieve more consistent results through the cross-sectional dependence and slope homogeneity tests used in the methodology section of the study. Based on the results obtained from these two tests, second-generation unit root, panel cointegration, long-term coefficient estimator, and causality tests were applied. First of all, it is expected that the method used in this study will contribute to literature. It uses two robust methods, such as FGLS and PCSE, and supports them with the MMQR method to enhance the reliability of the results. The MMQR method examines the effects of the dependent variable's distribution in different regions, providing more detailed results compared to traditional regression methods. Additionally, the examination of the impact of the environmental ICT variable on ecological footprint in the study is expected to contribute to the literature by providing a new perspective on the relationship between the ecological environment and the variable, as it is not widely covered in existing literature. Additionally, the fact that this study is conducted within the G7 country group and covers a long period from 1990 to 2021 is expected to contribute to the literature by allowing the observation of the effectiveness of environmentally sustainable development policies in a sample of developed countries like the G7. The next part of the study is planned as follows: In the second part, some studies from literature on the relationship between ICT, renewable energy, energy consumption, and economic growth with environmental degradation are included. In the third section, the data related to the study, the economic model, the method, and the analysis results are presented. In the fourth and final section, the results of the study are presented and interpreted, and policy recommendations are made in the conclusion section.

1. LITERATURE REVIEW

In the literature review section, studies examining the impact of ICT, renewable energy, energy consumption, and economic growth on environmental degradation are included. In this context, the literature review section is presented in tabular form. In the table below, the relationships between environmental degradation and independent variables are grouped separately. First, studies on the relationship between ICT and environmental degradation are presented. The impact of ICT on environmental degradation varies in the studies based on the environmental degradation indicators, samples, and the period of the dataset considered. The relationship between renewable energy and environmental degradation, on the other hand, presents findings in all the studies examined that an increase in renewable energy consumption improves environmental quality. In the relationship between energy consumption and environmental degradation, all studies present evidence that environmental quality deteriorates. Finally, it presents evidence that economic growth increases environmental degradation in all of the studies.

Table 1. Literature review

Author(s)	Period	Country/Region	Method	Findings
ICT and Environmental Degradation				
Ahmed et al. (2020)	1995-2017	Latin America and Caribbean	CUP-BC, CUP-FM	ICT reduces CO2 emissions; one-way causality from ICT to CO2.
Ahmed and Le (2021)	1990-2020	ASEAN-6	CUP-FM	ICT reduces emissions; one-way causality from ICT to emissions.
Charfeddine et al. (2024)	1995-2018	Top 10 polluting countries	Panel Vector Autoregression	ICT improves environmental quality.
Danish., (2019)	1990-2016	59 Belt and Road countries	Generalized Least Squares	ICT reduces CO2 emissions.
Huang et al. (2022)	1990-2020	G7 and E7	AMG, CCEMG	ICT reduces pollution in G7 but increases it in E7.
Kahouli et al. (2022)	1990-2018	Saudi Arabia	VECM	ICT reduces ecological footprint.
Ulucak et al. (2020)	1990-2015	BRICS	CUP BC, CUP FM	ICT reduces carbon emissions.
Kongbuamai et al. (2022)	1992-2015	N-11	FGLS	ICT increases ecological footprint; one-way causality from ICT to EF.
Lee and Brahmasrene (2014)	1991-2009	ASEAN-9	Cointegration Regression	ICT increases CO2 emissions.
Raheem et al. (2020)	1990-2014	G7	PMG	ICT increases carbon emissions in the long term.
Amri et al. (2019)	1975-2014	Tunisia	ARDL	ICT has an insignificant effect on CO2 emissions.
Dehghan et al. (2018)	2002-2013	Iran	DOLS	ICT reduces CO2 in service/transport but increases it in the industrial sector.
Higón et al. (2017)	1995-2010	142 developed and developing countries	POLS, Driscoll-Kraay	Inverted U-shaped relationship between ICT and CO2.

continued on following page

Table 1. Continued

Author(s)	Period	Country/Region	Method	Findings
Renewable Energy and Environmental Degradation				
Acheampong et al. (2019)	1980-2015	Sub-Saharan Africa	GMM	Renewable energy reduces CO2 emissions.
Alola et al. (2019)	1997-2014	16 EU countries	PMGARDL	Renewable energy improves environmental quality.
Aydın and Erdem (2024)	2000-2019	EU (Malta, Spain)	CCEMG, AMG	Renewable energy improves environmental quality.
Bozatli and Akca (2023)	1994-2018	OECD	AMG, CS-ARDL	Renewable energy reduces ecological footprint.
Celik et al. (2024)	1970-2019	Scandinavian countries	DOLS, FMOLS, GMM	Renewable energy reduces CO2 emissions.
Chen et al. (2023)	1990-2020	E7	MMQR	Renewable energy reduces CO2 emissions.
Cheng et al. (2021)	1991-2017	China	Quantile Regression	Renewable energy reduces CO2 emissions.
Hao et al. (2021)	1991-2017	G7	CS-ARDL	Renewable energy reduces CO2 emissions.
Hsu et al. (2021)	1990-2020	China	ARDL	Renewable energy reduces CO2 emissions.
Huang et al. (2022)	1990-2020	G7 and E7	AMG, CCEMG	Renewable energy reduces pollution in both groups.
Khan et al. (2020)	1990-2017	G7	AMG, CCEMG	Renewable energy reduces consumption-based CO2 emissions.
Inglesi-Lotz and Doğan (2018)	1980-2011	Sub-Saharan Africa	DOLS	Renewable energy reduces environmental degradation.
Obobisa and Ahakwa (2024)	1990-2019	25 European countries	GMM, Quantile Regression	Renewable energy reduces CO2 emissions.
Onwe et al. (2023)	1994-2020	G7	MMQR	Renewable energy reduces ecological footprint; bidirectional causality.
Rafique et al. (2022)	1996-2016	29 OECD countries	DOLS, FMOLS	Renewable energy reduces ecological footprint.
Wang et al. (2020)	1990-2017	N-11	AMG, CCEMG	Renewable energy reduces CO2 emissions.
Xie and Jamaani (2022)	1990-2020	G7	MMQR	Renewable energy reduces CO2 emissions.
Energy Consumption and Environmental Degradation				
Ahmed and Le (2021)	1990-2020	ASEAN-6	CUP-FM	Energy consumption increases emissions.
Ahmed et al. (2019)	1971-2021	Malaysia	ARDL	Energy consumption increases ecological footprint.

continued on following page

Table 1. Continued

Author(s)	Period	Country/Region	Method	Findings
Alola et al. (2019)	1997-2014	16 EU countries	PMGARDL	Non-renewable energy reduces environmental quality.
Chien et al. (2021)	1990-2017	Asian economies	CS-ARDL	Energy consumption increases CO2 emissions.
Dogan and Turkekul (2016)	1960-2010	USA	ARDL	Energy consumption increases CO2 emissions; bidirectional causality.
Jahanger et al. (2022)	1990-2018	73 developing countries	2-SLS	Energy consumption increases CO2 emissions.
Liu et al. (2023)	1995-2020	China	PMGARDL	Energy consumption increases CO2 emissions.
Inglesi-Lotz and Doğan (2018)	1980-2011	Sub-Saharan Africa	DOLS	Non-renewable energy increases environmental degradation.
Mensah et al. (2018)	1990-2014	28 OECD countries	ARDL	Non-renewable energy increases CO2 emissions.
Obobisa and Ahakwa (2024)	1990-2019	25 European countries	GMM, Quantile Regression	Non-renewable energy increases CO2 emissions.
Onwe et al. (2023)	1994-2020	G7	MMQR	Non-renewable energy increases ecological footprint; one-way causality.
Rafique et al. (2022)	1996-2016	29 OECD countries	DOLS, FMOLS	Energy consumption increases ecological footprint.
Sahoo and Sethi (2021)	1990-2016	Developing countries	FMOLS, DOLS	Non-renewable energy increases ecological footprint.
Shahbaz et al. (2017)	1970-2014	Japan	ARDL	Energy consumption increases CO2 emissions.
Shahzad et al. (2021)	1965-2017	USA	QARDL	Fossil fuel energy increases ecological footprint; causal relationship.
Yilanci and Pata (2020)	1995-2016	China	Fourier ARDL	Energy consumption increases ecological footprint.
Zafar et al. (2019)	1990-2014	OECD	CUP BC, CUP FM	Energy consumption increases environmental degradation; bidirectional causality.
Economic Growth and Environmental Degradation				
Ahmed and Le (2021)	1990-2020	ASEAN-6	CUP-FM	Economic growth increases emissions.
Ahmed et al. (2019)	1971-2021	Malaysia	ARDL	Economic growth increases ecological footprint.
Alola et al. (2019)	1997-2014	16 EU countries	PMGARDL	Economic growth reduces environmental quality.

continued on following page

Table 1. Continued

Author(s)	Period	Country/Region	Method	Findings
Chien et al. (2021)	1990-2017	Asian economies	CS-ARDL	Economic growth increases CO2 emissions.
Dogan and Turkekul (2016)	1960-2010	USA	ARDL	Economic growth increases CO2 emissions; bidirectional causality.
Jahanger et al. (2022)	1990-2018	73 developing countries	2-SLS	Economic growth increases CO2 emissions.
Liu et al. (2023)	1995-2020	China	PMGARDL	Economic growth increases CO2 emissions.
Inglesi-Lotz and Doğan (2018)	1980-2011	Sub-Saharan Africa	DOLS	Economic growth increases environmental degradation.
Mensah et al. (2018)	1990-2014	28 OECD countries	ARDL	Economic growth increases CO2 emissions.
Obobisa and Ahakwa (2024)	1990-2019	25 European countries	GMM, Quantile Regression	Economic growth increases CO2 emissions.
Onwe et al. (2023)	1994-2020	G7	MMQR	Economic growth increases ecological footprint; one-way causality.
Rafique et al. (2022)	1996-2016	29 OECD countries	DOLS, FMOLS	Economic growth increases ecological footprint.
Sahoo and Sethi (2021)	1990-2016	Developing countries	FMOLS, DOLS	Economic growth increases ecological footprint.
Shahbaz et al. (2017)	1970-2014	Japan	ARDL	Economic growth increases CO2 emissions.
Shahzad et al. (2021)	1965-2017	USA	QARDL	Economic growth increases ecological footprint; causal relationship.
Yilanci and Pata (2020)	1995-2016	China	Fourier ARDL	Economic growth increases ecological footprint.
Zafar et al. (2019)	1990-2014	OECD	CUP-BC, CUP FM	Economic growth increases environmental degradation; bidirectional causality.

2. METHODOLOGY

2.1. Cross-Section Dependence and Slope Homogeneity

In the context of panel data analysis, cross-sectional dependence may be a considerable difficulty; if it is not adequately addressed, it can result in estimates that are biased and conclusions that are erroneous. A substantial number of statistical tests have been developed in order to identify and solve the issue of cross-sectional dependency. There is a distinct collection of characteristics and application regions that are suitable for panel data formats that are possessed by each of these tests.

When it comes to identifying cross-sectional dependency in panels with N<T, the Breusch-Pagan LM test, which was initially proposed by Breusch and Pagan in 1980, is the major approach that is utilized. The pairwise total correlations of the squared residuals between the various cross-sections are evaluated as part of the study. The Breusch-Pagan LM test has a null hypothesis that indicates that the residuals between units are not associated with one another. This hypothesis shows that there is no cross-sectional dependency. This test's findings provide substantial evidence of cross-sectional dependency on the variables under investigation. In order to do this evaluation, the equation is as follows:

$$LM = \sum_{i=1}^{N-1}\sum_{j=i+1}^{N}\hat{\rho}_{ij}^{2} \tag{1}$$

Here $\hat{\rho}_{ij}$, is the correlation coefficient of the residuals between units i and j.

Pesaran (2004) developed the Pesaran Scaled LM test in response to the findings of Breusch-Pagan's research. This test was designed to handle situations in which N greatly transcends T. In this option, the usual LM statistic is modified by the use of a scaling factor, which makes it acceptable for use with big panels. According to the null hypothesis, there is no reliance on cross-sectional measurements between the units. The following is an expression of the scaled LM test statistic:

$$LM_{scaled} = \frac{1}{N(N-1)}\sum_{i=1}^{N-1}\sum_{j=i+1}^{N}T\hat{\rho}_{ij}^{2} - 1 \tag{2}$$

Because of this expansion, the power of the test is boosted when it is used to datasets that contain a considerable number of cross-sectional units. This is because the test is applicable to more data. Specifically in the case of finite samples, the Bias-Corrected Scaled LM test was developed by Pesaran et al. (2008). This test improves the accuracy of finding cross-sectional dependence. This examination was developed with the intention of refining the correctness of research. By incorporating a bias adjustment term into the analysis, this test is able to enhance the precision of the findings for reduced panel sizes. As a result, the equation may be obtained as follows:

$$LM_{bc} = \frac{1}{N(N-1)}\sum_{i=1}^{N-1}\sum_{j=i+1}^{N}T\hat{\rho}_{ij}^{2} - 1 - \frac{\hat{\mu}_{T}}{N} \tag{3}$$

where $\hat{\mu}_T$, is a bias correction factor that gives greater accuracy to the test statistic.

Furthermore, the CD test that was created by Pesaran is still another strategy that is utilized on a consistent basis. The objective of this method is to determine the average correlation between all of the error term pairs that are obtained from cross-sectional units. By utilizing this test, it is possible to discover cross-sectional dependence in big panels with surprising efficiency. Additionally, it provides a diagnostic tool that is not only basic but also efficient in its computation. According to this premise, the operation is carried out in line with the assumption that there is no cross-sectional dependency between the units. This is the process that involves the calculation of its statistics, and it is as follows:

$$CD = \sqrt{\frac{2}{N(N-1)}} \sum_{i=1}^{N-1} \sum_{j=i+1}^{N} \hat{\rho}_{ij} \tag{4}$$

One of the most important characteristics of cross-sectional dependency (CD) statistics is the fact that it indicates the presence of probable dependencies between cross-sections in panel data. This is one of the most essential aspects of CD statistics. In the panel data model that Pesaran and Yamagata (2008) developed, the test for slope homogeneity is utilized to investigate the consistency of slope coefficients across cross-sectional units. This is done in order to determine whether or not the slopes are consistent. By employing the Delta statistic, which may be defined as follows on its own, a comparison is conducted between the alternative hypothesis of heterogeneity and the null hypothesis of slope homogeneity. This comparison is made in order to determine whether hypothesis is more reliable.

$$\Delta = \sqrt{N}\left(\frac{\hat{S} - (K+1)}{\sqrt{2(K+1)}}\right) \tag{5}$$

Let N be the number of cross-sections; K be the number of regression variables; and $\hat{S}$, be any test statistics obtained from individual unit regressions. This alternative version, Δ^*, is superior in terms of small sample properties. A significant Delta statistic indicates the presence of heteroskedasticity in the slopes, while a nonsignificant value indicates homogeneity. This test effectively detects variation in slope coefficients across units, allowing robust inferences to be made in panel data models.

2.2. Unit Root

When there is an issue with cross-sectional dependency and slope heterogeneity, the unit root tests of the first generation are unable to work successfully as they were supposed to. In spite of the cross-sectional dependency and slope heterogeneity, the stationarity of the variables was investigated with the help of the second-generation unit root test of the CIPS (Pesaran, 2007). As a prerequisite for carrying out the CIPS computation, it is essential to compute the cross-sectional mean of t_i, which is represented by the equation that is presented below.

$$CIPS = \frac{1}{N}\sum_{i=1}^{N} t_i\left(N,T\right) \tag{6}$$

2.3. Westerlund Panel Cointegration

Westerlund cointegration tests are designed to examine cointegration links in panel data using a modular approach that takes into account heterogeneity and cross-sectional dependence. According to Westerlund (2007), these tests work within the framework of the Error Correction Model (ECM) and provide a collection of group and panel-based test data. The ECM framework facilitates the determination of long-term correlations between dependent and independent variables. The cointegration test is expressed by the following equation:

$$\Delta Y_{it} = \alpha_i + \phi_i\left(Y_{i,t-1} - \beta' X_{it-1}\right) + \sum_{j=1}^{p}\gamma_{ij}\Delta Y_{i,t-j} + \sum_{j=0}^{q}\delta_{ij}\Delta X_{it-j}$$

$$+\epsilon_{it} \tag{7}$$

The error correction term is denoted by ϕ_i and the individual specific intercept is denoted by α_i. Cointegration will be indicated by a return to equilibrium in the long run, which is indicated by a significantly negative ϕ_i. G_τ, G_α, P_τ, P_α are the basic statistics used in Westerlund tests. Panel statistics, such as P_τ and P_α, evaluate evidence of cointegration across the entire panel, while individual unit statistics, such as G_τ and G_α, evaluate evidence of cointegration for individual units. If the unit root hypothesis is rejected, a long-run relationship exists.

2.4. FGLS and PCSE

As the long-term coefficient estimator of this study, since the covariance matrices created with generalized least squares (GLS) estimators may contain unknown parameters, Parks (1967) applicable generalized least squares (FGLS) method, which corrects cross-sectional dependency and heterogeneity in the same way, was used to increase the consistency of the parameters in the applied method. Then, the panel corrected standard error (PCSE) approach suggested by Beck and Katz (1995) was used. This method is a modified version of the Least Squares approach that can provide consistent results while assuming the existence of heterogeneity and cross-sectional dependency problems in panel data. It obtains more efficient results by using Monte Carlo simulation in Least Squares estimation.

2.5. Method of Moment Quantile Regression

Machado and Silva (2019) developed the Method of Moment Quantile Regression (MMQR) estimator out of their research. Conventional regression models have several inherent faults; our estimation helps to avoid some of them. Koenker (2004) is used to investigate the impact of the independent factors used in the CO2 dependent variable research on. This approach is thus selected as it considers the asymmetric and nonlinear correlations between the variables, hence addressing heterogeneity and endogeneity. This yields conclusions more consistent and stronger than those generated by previous panel quantile regression techniques (Onwe et al. 2023). This generates more reliable results as a result. Hence, on the ecological footprint in G7 countries, the MMQR technique offers a more consistent observation of the conditional heterogeneous covariance effect of information and communication technology (ICT), gross domestic product (GDP), renewable energy (REN), and environmental net worth (ENR). This is so because the MMQR approach considers the fact that the ecological effect is influenced by all these elements. The complete conditional quantile estimate for the location-scale variant model is provided below under the symbol $\theta_{\rho}\left(\tau/X_{i,t}\right)$.

$$\theta_{\rho}\left(\tau/X_{i,t}\right) = \sigma_1 + X'_{i,t}\zeta + \left(\alpha_i + Z'_{i,t}\gamma\right) U_{i,t} \tag{8}$$

In the equation shown above, where the dependent variable (EF) θ_{ρ} is reported as the dependent variable (CO2), followed by $X_{i,t}$ representing the predictors (ICT, GDP, REN and ENR). The scale factor $(\alpha_i + Z'_{i,t}\gamma)$ indicates the fixed effect of quantile "i" on cross-sectional units "t". The above equation should be examined as $(\alpha_i + Z'_{i,t}\gamma >) = 1$ and $(\sigma, \zeta, \alpha, \gamma)$. (σ_1, α_i), $i = 1, \ldots.., n$, reports the fixed effect of individual. i, and Z denotes the k-element vector explained by X, separating the one explaining element i.

$$Z_i = Z_i(X), i = 1, \ldots.., k \tag{9}$$

For any fixed i. $X_{i,t}$ forms a similar and independent distribution and is independent at t. In contrast, $U_{i,t}$ is independent and identical among units (*i*) during t and is unrelated to $X_{i,t}$, which is a necessary condition to satisfy the moment defined by (Machado and Silva, 2019).

2.6. Dumitrescu-Hurlin Causality Test

To determine the nature of the causal connection that exists between the variables, the causality test that was established by Dumitrescu and Hurlin (2012) was utilized in this study. In comparison to the panel Granger causality test, this test is more favorable since it takes into account the cross-sectional dependency analysis. The following is one possible way to represent the causality of the D-H panel:

$$y_{it} = \theta_i + \sum_{j-1}^{j} \gamma_i^n P_{i,t-i} + \sum_{i=1}^{q} \lambda_i^n \beta_{i,t-i} + \varepsilon_t \quad (10)$$

The alternative hypothesis in this test asserts that there is a causal link between the variables, in contrast to the null hypothesis, which asserts that there is no causal relationship between the variables.

3. DATA AND ANALYSIS

In this study, annual data for the period 1990-2021 is used to examine the impact of environmental ICT, renewable energy, energy consumption, and economic growth on the ecological footprint in G7 (USA, Germany, UK, Italy, France, Japan, Canada) countries. The series used in the study were subjected to analysis by taking their logarithms. Additionally, the model used in the study is shown in the equation below:

$$EF_{it} = ICT_{it} + REN_{it} + ENR_{it} + GDP_{it} + \varepsilon_{it}$$

The term EF_{it}, used as the dependent variable in the equation, is explained by the ecological footprint of consumption (in global hectares) divided by the population of the country, and this dataset has been obtained from the Global Footprint Network database. The variable ICT_{it} is represented by the percentage of environment-related technologies by climate change mitigation in ICT and is obtained from the OECD database. Among the independent variables, the GDP_{it} variable is defined as GDP per capita (constant 2015 US$); the REN_{it} variable is expressed as renewable energy consumption (% of total final energy consumption), and these two variables are obtained from the World Bank database. ENR_{it}, on the other hand, has been obtained from the Energy Institute as primary energy: Consumption per capita. Additionally, ε_{it} represents the error term. The compiled version of the information provided above is shown in Table 2 below.

Table 2. Variables and data sources

Variables	Symbol	Units	Source
Ecological Footprint	EF	Ecological Footprint of consumption (in global hectares) divided by the population of the country)	Global Footprint Network
Environmental ICT	ICT	Percentage of environment-related technologies by climate change mitigation in ICT	OECD
Gross Domestic Product	GDP	GDP per capita (constant 2015 US$)	World Bank
Renewable Energy	REN	Renewable energy consumption (% of total final energy consumption)	World Bank
Energy Consumption	ENR	Primary energy: Consumption per capita	Energy Institute

Table 3 shows the descriptive statistics and correlation matrix for the series used in the study. As the average value, it is shown that GDP has the highest value, while ICT has the lowest value. It is observed that the maximum value is in the highest GDP variable, while the lowest is in the ICT series. While the minimum value is the lowest in ICT, the series with the highest value is GDP. According to the Kurtosis Values, which provide information about the distribution of the series, it is understood that the EF, GDP, and ENR series are more evenly distributed and have fewer outliers because their values are less than 3, while the ICT and REN variables have more outliers than normal because their values are greater than 3.

Table 3. Descriptive statistics and correlation matrix

	EF	ICT	GDP	REN	ENR
Mean	1.831322	-1.313836	10.51581	1.934947	5.277512
Median	1.757858	-1.177990	10.47299	2.041220	5.153891
Max.	2.407846	0.372966	11.02403	3.173878	6.061389
Min.	1.316408	-5.158662	10.22160	-0.510826	4.604667
Std. Dev.	0.258199	1.078564	0.184289	0.866109	0.410590
Skewness	0.629846	-0.897459	0.697790	-0.697495	0.676469
Kurtosis	2.353949	3.717142	2.809641	3.104004	2.187196
Jarque-Bera	18.12138	33.77987	17.93758	17.69288	22.52365

continued on following page

Table 3. Continued

	EF	ICT	GDP	REN	ENR
Probability	0.000116	0.000000	0.000127	0.000144	0.000013
Sum	397.3969	-285.1025	2281.932	419.8834	1145.220
Sum Sq. Dev.	14.39999	251.2728	7.335852	162.0314	36.41423
Observations	217	217	217	217	217
Correlation Matrix					
EF	1.000000				
ICT	0.255074	1.000000			
GDP	0.504594	0.685464	1.000000		
REN	0.168039	0.297062	0.157490	1.000000	
ENR	0.901525	0.399628	0.466034	0.350461	1.000000

After providing descriptive information about the variables and the data set, the findings obtained from the method used in the study are presented. First, the cross-sectional dependence test, which examines whether there is any interaction between countries, and the slope homogeneity tests, which investigate whether the variables in the model show homogeneous effects across country units, are included. Based on the results obtained from these tests, other tests to be applied are determined. Then, a second-generation unit root test was applied to examine the stationarity of the series. Then, a panel cointegration test was applied to examine whether the series move together in the long term. The panel quantile regression method was used for the long-term predictions of the coefficients in the model. Finally, a panel causality test was applied to determine the causal relationship between the variables.

Table 4. Cross-section dependence tests results

Variable	Breusch-Pagan LM	Peseran scaled LM	Bias-corrected scaled LM	Peseran CD
EF	396.4475***(0.000)	57.93280***(0.000)	57.81614***(0.000)	19.57975***(0.000)
ICT	426.6450***(0.000)	62.59239***(0.000)	62.47572***(0.000)	20.58822***(0.000)
GDP	505.5251***(0.000)	74.76385***(0.000)	74.64718***(0.000)	22.08224***(0.000)
REN	381.9582***(0.000)	55.69706***(0.000)	55.58039***(0.000)	18.49877***(0.000)
ENR	433.8737***(0.000)	63.70779***(0.000)	63.59112***(0.000)	20.57527***(0.000)

Note: ***, ** & * explain the level of significance at 1%, 5% and 10% respectively.

According to the results of the Peseran CD, Peseran scaled LM, Bias-corrected scaled LM, and Breusch-Pagan LM cross-sectional dependence tests shown in Table 4, the null hypothesis of no cross-sectional dependence is rejected for all four tests. According to the obtained results, it is concluded that there is an interaction or correlation among the countries used in the panel in terms of cultural, economic, or social aspects. According to the obtained result, the presence of cross-sectional dependence can lead to bias in other tests that will be applied.

Table 5. Result of slope heterogeneity analysis

Dependent Variable: EF	
Statistics	Test value (P-value)
Delta tilde	8.405***(0.000)
Delta tilde Adjusted	9.359***(0.000)

Note: ***, ** & * explain the level of significance at 1%, 5% and 10% respectively, whereas the values are in parentheses contains p-values.

The results of the slope homogeneity test are shown in Table 5. According to the obtained result, the slope coefficient is homogeneous, and the null hypothesis has been rejected. As a result of the findings for the model, it is concluded that the slope coefficients are heterogeneous. This situation indicates that the independent variables have a heterogeneous effect on the countries and should be taken into account in the tests to be applied thereafter.

Table 6. Unit root analysis

Variables	**Second-generation CIPS**	
	Level	Differences
EF	0.378 (0.647)	-3.469***(0.000)
ICT	-2.118** (0.017)	-7.754*** (0.000)
GDP	2.606 (0.995)	-3.227***(0.001)
REN	2.590 (0.752)	-3.167***(0.001)
ENR	-0.677 (0.249)	-4.779***(0.000)

Note: ***, ** & * explain the level of significance at 1%, 5% and 10% respectively.

In Table 6, which contains the unit root test results, the second-generation Pesaran (2006) CIPS test is included. In deciding on this test, the second-generation unit root test was used, taking into account the cross-sectional dependence and heterogeneity assumptions. According to the test results, the EF, GDP, REN, and ENR series contain a unit root at their level values, but they become stationary when first differences are taken, I (1). However, the ICT series does not contain a unit root at the level value I (0).

Table 7. Westerlund panel cointegration

Model, EF = ICT, GDP, REN, ENR				
Statistic	Value	Z-value	P-value	Robust P-value
Gt	-2.827	-2.182	0.015	0.010**
Ga	-11.593	-0.638	0.262	0.000***
Pt	-7.556	-2.588	0.005	0.020**
Pa	-11.185	-1.835	0.033	0.000***

Note: ***, ** & * explain the level of significance at 1%, 5% and 10% respectively.

The results of the Westerlund (2007) panel cointegration test, applied to examine how the variables with determined stationarity levels according to the unit root test results behave in the long term, are shown in Table 7. According to the analysis results, the null hypothesis of no cointegration is rejected based on the results of four different tests. According to these results, it is concluded that the variables move together in the long term.

Table 8. Result of FGLS and PCSE for panel

Variables	FGLS	PCSE
ICT	-0.0148*** (0.009)	-0.0151** (0.019)
GDP	0.2307*** (0.000)	0.2288*** (0.006)
REN	-0.0693*** (0.000)	-0.0687*** (0.000)
ENR	0.6187** (0.000)	0.6179*** (0.000)
Constant	-3.7458*** (0.000)	-3.7237*** (0.000)
Obs.	223	223
Number of Groups	7	7
Wald chi2	520.89	496.24
Prob>chi2	0.000	0.000

Note: ***, ** & * explain the level of significance at 1%, 5% and 10% respectively. Heteroskedastic with cross-sectional correlation and panel specific AR(1).

As the long-term coefficient estimator of this study, it has been concluded that the increases in ICT reduce EF according to both FGLS and PCSE results shown in Table 8. For both test groups, it is concluded that the increase in GDP raises EF, thereby increasing environmental degradation. Another variable, REN's effect on EF, appears to be negative. The increase in renewable energy consumption reduces environmental degradation. It is observed that an increase in the ENR variable raises the EF level and increases environmental degradation.

Table 9. Method of moment quantile regression test results

Variable			Quantile Regression								
	location	scale	10th	20th	30th	40th	50th	60th	70th	80th	90th
ICT	-0.0641 (0.000)	-0.018 (0.000)	-0.035 (0.002)	-0.043 (0.000)	-0.050 (0.000)	-0.055 (0.000)	-0.063 (0.000)	-0.069 (0.000)	-0.075 (0.000)	-0.085 (0.000)	-0.093 (0.000)
GDP	0.3895 (0.000)	0.097 (0.000)	0.235 (0.000)	0.281 (0.000)	0.314 (0.000)	0.342 (0.000)	0.387 (0.000)	0.419 (0.000)	0.449 (0.000)	0.501 (0.000)	0.545 (0.000)
REN	-0.035 (0.000)	0.005 (0.222)	-0.043 (0.000)	-0.041 (0.000)	-0.039 (0.000)	-0.037 (0.000)	-0.035 (0.000)	-0.033 (0.000)	-0.032 (0.000)	-0.029 (0.001)	-0.027 (0.009)
ENR	0.5790 (0.000)	-0.013 (0.157)	0.600 (0.000)	0.593 (0.000)	0.589 (0.000)	0.585 (0.000)	0.579 (0.000)	0.574 (0.000)	0.570 (0.000)	0.563 (0.000)	0.557 (0.000)
Constant	-5.3364 (0.000)	-0.911 (0.001)	-3.889 (0.000)	-4.318 (0.000)	-4.636 (0.000)	-4.898 (0.000)	-5.319 (0.000)	-5.621 (0.000)	-5.903 (0.000)	-6.389 (0.000)	-6.797 (0.000)
Obs.	217	217	217	217	217	217	217	217	217	217	217

The results of the Method of Moments Quantile Regression (MMQREG) are presented in Table 9. The ICT coefficient is negative and statistically significant at the 1% significance level across all quantiles. As the quantile value increases, the effect of ICT on EF also increases. At the extreme value of the 0.9 quantile, when ICT increases by 1%, EF shows a decrease of 0.093. It can be seen that the effect of economic growth on EF is positive and significant across all percentiles. It is observed that the effect of Economic Growth increases towards higher quantile values, and the highest effect is seen at the 0.9 quantile value, where a 1% increase in economic growth raises EF by 0.545. The REN coefficient is statistically significant and negative at all quantile values, with the effect decreasing as the quantile values increase, and it is observed that a 1% increase in REN at the 0.1 quantile reduces EF by 0.043. Finally, the ENR coefficient is statistically significant and has a positive effect at all quantiles. In lower quantile values, the effect of ENR on EF is greater, while in the 0.1 quantile, when ENR increases by 1%, EF increases by 0.6%.

Table 10. Dumitrescu & Hurlin (2012) granger non-causality test results

Pair	Null Hypothesis: No Causality	Statistics			Decision
		W-bar	Z-bar	Z-bar tilde	
1	EF and ICT	1.2579	0.4826 (0.6294)	0.2899 (0.7719)	Failed to reject
2	EF and GDP	4.6790	6.8827*** (0.000)	5.8636*** (0.000)	Reject
3	EF and REN	6.6532	10.5763*** (0.000)	9.0802*** (0.000)	Reject
4	EF and ENR	4.3061	6.1851*** (0.000)	5.2561*** (0.000)	Reject
5	ICT and EF	0.2848	-1.3380 (0.1809)	-1.2956 (0.1951)	Failed to reject
6	GDP and EF	0.6807	-0.5974 (0.5502)	-0.6506 (0.5153)	Failed to reject
7	REN and EF	1.6378	1.1932 (0.2328)	0.9088 (0.3635)	Failed to reject
8	ENR and EF	1.8013	1.4991 (0.1339)	1.1752 (0.2399)	Failed to reject

Note: ***, ** & * explain the level of significance at 1%, 5% and 10% respectively.

Table 10 shows the Dumitrescu-Hurlin panel causality test results, indicating the reciprocal relationship between ecological footprint and explanatory variables. The relationship between EF and ICT appears to be non-reciprocal, as there is no causal relationship from ICT to EF or from EF to ICT. In both relationships, the hypothesis that there is no causality cannot be rejected. The relationship between EF and GDP is concluded to have a strong causal relationship from GDP to EF, but no causal relationship from EF to GDP. In the other relationship between EF and REN, while there is no causal relationship from EF to REN, it is concluded that there is a strong causal relationship from REN to EF. Finally, for the relationship between EF and ENR, there is a strong causal relationship from EF to ENR, while there is no causal relationship from ENR to EF. This highlights the causal relationship from renewable energy consumption, energy consumption, and economic growth to ecological footprint in the examined context, emphasizing a highly interconnected ecological issue.

4. RESULTS AND POLICY IMPLICATIONS

The study generally examines the complex interaction between environmental ICT, renewable energy consumption, energy consumption, and the combined effects of per capita income on the ecological footprint. The study aims to provide a comprehensive analysis of this phenomenon by examining the empirical evidence among the aforementioned variables for the G7 countries during the period from 1990 to 2021. In the analysis section of the study, a cross-sectional dependence test was first applied to understand whether a shock occurring in one of the sections of the panel data set affects the others, and it was examined whether the regression coefficients of the countries in the panel data set are the same. As a result of the cross-sectional dependence and slope homogeneity tests, it was understood that there are cross-sectional dependence and heterogeneity conditions among the variables. Then, to increase the reliability of the tests to be applied, these two assumptions need to be taken into account. In this context, the stationarity of the variables was tested using the CIPS unit root test, which is a second-generation unit root test. The ICT variable is assumed to be stationary at level I(0), but the other variables are believed to be stationary at level I(1) when their starting differences are taken into consideration The gathered observations form the foundation for this judgment. After the integration degrees of the variables were ascertained, a cointegration test was conducted to ascertain whether or not there is a link between the variables over a significant length of time. This was carried out to ascertain whether the variables had any relationship at all. Long-range projections will produce consistent findings, and it is known from the panel cointegration test conducted by Westerlund (2007) that the variables move together over the long term. These are confirmed to be accurate both of them. Following that, employed were the long-term coefficient estimation tests, FGLS and PCSE. These tests respect heteroskeasticity and cross-sectional dependability accordingly. The Method of Moment Quantile Regression test was used to raise the validity of the acquired test outcomes even further. This was carried out to increase the consistency of the findings. The study that has been published in scholarly literature lends additional weight to the conclusions derived from the analysis. The results of this study line up with those of Charfeddine et al. (2024), Ahmed et al. (2020), Ulucak et al. (2020), all of which find that information and communication technology (ICT) helps to lower environmental degradation. 2020 saw this research carried out. Research by Chen et al. (2023), Celik et al. (2024), and Hsu et al. (2021) shows similar results implying that the use of renewable energy might help to lower the degree of environmental damage that results. The conclusion that the use of energy results in an increase in the deterioration of the environment is congruent with the results of the research of Alola et al. (2019) and Rafique et al. (2022) which investigate the relationship between energy consumption and the environment. 2019 saw both of these investigations finished. The research shown in the literature supports the conclusions showing that a rise in per capita income results in environmental degradation. All of these conclusions, then, fit one another. Those who conducted these studies were Mensah et al. (2018) and Zaidi et al. (2019).

Regarding the policies that must be carried out in information and communication technologies, the results of the research indicate that member countries of the G7 should increase the usage of environmentally friendly ICTs. One has to satisfy this need. Encouragement of the use of digitalization that is favorable to the environment, of broad use of green certification programs, and of digital support for recycling operations is very vital. Furthermore, assured should be the support of green digital transformation, the encouragement of artificial intelligence in this context, and the provision of research and development incentives for the development of green software that will help to lower the carbon footprint. At the same time, carbon emissions should be reduced through smart transportation systems, and the efficiency of water usage should be increased by developing digital agriculture technologies. In terms of energy policies, increasing energy efficiency in the residential and industrial sectors, transitioning to low-carbon production methods in heavy industry, and promoting electric transportation methods will be beneficial in this context. However, it is necessary to reduce fossil fuel consumption, eliminate subsidies provided for fossil fuel consumption, provide incentives and subsidies to increase renewable energy investments, and offer incentives to develop storage technologies, which are one of the biggest problems hindering the use of renewable energy. Since one of the drivers of economic growth and environmental degradation is the promotion of environmentally friendly production processes, it is necessary to encourage firms that lead these processes, implement sustainable development policies with a green economy focus, and enforce and apply emission trading systems and environmental tax mechanisms.

REFERENCES

Acheampong, A. O., Adams, S., & Boateng, E. (2019). Do globalization and renewable energy contribute to carbon emissions mitigation in Sub-Saharan Africa? *The Science of the Total Environment*, *677*, 436–446. DOI: 10.1016/j.scitotenv.2019.04.353 PMID: 31059886

Acquaye, A., Feng, K., Oppon, E., Salhi, S., Ibn-Mohammed, T., Genovese, A., & Hubacek, K. (2016). Measuring the environmental sustainability performance of global supply chains: A multi-regional input-output analysis for carbon, sulphur oxide and water footprints. *Journal of Environmental Management*, *187*, 571–585. DOI: 10.1016/j.jenvman.2016.10.059 PMID: 27876164

Afridi, F. E. A., Afridi, S. A., Zahid, R. M. A., Khan, W., & Anwar, W. (2023). Embracing green banking as a mean of expressing green behavior in a developing economy: Exploring the mediating role of green culture. *Environmental Science and Pollution Research International*, *31*(34), 46137–46147. DOI: 10.1007/s11356-023-25449-z PMID: 36694066

Ağan, B. (2024). Sustainable development through green transition in EU countries: New evidence from panel quantile regression. *Journal of Environmental Management*, *365*, 121545. DOI: 10.1016/j.jenvman.2024.121545 PMID: 38917545

Ahmed, Z., & Le, H. P. (2021). Linking Information Communication Technology, trade globalization index, and CO2 emissions: Evidence from advanced panel techniques. *Environmental Science and Pollution Research International*, *28*(7), 8770–8781. DOI: 10.1007/s11356-020-11205-0 PMID: 33070290

Ahmed, Z., Nathaniel, S., & Shahbaz, M. (2020). The criticality of information and communication technology and human capital in environmental sustainability: Evidence from Latin American and Caribbean countries. *Journal of Cleaner Production*, *286*, 125529. Advance online publication. DOI: 10.1016/j.jclepro.2020.125529

Ahmed, Z., Wang, Z., Mahmood, F., Hafeez, M., & Ali, N. (2019). Does globalization increase the ecological footprint? Empirical evidence from Malaysia. *Environmental Science and Pollution Research International*, *26*(18), 18565–18582. DOI: 10.1007/s11356-019-05224-9 PMID: 31054053

Alola, A. A., Bekun, F. V., & Sarkodie, S. A. (2019). Dynamic impact of trade policy, economic growth, fertility rate, renewable and non-renewable energy consumption on ecological footprint in Europe. *The Science of the Total Environment*, *685*, 702–709. DOI: 10.1016/j.scitotenv.2019.05.139 PMID: 31203164

Amri, F., Ben Lahouel, B., & Ben Zaied, Y. (2019). ICT, total factor productivity, and carbon dioxide emissions in Tunisia. *Technological Forecasting and Social Change*, *146*, 212–217. DOI: 10.1016/j.techfore.2019.05.028

Aydin, M., & Erdem, A. (2024). Analyzing the impact of resource productivity, energy productivity, and renewable energy consumption on environmental quality in EU countries: The moderating role of productivity. *Resources Policy*, *89*, 104613. DOI: 10.1016/j.resourpol.2023.104613

Aziz, G., Strielkowski, W., Sarwar, S., & Tiwari, A. K. (2024). Implications of circular economy, digitalization and technological innovation to achieve sustainable environmental goal: Pre and post-vision 2030. *Heliyon*, *10*(10), e30978. Advance online publication. DOI: 10.1016/j.heliyon.2024.e30978 PMID: 38770279

Beck, N., & Katz, J. N. (1995). What to do (and not to do) with time-series cross-section data. *The American Political Science Review*, *89*(3), 634–647. DOI: 10.2307/2082979

Bocken, N., Short, S. W., Rana, P., & Evans, S. (2013). A literature and practice review to develop sustainable business model archetypes. *Journal of Cleaner Production*, *65*, 42–56. DOI: 10.1016/j.jclepro.2013.11.039

Bozatli, O., & Akca, H. (2023). The effects of environmental taxes, renewable energy consumption and environmental technology on the ecological footprint: Evidence from advanced panel data analysis. *Journal of Environmental Management*, *345*, 118857. DOI: 10.1016/j.jenvman.2023.118857 PMID: 37657289

Celik, A., Kostekci, A., & Alola, A. A. (2024). Carbon neutrality implication of material productivity, total factor productivity and renewable energy uptake in the Nordics. *Ecological Indicators*, *160*, 111813. DOI: 10.1016/j.ecolind.2024.111813

Charfeddine, L., Hussain, B., & Kahia, M. (2024). Analysis of the impact of information and communication technology, digitalization, renewable energy and financial development on environmental sustainability. *Renewable & Sustainable Energy Reviews*, *201*, 114609. DOI: 10.1016/j.rser.2024.114609

Chen, F., Wang, Q., Umar, M., & Zheng, L. (2023). Towards sustainable resource management: The role of governance, natural resource rent and energy productivity. *Resources Policy*, 85(A), 104026. DOI: 10.1016/j.resourpol.2023.104026

Chen, Y., Subhan, M., Ahmad, G., Adil, M., & Zamir, M. N. (2024). Unveiling the linkages among digital technology, economic growth, and carbon emissions: A resource management perspective. *Resources Policy*, *91*, 104868. DOI: 10.1016/j.resourpol.2024.104868

Cheng, G., Zhao, C., Iqbal, N., Gülmez, Ö., Işik, H., & Kirikkaleli, D. (2021). Does energy productivity and public-private investment in energy achieve carbon neutrality target of China? *Journal of Environmental Management, 298*, 113464. DOI: 10.1016/j.jenvman.2021.113464 PMID: 34385111

Cheng, N., Stambula, S., Wang, D., Banis, M. N., Liu, J., Riese, A., Xiao, B., Li, R., Sham, T.-K., Liu, L., Botton, G. A., & Sun, X. (2016). Platinum single-atom and cluster catalysis of the hydrogen evolution reaction. *Nature Communications, 7*(1), 13638. Advance online publication. DOI: 10.1038/ncomms13638 PMID: 27901129

Chien, F., Sadiq, M., Nawaz, M. A., Hussain, M. S., Tran, T. D., & Thanh, T. L. (2021). A step toward reducing air pollution in top Asian economies: The role of green energy, eco-innovation, and environmental taxes. *Journal of Environmental Management, 297*, 113420. DOI: 10.1016/j.jenvman.2021.113420 PMID: 34333309

Danish. (2019). Effects of information and communication technology real income, and CO2 emissions: The experience of countries along Belt and Road. *Telematics and Informatics*, 45. 101300. DOI: 10.1016/j.tele.2019.101300

Dehghan, Z., & Shahnazi, R. (2018). Energy consumption, carbon dioxide emissions, information and communications technology, and gross domestic product in Iranian economic sectors: A panel causality analysis. *Energy, 169*, 1064–1078. Advance online publication. DOI: 10.1016/j.energy.2018.11.062

Dogan, E., & Turkekul, B. (2016). CO2 emissions, real output, energy consumption, trade, urbanization and financial development: Testing the EKC hypothesis for the USA. *Environmental Science and Pollution Research International, 23*(2), 1203–1213. DOI: 10.1007/s11356-015-5323-8 PMID: 26351068

Dumitrescu, E. I., & Hurlin, C. (2012). Testing for Granger non-causality in heterogeneous panels. *Economic Modelling, 29*(4), 1450–1460. DOI: 10.1016/j.econmod.2012.02.014

Duran, D. C., Artene, A. E., Gogan, L. M., & Duran, V. (2015). The objectives of sustainable development - ways to achieve welfare. *Procedia Economics and Finance, 26*, 812–817. DOI: 10.1016/S2212-5671(15)00852-7

Elfaki, K. E., & Ahmed, E. M. (2024). Digital technology adoption and globalization innovation implications on Asian Pacific green sustainable economic growth. *Journal of Open Innovation, 10*(1), 100221. DOI: 10.1016/j.joitmc.2024.100221

Eregha, P. B., Nathaniel, S. P., & Vo, X. V. (2024). COP28 Roadmap: Do ICT, education, and renewable energy consumption matter for environmental quality? Quantile evidence for emerging economies. *Journal of Cleaner Production, 143744*, 143744. Advance online publication. DOI: 10.1016/j.jclepro.2024.143744

Ghosh, P., Jha, A., & Sharma, R. (2020). Managing carbon footprint for a sustainable supply chain: A systematic literature review. *Modern Supply Chain Research and Applications*, *2*(3), 123–141. DOI: 10.1108/MSCRA-06-2020-0016

Gopinath, G. R., Sattanathan, M., Muthukannan, M., Pandian, R. S., & Manokar, A. M. (2021). Experimental study of solar dryer with thermal energy storage system for drying of agro products. *Research Square*. DOI: 10.21203/rs.3.rs-673328/v1

Gupta, U., Kim, Y. G., Lee, S., Tse, J., Lee, H.-H. S., Wei, G.-Y., Brooks, D., & Wu, C.-J. (2021). Chasing carbon: The elusive environmental footprint of computing. DOI: 10.1109/HPCA51647.2021.00076

Hao, L., Umar, M., Khan, Z., & Ali, W. (2021). Green growth and low carbon emission in G7 countries: How critical the network of environmental taxes, renewable energy and human capital is? *The Science of the Total Environment*, *752*, 141853. DOI: 10.1016/j.scitotenv.2020.141853 PMID: 32889278

Higón, D. A., Gholami, R., & Shirazi, F. (2017). ICT and environmental sustainability: A global perspective. *Telematics and Informatics*, *34*(4), 85–95. DOI: 10.1016/j.tele.2017.01.001

Hsu, C., Zhang, Y. Ch. P., Aqdas, R., Chupradit, S., & Nawaz, A. (2021). A step towards sustainable environment in China: The role of eco-innovation renewable energy and environmental taxes. *Journal of Environmental Management, 299*, 113609. DOI: 10.1016/j.jenvman.2021.113609 PMID: 34467863

Huang, Y., Haseeb, M., Usman, M., & Ozturk, I. (2022). Dynamic association between ICT, renewable energy, economic complexity and ecological footprint: Is there any difference between E-7 (developing) and G-7 (developed) countries? *Technology in Society*, *68*, 101853. DOI: 10.1016/j.techsoc.2021.101853

Inglesi-Lotz, R., & Dogan, E. (2018). The role of renewable versus non-renewable energy to the level of CO2 emissions a panel analysis of sub- Saharan Africa's Big 10 electricity generators. *Renewable Energy*, *123*, 36–43. DOI: 10.1016/j.renene.2018.02.041

Jahanger, A., Usman, M., & Ahmad, P. (2022). Investigating the effects of natural resources and institutional quality on CO2 emissions during globalization mode in developing countries. *International Journal of Environmental Science and Technology*, *20*(9), 9663–9682. Advance online publication. DOI: 10.1007/s13762-022-04638-2

Kahouli, B., Hamdi, B., Nafla, A., & Chabaane, N. (2022). Investigating the relationship between ICT, green energy, total factor productivity, and ecological footprint: Empirical evidence from Saudi Arabia. *Energy Strategy Reviews*, *42*, 100871. DOI: 10.1016/j.esr.2022.100871

Khan, A., & Wu, X. (2022). Digital economy and environmental sustainability: Do information communication and technology (ICT) and economic complexity matter? *International Journal of Environmental Research and Public Health*, *19*(19), 12301. DOI: 10.3390/ijerph191912301 PMID: 36231602

Khan, Z., Ali, S., Umar, M., Kirikkaleli, D., & Jiao, Z. (2020). Consumption-based carbon emissions and international trade in G7 countries: The role of environmental innovation and Renewable energy. *The Science of the Total Environment*, *730*, 138945. DOI: 10.1016/j.scitotenv.2020.138945 PMID: 32416502

Khumalo, P. (2014). Leadership will remain a key factor in attaining global environmental sustainability: What are the stakes? *Journal of Human Ecology (Delhi, India)*, *46*(2), 187–194. DOI: 10.1080/09709274.2014.11906719

Koenker, R. (2004). Quantile regression for longitudinal data. *Journal of Multivariate Analysis*, *91*(1), 74–89. DOI: 10.1016/j.jmva.2004.05.006

Kongbuamai, N., Bui, Q., Adedoyin, F., & Victor Bekun, F. (2022). Developing environmental policy framework for sustainable development in Next-11 countries: The impacts of information and communication technology and urbanization on the ecological footprint. *Environment, Development and Sustainability*. Advance online publication. DOI: 10.1007/s10668-022-02528-8

Kumar, R., & Kumar, S. (2017). Different technologies and components of renewable energy sources based power system: A Review. *International Journal of Trend in Scientific Research and Development*, DOI: 10.31142/ijtsrd165

Latif, Z., Xin, W., Khan, D., Iqbal, K., Pathan, Z. H., Salam, S., & Jan, N. (2017). ICT and sustainable development in South Asian countries. *Human Systems Management*, *36*(4), 353–362. DOI: 10.3233/HSM-17166

Lee, J. W., & Brahmasrene, T. (2014). ICT, CO2 emissions and economic growth: Evidence from a panel of ASEAN. *Global Economic Review*, *43*(2), 93–109. Advance online publication. DOI: 10.1080/1226508X.2014.917803

Li, J., Irfan, M., Samad, S., Ali, B., Zhang, Y., Bădulescu, D., & Bădulescu, A. (2023). The relationship between energy consumption, CO2 emissions, economic growth, and health indicators. *International Journal of Environmental Research and Public Health*, *20*(3), 2325. DOI: 10.3390/ijerph20032325 PMID: 36767690

Liu, H., Wong, W. K., Cong, P., Nassani, A. A., Haffar, M., & Abu-Rumman, A. (2023). Linkage among urbanization, energy consumption, economic growth and carbon Emissions: Panel data analysis for China using ARDL model. *Fuel*, *332*(2), 126122. DOI: 10.1016/j.fuel.2022.126122

Lopez, A., Mai, T., Lantz, E., Harrison-Atlas, D., Williams, T., & Maclaurin, G. (2021). Land use and turbine technology influences on wind potential in the United States. *Energy*, *223*, 120044. DOI: 10.1016/j.energy.2021.120044

Machado, J. A., & Silva, J. S. (2019). Quantiles via moments. *Journal of Econometrics*, *213*(1), 145–173. DOI: 10.1016/j.jeconom.2019.04.009

Mensah, C. N., Long, X., Boamah, K. B., Bediako, I. A., Dauda, L., & Salman, M. (2018). The effect of innovation on CO2 emissions of OCED countries from 1990 to 2014. *Environmental Science and Pollution Research International*, *25*(29), 29678–29698. DOI: 10.1007/s11356-018-2968-0 PMID: 30144011

Mngumi, F., Huang, L., Geng, X., & Ayub, B. (2024). Financial efficiency and CO2 emission in BRICS: Does digital economy development matter? *Heliyon*, *10*(2), e24321. Advance online publication. DOI: 10.1016/j.heliyon.2024.e24321 PMID: 38304825

Moyer, J. D., & Hughes, B. B. (2012). ICTs: Do they contribute to increased carbon emissions? *Technological Forecasting and Social Change*, *79*(5), 919–931. DOI: 10.1016/j.techfore.2011.12.005

Obobisa, E. S., & Ahakwa, I. (2024). Stimulating the adoption of green technology innovation, clean energy resources, green finance, and environmental taxes: The way to achieve net zero CO2 emissions in Europe? *Technological Forecasting and Social Change*, *205*, 123489. DOI: 10.1016/j.techfore.2024.123489

Ollo-López, A., & Aramendia-Muneta, M. E. (2011). ICT impact on competitiveness, innovation and environment. *Telematics and Informatics*, *29*(2), 204–210. DOI: 10.1016/j.tele.2011.08.002

Onwe, J. C., Bandyopadhyay, A., Hamid, I., Rej, S., & Hossain, E. (2023). Environment sustainability through energy transition and globalization in G7 countries: What role does environmental tax play? *Renewable Energy*, *218*, 119302. DOI: 10.1016/j.renene.2023.119302

Parks, G. A. (1967). Aqueous surface chemistry of oxides and complex oxide minerals: Isoelectric point and zero point of charge.

Pesaran, M. (2007). A simple panel unit root test in the presence of cross-section dependence. *Journal of Applied Econometrics*, *22*(2), 265–312. DOI: 10.1002/jae.951

Pesaran, M. H. (2004). General diagnostic tests for cross section dependence in panels. Cambridge Working Papers in Economics, 0435, Faculty of Economics, University of Cambridge.

Pesaran, M. H., Ullah, A., & Yamagata, T. (2008). A bias-adjusted Lm test of error cross-section independence. *The Econometrics Journal*, *11*(1), 105–127. DOI: 10.1111/j.1368-423X.2007.00227.x

Pesaran, M. H., & Yamagata, T. (2008). Testing slope homogeneity in large panels. *Journal of Econometrics*, *142*(1), 50–93. DOI: 10.1016/j.jeconom.2007.05.010

Pu, Z., Liu, J., & Yang, M. (2024). The effect of digital technology on residential and non-residential carbon emission. *International Review of Economics & Finance*, *95*, 103495. DOI: 10.1016/j.iref.2024.103495

Rafique, M. Z., Fareed, Z., Ferraz, D., Ikram, M., & Huang, S. (2022). Exploring the heterogenous impacts of environmental taxes on environmental footprints: An empirical assessment from developed economies. *Energy*, 238(A), 121753. DOI: 10.1016/j.energy.2021.121753

Raheem, I., Tiwari, A., & Lorente, D. (2020). The role of ICT and financial development in CO2 emissions and economic growth. *Environmental Science and Pollution Research International*, *27*(2), 1912–1922. Advance online publication. DOI: 10.1007/s11356-019-06590-0 PMID: 31760620

Sahoo, M., & Sethi, N. (2021). The intermittent effects of renewable energy on ecological footprint: Evidence from developing countries. *Environmental Science and Pollution Research International*, *28*(40), 56401–56417. DOI: 10.1007/s11356-021-14600-3 PMID: 34053045

Shahbaz, M., Shahzad, S. J. H., & Mahalik, M. K. (2017). Is globalization detrimental to CO2 emissions in Japan? New threshold analysis. *Environmental Modeling and Assessment*. Advance online publication. DOI: 10.1007/s10666-017-9584-0

Shahzad, U., Fareed, Z., Shahzad, F., & Shahzad, K. (2021). Investigating the nexus between economic complexity, energy consumption and ecological footprint for the United States: New insights from quantile methods. *Journal of Cleaner Production*, *279*, 123806. DOI: 10.1016/j.jclepro.2020.123806

Tran, T. T. D., & Smith, A. D. (2017). Evaluation of renewable energy technologies and their potential for technical integration and cost-effective use within the U.S. energy sector. https://ideas.repec.org/a/eee/rensus/v80y2017icp1372-1388.html

Ukoba, K., Olatunji, K. O., Adeoye, E., Jen, T., & Madyira, D. M. (2024). Optimizing renewable energy systems through artificial intelligence: Review and future prospects. *Energy & Environment, 35*(7), 3833–3879. DOI: 10.1177/0958305X241256293

Ulucak, R., Danish, , & Khan, S. U.-D. (2020). Does information and communication technology affect CO2 mitigation under the pathway of sustainable development during the mode of globalization? *Sustainable Development (Bradford), 28*(4), 857–867. Advance online publication. DOI: 10.1002/sd.2041

Wang, R., Mirza, N., Vasbieva, D. G., Abbas, Q., & Xiong, D. (2020). The nexus of carbon emissions, financial development, renewable energy consumption, and technological innovation: What should be the priorities in light of COP 21 Agreements? *Journal of Environmental Management, 271*, 111027. DOI: 10.1016/j.jenvman.2020.111027 PMID: 32778307

Waters, J. (2008). *Sustainable development.* The International Encyclopedia of Communication., DOI: 10.1002/9781405186407.wbiecs122

Wei, Y., Zhang, H., & Zhao, Z. (2024). Digitalization and pollution: Evidence from South Africa. *China Economic Review, 102266*, 102266. Advance online publication. DOI: 10.1016/j.chieco.2024.102266

Westerlund, J. (2007). Testing for error correction in panel data. *Oxford Bulletin of Economics and Statistics, 69*(6), 709–748. DOI: 10.1111/j.1468-0084.2007.00477.x

Xie, J., Ahmed, Z., Zhang, P., Khan, S., & Alvarado, R. (2023). Financial expansion and CO2 mitigation in top twenty emitters: Investigating the direct and moderating effects of the digital economy. *Gondwana Research, 125*, 1–14. DOI: 10.1016/j.gr.2023.07.013

Xie, P., & Jamaani, F. (2022). Does green innovation, energy productivity and environmental taxes limit carbon emissions in developed economies: Implications for sustainable development. *Structural Change and Economic Dynamics, 63*, 66–78. DOI: 10.1016/j.strueco.2022.09.002

Xing, C., & Ding, Y. (2019). Research on urban eco-environmental problems and sustainable development. *IOP Conference Series. Earth and Environmental Science, 300*(3), 32103. DOI: 10.1088/1755-1315/300/3/032103

Yilanci, V., & Pata, U. K. (2020). Investigating the EKC hypothesis for China: The role of economic complexity on ecological footprint. *Environmental Science and Pollution Research International*, *27*(26), 32683–32694. DOI: 10.1007/s11356-020-09434-4 PMID: 32514916

Zafar, M. W., Saud, S., & Hou, F. (2019). The impact of globalization and financial development on environmental quality: Evidence from selected countries in the Organization for Economic Co-operation and Development (OECD). *Environmental Science and Pollution Research International*, *26*(13), 13246–13262. DOI: 10.1007/s11356-019-04761-7 PMID: 30900127

Zaidi, S. A., Zafar, M. W., Shahbaz, M., & Hou, F. (2019). Dynamic linkages between globalization, financial development and carbon emissions: Evidence from Asia Pacific Economic Cooperation countries. *Journal of Cleaner Production*, *228*, 533–543. Advance online publication. DOI: 10.1016/j.jclepro.2019.04.210

Zeghlouli, J., Christophe, G., Guendouz, A., Modafar, C. E., Belkamel, A., Michaud, P., & Delattre, C. (2021). Optimization of Bioethanol Production from Enzymatic Treatment of Argan Pulp Feedstock. In Molecules (Vol. 26, Issue 9, p. 2516). Multidisciplinary Digital Publishing Institute. DOI: 10.3390/molecules26092516

Zhu, J. Z., & Cheung, K. W. (2012). Summary of environment impact of renewable energy resources. *Advanced Materials Research*, *1133*, 1133–1136. Advance online publication. . DOI: 10.4028/www.scientific.net/AMR.616-618.1133

Ziaul, I., & Shuwei, W. (2023). Environmental sustainability: A major component of sustainable development. *International Journal of Environmental Sustainability and Social Science*, *4*(2), 620–627. DOI: 10.38142/ijesss.v4i2.296

Chapter 8
Exploring Sustainability Practices of SMEs in an Emerging Market Economy:
Opportunities, Impact, Support, and Challenges

Peter Ansu-Mensah
https://orcid.org/0000-0002-3961-3109
Sunyani Technical University, Ghana

Paul Adjei Kwakwa
https://orcid.org/0000-0002-6516-217X
University of Energy and Natural Resources, Ghana

Nicholas Appiah
Sunyani Technical University, Ghana

ABSTRACT

This research explores the levels, opportunities, impact, support systems, and challenges associated with sustainability practices among SMEs within the Sunyani Municipality of Ghana. A descriptive and cross-sectional research design was employed, utilizing quantitative methods to collect data from SMEs across sectors such as manufacturing, services, retail, and agriculture. The results indicate that SMEs prefer energy-efficient processes as a sustainability practice. Again, the results suggest implementing sustainable practices can enhance market competitiveness, increased customer loyalty and improved brand reputation of SMEs. Moreover,

DOI: 10.4018/979-8-3373-1822-6.ch008

financial support, training and education were the support systems needed to practice sustainability. Similarly, high initial costs, insufficient funding, and inadequate infrastructure were the most challenges businesses faced. Finally, there includes discussion of the study's shortcomings, implications, and potential future research.

1. INTRODUCTION

Generally, businesses are increasingly taking sustainability initiatives into account to realize long-term advantages (Cantele & Zardini, 2018; Thorlakson et al., 2018). As a result, the significance of sustainability in today's globalized world has expanded beyond big businesses to include small and medium-sized businesses (SMEs) as vital change agents (Afolabi et al., 2023; Kiesnere & Baumgartner, 2019; Tomšič et al., 2015). Both developed and developing nations' economies are made up of SMEs and these companies are essential to promoting innovation, job creation, and economic growth because of their agility and spirit of entrepreneurship (Mugano & Dorasamy, 2024; Enaifoghe & Vezi-Magigaba 2023; Enaifoghe, 2024; Dabić et al., 2021; Heinicke, 2018). It is estimated that over 90% of global businesses are SMEs, responsible for more than half of all jobs (Enaifoghe, 2024). Yet, many SMEs lack the knowledge and expertise to recognize the benefits of environmental management systems (Aghelie, 2017).

Ideally, SMEs should actively adopt sustainability practices to positively impact environmental preservation while boosting their business performance (Franco & Rodrigues, 2021). By implementing socially and environmentally responsible practices, these businesses not only become more competitive but also spur societal advancement and prosperity (Islam et al., 2024; Yakymchuk, & Rataj, 2024; Elpisah, 2023; Veleva, 2021; Sommer, 2017).

Implementing and adopting sustainable practices by SMEs can bring numerous advantages, including cost savings, enhanced brand reputation and public image, improve business processes, customer loyalty, and competitive positioning (Okeke, 2024; Hopkins et al., 2011; Danciu, 2018). Similarly, blending in sustainability is not only vital for the preservation of the environment but also improves business performance and promotes social prosperity (Okeke, 2024), and these agree with SDGs of the United Nations Development Program (Fonseca et al., 2024; Omowole et al., 2024; Smith et al., 2022).

Notwithstanding their potential and substantial economic impact most SMEs show limited interest in environmental issues and often prioritize financial performance over social and environmental aspects due to challenges such as time constraints, limited expertise, and lack of resources which makes it difficult for SMEs to adopt sustainable practices (Antwi-Boateng et al., 2025; Journeault et al., 2021; Malesios et al., 2021 Caldera, 2019; Hoogendoorn et al., 2015).

Furthermore, SMEs contribute substantially to environmental pollution, and that about 60-70% of total pollution is linked to SMEs (Hoogendoorn et al., 2015), with those in manufacturing alone accounting for 64% of air pollution, although only 0.4% comply with environmental management standards (Behjati, 2017), while rapid urbanization also adds to these challenges (Liang et al., 2019).

Emerging market economies, which are marked by their fast industrialization and changing consumer dynamics (Rubaj, 2023; Sekarini & Selvabaskar, 2024), offer a special environment in which SMEs can significantly contribute to sustainable development (Oppong, 2022). The IMF's regional economic outlook for sub-Saharan Africa (SSA) emphasizes that SMEs provide most jobs in the region, thereby highlighting their economic significance (Nyika et al., 2024). Ghana, a developing market country in SSA region has a population of 31 million, whiles SMEs contribute 60-70% of its GDP (Ansu-Mensah et al., 2021; Adongo et al., 2020). Sunyani is also located in Ghana's Bono Region, and it is a vibrant commercial hub where various SMEs play an essential and solid role in the local economy (Adongo et al., 2020; Sunyani Municipal Assembly, 2019).

The lack of research on SMEs sustainability practices, particularly in areas like Ghana, highlights a significant gap in understanding how these businesses can shift toward sustainable operations. Again, existing studies lack detailed analysis of SMEs sustainability practices, especially in SSA regions like Ghana, where these enterprises are vital to economic growth. Also, despite their critical role, SMEs in Ghana lack opportunities that sustainability practices present. Besides, SMEs are unaware of how sustainability practices impact them. What is more, SMEs are ignorant of the support needed to adopt sustainability practices. Finally, SMEs face considerable challenges in adopting sustainable practices in an emerging market like Ghana. It is therefore important for SMEs to navigate a complicated web of possibilities, impacts, support systems, and obstacles as they move toward sustainable practices.

Against this backdrop, this study delves into the levels, opportunities, impact, support systems, and challenges associated with sustainability practices among SMEs in emerging markets. Consequently, the following research questions would be addressed:

RQ1. What levels of sustainability practices are performed by SMEs?
RQ2. How can SMEs leverage the opportunities presented by sustainability practices?
RQ3. To what extent do sustainability practices impact on SMEs?
RQ4. What support do SMEs need towards executing sustainability practices?

RQ5. What are the challenges SMEs face in implementing sustainability practices?

This paper provides valuable insights for both SMEs stakeholders and managers by exploring the practical implementation of sustainable management practices within SMEs, examining how these practices contribute to improved performance, and suggesting additional strategies. Also, for a number of factors that include social justice, environmental conservation, and economic growth, SMEs' adoption of sustainable practices is essential in emerging market economies. Besides, in emerging market economies, SMEs' adoption of sustainable practices is not merely a fad; rather, it is essential to attaining long-term social welfare, environmental preservation, and economic stability. Moreover, by prioritizing sustainability, SMEs can drive innovation, improve their competitiveness, and contribute to a more sustainable future for their communities and beyond. Furthermore, the findings from this research can guide policy and practice, fostering sustainable development and economic growth in emerging markets.

2. LITERATURE REVIEW

2.1 SMEs and Sustainability

All countries, whether developed or developing countries, benefit immensely from SMEs through job creation and income generation (Enaifoghe, 2024). The definitions of SMEs are varied in the literature (Abor & Quartey, 2010). Based on the size of workers and revenue the World Bank defines SMEs as s enterprises with up to 300 employees with a total annual sale of US$15 million (Mohammed & Bunyaminu, 2021).

According to Brundtland (1987), sustainable development focuses on meeting the needs of current generation without disadvantaging future generations to meet their needs. The principle behind this definition has shaped discussions in various disciplines. From the business perspective, some have hinted that sustainable development should incorporate a consideration of people, the planet and profit (Hoogendoorn, 2020). This has been extended to include that companies should engage in actions by taking the social and environmental concerns into consideration even though they may have profit as their main motive (Van Marrewijk & Werre, 2003). In order for firms to achieve their economic goals alongside the social and environmental concerns, scholars have shown that firm's ability to incorporate innovative means of production which enhances efficiency is crucial (Heenkenda et al., 2022, Toromade & Chiekezie, 2024).

2.2 Theories of Sustainability

2.2.1 Institutional Theory

Institutional theory argues that firms engage in sustainable behavior because it aligns with cultural, social, legal, and economic expectations (Glover, 2014; Ervin et al., 2013). The theory showcases norms, rules and some practices in a society shape the actions of firms towards sustainability (Scott, 2004). Two key concepts related to this theory are 'legitimacy' and 'isomorphism'. Legitimacy is about firms' adoption of sustainable practices that are deemed as appropriate and fall in line with organizational norms while Isomorphism is the adopted processes by organizations to achieve rationality and uniformity, that make them gain (Scott, 2004).

2.2.2 Natural Resource-Based View Theory

Resource-Based View argues that firms that manage their natural resources well and minimize the environmental impact of their activities have competitive advantage (Barney, 1991). This implies that firms need to be efficient with their resource utilization and have mechanism of addressing environmental impact of their activities to gain competitive advantage (Hart, 1995; Fowler and Hope (2007). Thus, companies are encouraged to adopt strategies including pollution reduction/ prevention, product stewardship, and promoting sustainable development. Pollution reduction/prevention highlights that firms should improve their operational efficiency by minimizing waste, emissions, and effluents (Hart, 1995). Product stewardship requires that companies involve the views of stakeholders when analyzing the environmental impacts of their products (Hart, 1995). According to Hart (1995), sustainable development prioritizes long-term market growth using low-impact products and technologies while actively engaging stakeholders.

2.2.3 Stakeholder Theory

Stakeholder theory posits it is important to incorporate the interests of all stakeholders, not solely those of shareholders in order to achieve sustainability (Freeman, 1984). Thus, involving the views of customers, suppliers, and local communities can enhance sustainability efforts. This holistic approach encourages businesses to generate value for all stakeholders while balancing economic, social, and environmental goals (Freeman, 1984). This theory is relevant for SMEs since there is a strong bond between them and the local communities where they operate. By engaging their stakeholder, such SMEs are able to address environmental issues and enhance their reputation as well (Pham et al., 2021).

2.3 Sustainability Practices in SMEs

2.3.1 Economic Sustainability Practices

As noted by Almaleeh (2019), economic sustainability involves achieving long-term growth through the responsible and efficient use of resources without compromising the well-being of the environment or society. This form of sustainability relies heavily on assets, equity, and returns on investment, which are generally more favorable for businesses that adopt economic, environmental, and social strategies (Shad et al., 2019). Arend (2014) identifies three key components: innovation, cost control, and technological investment. Cost containment and effective resource management are crucial for economic sustainability and these efforts often lead to increased customer loyalty, an enhanced brand reputation, and higher sales (Pham et al., 2021).

2.3.2 Social Sustainability Practices

The social aspect of sustainability involves an organization's influence on the community, encompassing community engagement, labor rights, human rights, and working conditions (Sy, 2016). Although implementing social sustainability practices can pose challenges such as costs and difficulties in measuring returns, they can also improve business processes, public image, employee morale, and brand recognition (Pham et al., 2021). According to Ciasullo and Troisi (2013), businesses that embrace social sustainability actively support their communities by promoting equitable access to opportunities and resources while enhancing health, well-being, and educational prospects. Examples of social sustainability practices for SMEs include fair labor practices, corporate social responsibility initiatives, and community engagement (Jenkins, 2009).

2.3.3 Environmental Sustainability Practices

Practices that foster environmental sustainability ensure that future generations have access to the natural resources necessary for healthy living. These strategies involve minimizing the environmental impact of business activities and are defined as improving human quality of life while staying within the planet's ecological limits (Brundtland, 1987). Incorporating environmental sustainability into business operations offers numerous benefits, including cost savings, opportunities to enter new markets, and improved brand reputation (Jenkins, 2009). The three primary pillars of environmental sustainability are waste management, the utilization of renewable resources, and the reduction of energy consumption and implementing these initiatives can lead to lower operating costs and diminish environmental harm (Agan et al., 2013). Efforts to decrease energy and fuel consumption are critical for reducing greenhouse gas emissions and pollutants (Ansu-Mensah & Kwakwa, 2022; Ansu-Mensah & Bein, 2019). Compliance with environmental regulations, along with the potential for cost savings through energy efficiency, are significant motivators for SMEs pursuing environmental sustainability (Cantele & Zardini, 2018).

3. METHODOLOGY

3.1 Study Area

The study took place in the Sunyani municipality. This is one of the twelve districts in the Bono Region of Ghana. The municipality was established on 10th March, 1989 by a legislative instrument (LI) 1473. This was the period Ghana adopted the District Assembly concept. The overall goal is to accelerate growth and development in the Municipality. The Sunyani Municipality can be located between 7°20' N, 2°20'W, and 7°33'N 2°33'W at an altitude of between 229 to 376 m. It has a total land area of 2,488 km2 with a population of 123,224 (GSS, 2012) and is mainly urban with more than 8 out of every 10 persons living in urban areas. The Municipality exhibits feature of an urban, peri-urban and rural area.

In the past, agricultural was the main economic activity in Sunyani. However, this has changed in recent times as a result of the pace of commercial, services and industrial activities. Currently, the service sector employs majority (58.3%) of the population in the Municipality. Many of the businesses in Sunyani like it is in the entire country are classified as SMEs. The location of the Sunyani Township in the middle belt makes it attractive to people from the Northern and Southern part of the country. The increased SMEs in the town necessitates they undertake sustainability practices. However, the current sustainability practices, opportunities and challenges of SMEs in Sunyani are not known.

3.2 Research Design and Sampling Issues

A descriptive and cross-sectional research design was chosen for this study. A quantitative research method was employed to gather numerical data and perform statistical analysis to understand the current state of sustainability practices among SMEs, the challenges they face, and the opportunities available to them. This design allows for a systematic collection of data and provides a snapshot of the phenomena at a specific point in time (Baker, 2000).

The target population for this study comprises all SMEs operating within the Sunyani municipality in the Bono Region of Ghana. SMEs are defined according to the criteria set by the Ghanaian government, typically characterized by the number of employees and annual turnover (Ghana Statistical Service, 2020). Determining the appropriate sample size is crucial for the reliability and validity of the study. The sample size was calculated using the formula for determining sample sizes in survey research.

$n = \frac{N}{1 + N(e^2)}$ where: (n) is the sample size, (N) is the population size, and (e) is the margin of error for this study (0.05). Given the estimated population of SMEs in Sunyani is 456, the sample size was calculated as follows:

$$n = \frac{N}{1 + N(e^2)}$$

$$n = \frac{456}{1 + 456(0.05^2)}$$

$n = 213$ respondents. Therefore, the sample size for this study is 213 SMEs.

A stratified random sampling technique was employed to ensure that the sample is representative of the diverse sectors within the SMEs population in Sunyani. The population was divided into strata based on the type of industry (e.g., manufacturing, services, retail), and a random sample was drawn from each stratum proportionally (Saunders et al., 2012).

3.3 Data Collection Methods

The primary data collection instrument used in this study was a structured questionnaire. The questionnaire was developed based on the research objectives and reviewed literature, ensuring it covers all relevant aspects of sustainability practices, opportunities, impacts, support systems, and challenges. The questionnaire consisted of closed-ended questions to facilitate quantitative analysis. The structured questionnaire included questions related to sustainability practices and the corresponding performance of the SMEs, attempting to measure the sustainability performance of the company according to the views of the owner/manager by providing ratings using a suitable Likert scale from 1-10 (10 being the highest and 1 lowest), with managers/owners ranking their company's practices and performance.

Data collection was carried out through face-to-face interviews and online surveys to maximize response rates. Two (2) members conducted the interviews, ensuring consistency and accuracy in data collection. The online surveys were distributed via email and social media platforms to reach a broader audience. The closed-ended questions comprised 20 items. Shown in Table 1 are the demographic information of respondents.

The questionnaires were self-designed and self-administered, consisting of both closed-ended questions and Likert-scale items. It was organized into five sections. Section "A" collected basic information about the respondents, including their age, gender, education level, business type, and years in operation. Section "B" inquired about the understanding of sustainability in SMEs. Section "C" considered the current sustainability practices implemented in various SMEs. Section "D" also considered the challenges in implementing sustainability practices in SMEs operations. Section "E" reflected on the impact of sustainability practices in SMEs operations. Section "F" identified the opportunities for SMEs in embracing sustainability practices. In sum, the questionnaire contained 20 items, formulated based on an extensive literature review to ensure content validity and comprehensiveness in covering the research objectives.

The questionnaire was pre-tested with a small group of SMEs (30 respondents) to identify any ambiguities and ensure the reliability and validity of the instrument. Feedback from the pre-test was used to make necessary adjustments to the questionnaire.

3.4 Data Analysis

The collected data were first checked for completeness and consistency. Any incomplete or inconsistent responses were discarded. The valid responses were then coded and entered in an SPSS Version 27.0 software for analysis. Descriptive statistics such as frequencies, percentages, means, and standard deviations were used to summarize and describe the data. These statistics provided an overview of the current state of sustainability practices, support systems, opportunities available and the challenges faced by SMEs in Sunyani, Ghana.

4. RESULTS AND DISCUSSIONS

4.1 Results

4.1.1 Demographics of Respondents

Most respondents were males, 90 (58.1%) and females 65 (41.9%). Respondents' ages ranged under 25 years (82) representing 53.2%, 25-34 years (55) with 35.7%, 35-44 years (13) with 8.4%, 45-54 years (3) with 1.9%, and over 55 years (1) representing 0.6%. Regarding educational level, 136 (88.3) were with tertiary qualifications, 17 (11%) had high school qualifications, and 1 (0.6%) had primary education. There were 90 (58.4%) employees, 19 (12.3%) were managers, and 45 (29.2%) were business owners. The businesses represented are predominantly SMEs, with 51.9% employing 1-10 people, and 28.2% employing 11-50 people. Most businesses have been operational for 1-5 years (50%), and 22.4% for less than a year, suggesting a predominance of relatively young firms that may be more flexible in adopting new sustainability practices. The presence of businesses operational for more than five years (27.5%) reflects some stability within the sector. Most respondents are from the services sector with 39.7% of respondents, followed by retail (28.5%) and manufacturing (19.2%).

Table 1. Demographics of respondents

Variable	**Frequency**	**Percentage**
Gender		
Male	90	58.1
Female	65	41.9
Age		
Under 25	82	53.2
25 – 34	55	35.7
35 – 44	13	8.4
45 – 54	3	1.9
Over 55	1	0.6
Education		
Primary	1	0.6
High School	17	11.0
Tertiary	136	88.3
Position in Business		
Employee	90	58.4
Manager	19	12.3
Owner	45	29.2
Business Size		
1 – 10	81	51.9
11 – 50	44	28.2
51 – 250	17	10.9
251+	9	5.8

continued on following page

Table 1. Continued

Variable	Frequency	Percentage
Years in Operation		
Less than 1 year	35	22.4
1 – 5 years	78	50.0
6 – 10 years	25	16.0
More than 10 years	18	11.5
Industry Business Operate in		
Manufacturing	29	19.2
Retail	43	28.5
Services	60	39.7
Agriculture	19	12.6

4.1.2 Sustainability Practices of SMEs

Results on the sustainability practices of SMEs are shown in Table 2. Majority of businesses in this study implement sustainability practices, indicating widespread adoption of sustainability practices. The practices include energy-efficient processes, sustainable sourcing, employee training, waste reduction, and use of renewable energy. In interpreting the results presented, it revealed the specific practices and how they were ranked using the Kendall's Coefficient of Concordance (KCC). What the KCC does is to simplify and summarize the rankings. Energy-efficient processes which was ranked first with a mean of 4.92 is the most common practice, and the following; sustainable sourcing (4.89), employee training (4.76), and waste reduction (4.63) in that order respectively. Eco-friendly product design (4.28), water reuse (4.27), use of renewable energy (4.25), and waste recycling (4.00) were fewer common practices.

4.1.3 Sustainability Practices Opportunities of SMEs

The opportunities for sustainability practices are shown in Table 2. Having ranked the opportunities in sustainability practices, using the Kendall's Coefficient of Concordance, it was discovered that the significant opportunities that sustainability practices create for SMEs include enhanced brand reputation with a mean rank of (5.92), followed by access to new markets (5.77), and long-term viability (5.66). However, regulatory compliance with (5.05) and government incentives/grants (5.27) had the lowest mean scores and were seen as not valuable.

Table 2. Results on sustainability practices and opportunities for SMEs

Variables	Items	Mean Rank	Rank	Kendall's Coefficient of Concordance
Sustainability practices				0.027***
	Energy-efficient processes	4.92	1st	
	Sustainable sourcing	4.89	2nd	
	Employee training on sustainability	4.76	3rd	
	Waste reduction	4.63	4th	
	Eco-friendly product design	4.28	5th	
	Waste reuse	4.27	6th	
	Use of renewable energy	4.25	7th	
	Waste recycling	4.00	8th	
Opportunities				0.009***
	Enhanced brand reputation	5.92	1st	
	Access to new markets	5.77	2nd	
	Long-term viability	5.66	3rd	
	Attracting investments	5.58	4th	
	Partnerships with other businesses	5.54	5th	
	Improved employee morale and productivity	5.44	6th	
	Innovations and new technologies	5.42	7th	
	Product development	5.36	8th	
	Government incentives and grants	5.27	9th	
	Regulatory compliance	5.05	10th	

*** p<1%

4.1.4 Impact of Sustainability Practices on SMEs

In Table 3 there is evidence of some impact of sustainability practices on SMES. Many businesses reported that enhanced market competitiveness (6.45), increased customer loyalty (6.27), improved brand reputation (6.02), and cost savings were notable impacts sustainable practices have on their businesses. Conversely, access to funding (4.92), market differentiation (5.07), and risk management (5.09) did not have any significant impact of sustainable practices on their businesses.

4.1.5 Challenges of Sustainability Practices of SMEs

The major challenges identified that impede sustainability efforts included high initial costs (6.68), insufficient funding (6.32), and inadequate infrastructure (5.66). however, regulatory and return on investment uncertainties (4.83, 4.89) are the least obstacles faced by SMEs (Table 3).

4.1.6 Sustainability Practices Support for SMEs

The most sought-after forms of support are financial (6.10), training and education (5.82) and access to information and resources. These preferences highlight the need for resources that can help alleviate the financial burden and improve the knowledge base for implementing sustainability practices (Table 3).

Table 3. Results on impact, support and challenges of sustainability practices

Variables	Items	Mean Rank	Rank	Kendall's Coefficient of Concordance
Impact				**0.060*****
	Enhanced market competitiveness	6.45	1st	
	Increased customer loyalty	6.27	2nd	
	Improved brand reputation	6.02	3rd	
	Cost savings	5.78	4th	
	Access to new markets	5.70	5th	
	Innovative and competitive	5.12	6th	
	Risk management	5.09	7th	
	Market differentiation	5.07	8th	
	Access to funding	4.92	9th	
	No noticeable impact	4.59	10th	
Support				**0.027*****
	Financial support	6.10	1st	
	Training and education	5.82	2nd	
	Access to information and resources	5.77	3rd	
	Networking opportunities	5.73	4th	
	Collaborations with NGOs	5.55	5th	
	Sustainability certifications	5.28	6th	
	Customer engagement programs	5.27	7th	
	Innovation grants	4.95	8th	

continued on following page

Table 3. Continued

Variables	Items	Mean Rank	Rank	Kendall's Coefficient of Concordance
	Policy and regulatory support	4.88	9th	
Challenges				**0.061*****
	High initial costs	6.68	1st	
	Insufficient funding	6.38	2nd	
	Inadequate infrastructure	5.66	3rd	
	Lack of knowledge & expertise	5.64	4th	
	Measurement & reporting	5.40	5th	
	Balancing profitability & sustainability	5.29	6th	
	Resistance to change from employees	5.24	7th	
	Short-term focus	4.98	8th	
	Uncertainty about returns on investment	4.89	9th	
	Regulatory uncertainties	4.83	10th	

*** p<1%

4.2 Discussion

The results indicate that among the SMEs in Sunyani, especially when it comes to sustainable practices, what they often do or prefer is energy-efficient processes, which means it is either they go in for efficient energy bulbs/appliances or they choose modern machines and not old machines (Reem et al., 2022, Kosyak & Popov 2020). By choosing sustainable sourcing which ranked second, also show that they look for places where their source of materials is sustainable (Wilhelm et al., 2016). On the contrary, waste recycling and waste reuse were the least used sustainability practices. This result bodes well with what goes on in an emerging market economy like Ghana. This is because as a sub-Sahara African country, recycling and reuse are not on the agenda of SMEs. Thus, firms in this part of the world are not overly concerned about recycling (Kassahun et al., 2025; Ayeleru et al., 2020).

The results of this study point out that enhanced market competitiveness, increased customer loyalty and improved brand reputation are distinguished impacts that sustainable practices have on SME businesses (Pham et al., 2021). In effect, firms in Ghana, Sunyani, feel that they are impacted greatly when they practice sustainability. SMEs become competitive in the marketplace, their customers become loyal and are retained and there is better-quality brand, and this supports the notion that sustainable practices not only benefit the environment but can also enhance competitive advantage in the marketplace (Kusumaryoko, 2024; Ahmed & Halil, 2023; Hendratmoko, 2023; Ali et al., 2021). Contrariwise, access to funding had little impact on firm's sustainable practices (Du et al., 2022). To the firms in Sunyani, Ghana, it is not about cost savings, because financial institutions do not regard sustainability practice as a condition to give them loans. To them, if they practice sustainability, it does not give them the chance to get loans from financial institutions. Thus, financial institutions do not regard sustainability practice as a condition to give them loans.

Besides, the results revealed that high initial costs, insufficient funding, and inadequate infrastructure were the most challenges businesses faced if they want to practice sustainability. (Álvarez Jaramillo et al., 2019). High initial cost because when one goes in for new machines which are efficient, their prices are very high. Again, going for renewable energy is equally expensive. Insufficient funding in the sense that it is difficult to secure funds to purchase new machines. In this part of the world, financial institutions are very skeptical about supporting SMEs. Thus, the above challenges are common barriers across emerging market economies (Msomi & Olarewaju, 2024; Moşteanu, 2023; Álvarez-Jaramillo et al., 2019; Sommer, 2017; Abdullahi et al., 2016). This aligns with existing literature that emphasizes how limited resources can hinder SMEs from adopting more advanced sustainable initiatives (Hossain, 2018). Actually, SMEs often struggle with the upfront costs and expertise needed to implement sustainable initiatives (Revell et al. 2009). Regulatory uncertainty was not perceived much as a challenge because of adaptability, entrepreneurial mindset, and community support (Kuratko et al., 2021). In contrast, studies show that regulatory pressure, entrepreneurial mindset, and community support play a vital role in SMEs sustainability and performance, mainly in uncertain environments and they can grow and compete efficiently in the digital market (Metris et al., 2024, Tyler et al., 2024, Akomea et al., 2023).

Additionally, when the opportunities in sustainability practices were ranked, there was no confirmation of any differences in the items. What it means is that the SMEs see all the items to be equal. The KCC of 0.009 is not significant, so they do not have differences in the opportunities. In contrast, various research suggests that sustainability practices can create opportunities for SMEs (Indriastuti et al., 2024; Abulibdeh et al., 2024; Pham et al., 2021; Onu & Mbohwa, 2021; Hur & Cassidy 2019; Webb et al., 2012).

Similarly, for support that they will need to help them practice sustainability, the rankings revealed financial support, training and education, and access to information and resources respectively. In short, financial support, training, education, and innovation are positively correlated with the sustainability of SMEs (Burlea-Schiopoiu & Mihai, 2019). Moreover, training, collaboration, and mentoring are important support systems to help SMEs practice sustainability (Brien & Hamburg, 2014).

Consequently, SMEs require more government incentives and financial support to overcome the barriers they face and to facilitate sustainability transitions (Alayón et al., 2022). The indication is that they need finance so as to improve their sustainability practices. Important facilitators, such as funding and information availability, can help remove several obstacles at once (Alayón et al., 2022). SMEs can improve their sustainability practices and even obtain a competitive edge in the market by concentrating on these support systems. Likewise, policy and regulatory support were very low. This indicates that SMEs are not keen on regulatory support but only on financial support that will enable them to practice sustainability (Burlea-Schiopoiu & Mihai, 2019).

5. CONCLUSION

The study examined the levels, opportunities, impact, support systems, and challenges associated with sustainability practices among SMEs in Sunyani, Ghana, an emerging market economy. The findings indicate a growing recognition among SMEs in Sunyani regarding the significance of sustainability. Again, among the SMEs in Sunyani, especially when it comes to sustainable practices, what they prefer is energy-efficient processes, and the choice of sustainable sourcing. In contrast, waste recycling and waste reuse were the least used sustainability practices. Similarly, the significant opportunities that sustainability practices create for SMEs include enhanced brand reputation, and access to new markets. However, regulatory compliance and government incentives/grants were not valued. Also, enhanced market competitiveness, increased customer loyalty, and improved brand reputation were outstanding impacts sustainable practices have on SMEs. Conversely, access to funding and market differentiation had insignificant impact on sustainable practices of SMEs. Moreover, financial support, and training and education were the coveted strong support systems SMEs need. Furthermore, though SMEs are actively striving to integrate sustainable practices into their operations, yet they are mired by significant challenges such as high initial costs, insufficient funding, and inadequate infrastructure.

6. RECOMMENDATIONS

Based on the findings, the following recommendations are proposed to facilitate the adoption of sustainability practices in SMEs. Policymakers should introduce financial incentives, such as grants, subsidies, or low-interest loans, specifically earmarked for sustainability initiatives. Additionally, regulatory bodies should endeavor to simplify and clarify environmental regulations to make them more accessible and understandable for small businesses. This will reduce the compliance burden and encourage more firms to adopt sustainable practices.

SMEs should prioritize enhancing their internal capacity by providing sustainability training for their employees. This will help improve the knowledge and skills necessary to effectively implement sustainable practices. Business owners should consider taking a phased approach to sustainability, beginning with cost-effective initiatives such as energy-efficient lighting and gradually advancing to more complex practices like eco-design and the adoption of renewable energy. SMEs can also benefit from collaborating with other businesses to exchange knowledge and resources, which can reduce costs and promote the adoption of best practices.

For SMEs to fully leverage the benefits of sustainability, a concerted effort from all stakeholders is required, including policymakers, business owners, NGOs, and support organizations. For SMEs in developing market economies to truly benefit from sustainability, a multi-stakeholder strategy is required. While policymakers should increase funding and streamline regulatory frameworks, industry groups may play a critical role in fostering peer learning and collaboration. Additionally, promoting a sustainable culture through public awareness campaigns can help create an environment that is conducive to the growth of SMEs. Ultimately, sustainable practices give SMEs a path to resilience, creativity, and long-term success. By overcoming challenges and grabbing opportunities, SMEs may significantly contribute to the sustainable growth of emerging economies, benefiting businesses, society, and the environment.

7. LIMITATIONS AND FUTURE RESEARCH

First, the use of non-random sampling introduced selection bias which may affect the generalizability of the findings and restrict conclusions about the sustainability practices of SMEs in Sunyani. Second, social desirability bias may have led to an overestimation of sustainability practices as respondents could have provided socially acceptable answers rather than accurate ones. Additionally, the focus was limited to the Sunyani Municipality, potentially limiting the applicability of the results to other regions in Ghana. Furthermore, time and financial constraints limited the depth and breadth of data collection and analysis. Lastly, focusing solely on the Sunyani Municipality within the Bono region may introduce location bias, as the sustainability practices and challenges faced by SMEs in this area may not fully represent those of SMEs in other parts of the Bono region or across Ghana.

REFERENCES

Abdullahi, M. S., Jakada, B. A., & Kabir, S. (2016). Challenges affecting the performance of small and medium scale enterprises (SMEs) in Nigeria. [JHCD]. *Journal of Human Capital Development, 9*(2), 21–46.

Abor, J., & Quartey, P. (2010). Issues in SME development in Ghana and South Africa. *International Research Journal of Finance and Economics, 39*(39), 218–228.

Abulibdeh, A., Zaidan, E., & Abulibdeh, R. (2024). Navigating the confluence of artificial intelligence and education for sustainable development in the era of industry 4.0: Challenges, opportunities, and ethical dimensions. *Journal of Cleaner Production, 437*, 140527. DOI: 10.1016/j.jclepro.2023.140527

Adongo, J. A., Atugeba, I. L. A., & Insah, B. (2020). An assessment of the relationship between capital investments and financial performance of selected SMEs in the Upper East Region of Ghana. *International Journal of Innovation and Applied Studies, 29*(3), 760–770.

Afolabi, H., Ram, R., Hussainey, K., Nandy, M., & Lodh, S. (2023). Exploration of small and medium entities' actions on sustainability practices and their implications for a greener economy. *Journal of Applied Accounting Research, 24*(4), 655–681. DOI: 10.1108/JAAR-09-2022-0252

Agan, Y., Acar, M. F., & Borodin, A. (2013). Drivers of environmental processes and their impact on performance: A study of Turkish SMEs. *Journal of Cleaner Production, 51*, 23–33. DOI: 10.1016/j.jclepro.2012.12.043

Aghelie, A. (2017). Exploring drivers and barriers to sustainability green business practices within small medium sized enterprises: Primary findings. *International Journal of Business & Economic Development, 5*(1), 41–48.

Ahmed, B., & Halil, M. (2023). Environmental Sustainability and Competitive Advantage: A Meta-Analysis. *Competitive Research Journal Archive, 1*(01), 80–90.

Akomea, S. Y., Agyapong, A., Ampah, G., & Osei, H. V. (2023). Entrepreneurial orientation, sustainability practices and performance of small and medium enterprises: Evidence from an emerging economy. *International Journal of Productivity and Performance Management, 72*(9), 2629–2653. DOI: 10.1108/IJPPM-06-2021-0325

Alayón, C. L., Säfsten, K., & Johansson, G. (2022). Barriers and enablers for the adoption of sustainable manufacturing by manufacturing SMEs. *Sustainability (Basel), 14*(4), 2364. DOI: 10.3390/su14042364

Ali, H., Chen, T., & Hao, Y. (2021). Sustainable manufacturing practices, competitive capabilities, and sustainable performance: Moderating role of environmental regulations. *Sustainability (Basel), 13*(18), 10051. DOI: 10.3390/su131810051

Almaleeh, N. (2019). Are Sustainable Firms More Profitable? Evidence from Egypt. *International Journal of Accounting and Financial Reporting*.

Álvarez Jaramillo, J., Zartha Sossa, J. W., & Orozco Mendoza, G. L. (2019). Barriers to sustainability for small and medium enterprises in the framework of sustainable development-Literature review. *Business Strategy and the Environment, 28*(4), 512–524. DOI: 10.1002/bse.2261

Ansu-Mensah, P., & Bein, M. A. (2019). Towards sustainable consumption: Predicting the impact of social-psychological factors on energy conservation intentions in Northern Cyprus. *Natural Resources Forum, 43*(3), 181–193. DOI: 10.1111/1477-8947.12174

Ansu-Mensah, P., & Kwakwa, P. A. (2022). Modelling electricity consumption in Ghana: The role of financial development indicators. *Green Finance, 4*(1), 54–70. DOI: 10.3934/GF.2022003

Ansu-Mensah, P., Kwakwa, P. A., & Nyamedor, S. T. M. (2021). Determinants of the business performance of small and medium enterprises in an emerging market economy. *International Journal of Management. Knowledge and Learning, 10*(1), 219–233.

Antwi-Boateng, C., Asumah, S., Kumi, E., & Ansu-Mensah, P. (2025). Examining the triple impact of eco-intellectual capital on the financial, social and environmental sustainability performance of SMEs in emerging economies. *Journal of Tropical Futures, 2*(1), 10–25. DOI: 10.1177/27538931241288959

Arend, R. J. (2014). Entrepreneurship and dynamic capabilities: How firm age and size affect the capability enhancement–SME performance relationship. *Small Business Economics, 42*(1), 33–57. DOI: 10.1007/s11187-012-9461-9

Ayeleru, O. O., Dlova, S., Akinribide, O. J., Ntuli, F., Kupolati, W. K., Marina, P. F., Blencowe, A., & Olubambi, P. A. (2020). Challenges of plastic waste generation and management in sub-Saharan Africa: A review. *Waste Management (New York, N.Y.), 110*, 24–42. DOI: 10.1016/j.wasman.2020.04.017 PMID: 32445915

Baker, M. J. (2000). Selecting a research methodology. *The Marketing Review, 1*(3), 373–397. DOI: 10.1362/1469347002530736

Barney, J. (1991). Firm resources and sustained competitive advantage. *Journal of Management, 17*(1), 99–120. DOI: 10.1177/014920639101700108

Behjati, S. (2017). Critical remarks about environmentalism implication by Iranian SMEs. *European Journal of Sustainable Development, 6*(3), 209–219. DOI: 10.14207/ejsd.2017.v6n3p209

Brien, E. O., & Hamburg, I. (2014). Supporting Sustainable Strategies for SMEs through Training, Cooperation and Mentoring. *Higher Education Studies, 4*(2), 61–69. DOI: 10.5539/hes.v4n2p61

Brundtland, G. H. (1987). *Report of the World Commission on Environment and Development: Our Common Future*. United Nations.

Burlea-Schiopoiu, A., & Mihai, L. S. (2019). An integrated framework on the sustainability of SMEs. *Sustainability (Basel), 11*(21), 6026. DOI: 10.3390/su11216026

Caldera, H. T. S., Desha, C., & Dawes, L. (2019). Evaluating the enablers and barriers for successful implementation of sustainable business practice in 'lean' SMEs. *Journal of Cleaner Production, 218*, 575–590. DOI: 10.1016/j.jclepro.2019.01.239

Cantele, S., & Zardini, A. (2018). Is sustainability a competitive advantage for small businesses? An empirical analysis of possible mediators in the sustainability–financial performance relationship. *Journal of Cleaner Production, 182*, 166–176. DOI: 10.1016/j.jclepro.2018.02.016

Ciasullo, M. V., & Troisi, O. (2013). Sustainable value creation in SMEs: A case study. *The TQM Journal, 25*(1), 44–61. DOI: 10.1108/17542731311286423

Dabić, M., Stojčić, N., Simić, M., Potocan, V., Slavković, M., & Nedelko, Z. (2021). Intellectual agility and innovation in micro and small businesses: The mediating role of entrepreneurial leadership. *Journal of Business Research, 123*, 683–695. DOI: 10.1016/j.jbusres.2020.10.013

Danciu, V. (2018). The Changing Focus of Green Marketing: From Ecological to Sustainable Marketing (III). *Romanian Economic Journal, 21*(68), 121–144.

Du, L., Razzaq, A., & Waqas, M. (2022). The impact of COVID-19 on small-and medium-sized enterprises (SMEs): Empirical evidence for green economic implications. *Environmental Science and Pollution Research International, 30*(1), 1540. DOI: 10.1007/s11356-022-22221-7 PMID: 35917071

Elpisah, E. (2023). Towards Inclusive Growth: Community-Centered Management Strategies for SMEs. *Golden Ratio of Community Services and Dedication, 3*(1), 29–39. DOI: 10.52970/grcsd.v3i1.606

Enaifoghe, A. (2024). Enhancing Small and Medium Enterprises (SMEs) in a Globalized and Innovative Economy: Challenges and Opportunities. *International Journal of Business. Economics and Social Development*, *5*(2), 130–138.

Enaifoghe, A., & Vezi-Magigaba, M. F. (2023). Conceptualizing the role of entrepreneurship and SME in fostering South Africa's local economic development. *International Journal of Research in Business and Social Science (2147-4478)*, *12*(4), 96-105.

Ervin, D., Wu, J., Khanna, M., Jones, C., & Wirkkala, T. (2013). Motivations and barriers to corporate environmental management. *Business Strategy and the Environment*, *22*(6), 390–409. DOI: 10.1002/bse.1752

Fonseca, L., Oliveira, E., Pereira, T., & Sá, J. C. (2024). Leveraging ChatGPT for Sustainability: A Framework for SMEs to Align with UN Sustainable Development Goals and tackle sustainable development challenges. *Management & Marketing*, *19*(3).

Fowler, S. J., & Hope, C. (2007). Incorporating sustainable business practices into company strategy. *Business Strategy and the Environment*, *16*(1), 26–38. DOI: 10.1002/bse.462

Franco, M., & Rodrigues, M. (2021). Sustainable practices in SMEs: Reducing the ecological footprint. *The Journal of Business Strategy*, *42*(2), 137–142. DOI: 10.1108/JBS-07-2019-0136

Freeman, R. E. (1984). *Strategic Management: A Stakeholder Approach*. Pitman.

Ghana Statistical Service. (2020) Population and Housing Census Report. Regional Office, Accra. https://statsghana.gov.gh

Glover, J., Champion, D., Daniels, K., & Dainty, A. (2014). An Institutional Theory perspective on sustainable practices across the dairy supply chain. *International Journal of Production Economics*, *152*, 152. DOI: 10.1016/j.ijpe.2013.12.027

Hart, S. L. (1995). A natural-resource-based view of the firm. *Academy of Management Review*, *20*(4), 986–1014. DOI: 10.2307/258963

Heenkenda, H. M. J. C. B., Xu, F., Kulathunga, K. M. M. C. B., & Senevirathne, W. A. R. (2022). The role of innovation capability in enhancing sustainability in SMEs: An emerging economy perspective. *Sustainability (Basel)*, *14*(17), 10832. DOI: 10.3390/su141710832

Heinicke, A. (2018). Performance measurement systems in small and medium-sized enterprises and family firms: A systematic literature review. *Journal of Management Control, 28*(4), 457–502. DOI: 10.1007/s00187-017-0254-9

Hendratmoko, S. (2023). Developing Green Sustainable Entrepreneurship for Competitive Advantages through Green Management. *Indonesian Journal of Business Analytics, 3*(6), 2163–2176. DOI: 10.55927/ijba.v3i6.5972

Hoogendoorn, B., Guerra, D., & Van Der Zwan, P. (2015). What drives environmental practices of SMEs? *Small Business Economics, 44*(4), 59–781. DOI: 10.1007/s11187-014-9618-9

Hopkins, M. S., Kruschwitz, N., Haanaes, K., Kong, M. T., Arthur, D., & Reeves, M. (2011). Sustainability: The 'embracers' seize advantage. *MIT Sloan Management Review, 52*(3), 3.

Hur, E., & Cassidy, T. (2019). Perceptions and attitudes towards sustainable fashion design: Challenges and opportunities for implementing sustainability in fashion. *International Journal of Fashion Design, Technology and Education, 12*(2), 208–217. DOI: 10.1080/17543266.2019.1572789

Indriastuti, N., Prilliandani, N. M. I., Sutadji, I. M., Setiyaningsih, T. A., & Gunawan, A. (2024). *Opportunities and Challenges: Implementation of Sustainable Business Practices in MSME's. Proceeding*. Islamic University of Kalimantan.

Islam, A., Wahab, S. A., & Tehseen, S. (2024). Rethinking holistic sustainable growth of Malaysian SMEs: A University Helix-Quadruple Bottom Line perspective. *International Journal of Innovation Science*. Advance online publication. DOI: 10.1108/IJIS-07-2023-0158

Jenkins, H. (2009). A 'business opportunity' model of corporate social responsibility for small- and medium-sized enterprises. *Business Ethics (Oxford, England), 18*(1), 21–36. DOI: 10.1111/j.1467-8608.2009.01546.x

Journeault, M., Perron, A., & Vallières, L. (2021). The collaborative roles of stakeholders in supporting the adoption of sustainability in SMEs. *Journal of Environmental Management, 287*, 112349. DOI: 10.1016/j.jenvman.2021.112349 PMID: 33735683

Kassahun, S. K., Assefa, B., Henzler, K., Weißert, J., Oteng-Ababio, M., Admassu, M., Mohammed-Amin, I., & Tesfahun, G. (2025). Firsthand report on solid waste management practice in the major town of Addis Ababa-Adama Economic Corridor, Ethiopia. *Heliyon, 11*(2), e41699. DOI: 10.1016/j.heliyon.2025.e41699 PMID: 39882454

Kiesnere, A. L., & Baumgartner, R. J. (2019). Sustainability management in practice: Organizational change for sustainability in smaller large-sized companies in Austria. *Sustainability (Basel), 11*(3), 572. DOI: 10.3390/su11030572

Kosyak, A., & Popov, S. (2020). Analyzing Consumer Preferences for Green Products and Environmental Impact. *Journal of Energy and Environmental Policy Options, 3*(4), 134–140.

Kuratko, D. F., Fisher, G., & Audretsch, D. B. (2021). Unraveling the entrepreneurial mindset. *Small Business Economics, 57*(4), 1681–1691. DOI: 10.1007/s11187-020-00372-6

Kusumaryoko, P. (2024). Exploring the Impact of Green Entrepreneurship on the Competitive Advantage of SMEs: A Sustainable Management Approach. *The Journal of Academic Science, 1*(7), 805–816.

Liang, L., Wang, Z., & Li, J. (2019). The effect of urbanization on environmental pollution in rapidly developing urban agglomerations. *Journal of Cleaner Production, 237*, 117649. DOI: 10.1016/j.jclepro.2019.117649

Malesios, C., De, D., Moursellas, A., Dey, P. K., & Evangelinos, K. (2021). Sustainability performance analysis of small and medium-sized enterprises: Criteria, methods and framework. *Socio-Economic Planning Sciences, 75*, 100993. DOI: 10.1016/j.seps.2020.100993

Metris, D., Kraugusteeliana, K., Amory, J. D. S., Mustafa, F., & Risdwiyanto, A. (2024). The Influence of Community Support on the Sustainability of MSMEs in the Digital Era. [TIRAKAT]. *Jurnal Terobosan Peduli Masyarakat, 1*(1), 21–29. DOI: 10.61100/j.tirakat.v1i1.102

Mohammed, I., & Bunyaminu, A. (2021). Major obstacles facing business enterprises in an emerging economy: The case of Ghana using the World Bank Enterprise Survey. *Journal of Small Business and Enterprise Development, 28*(3), 475–487. DOI: 10.1108/JSBED-04-2020-0110

Moşteanu, N. R. (2023). Navigating the intersection of sustainability and intelligent automation: Opportunities and challenges for entrepreneurs. *Human Resources Management and Services, 5*(1).

Msomi, T. S., & Olarewaju, O. M. (2024). Navigating the Green Frontier and Sustainable Innovation in the Dynamic Business Landscape. In *Waste Management and Life Cycle Assessment for Sustainable Business Practice*, 241-268. DOI: 10.4018/979-8-3693-2595-7.ch011

Mugano, G., & Dorasamy, N. (2024). *SMEs Perspective in Africa*. Springer Books. DOI: 10.1007/978-3-031-69103-4

Nyika, F., Muzekenyi, M., Akbar, K., Moodley, M., & Nzimande, S. (2024). Economic inclusion of rural small businesses in policy formulation: Strategies for sustainable development in Africa. *International Journal of Development and Sustainability, 13*(1), 51–67.

OECD. (2022). *Australia in Financing SMEs and Entrepreneurs 2022: An OECD Scoreboard*. OECD Publishing.

Okeke, A. (2024). Evaluating sustainable practices and supply chain management effectiveness in African small and medium-sized enterprises (SMEs). *Journal of Sustainability Research, 6*(2), e240033.

Omowole, B. M., Olufemi-Philips, A. Q., Ofodili, O. C., Eyo-Udo, N. L., & Ewim, S. E.Bamidele Micheal OmowoleAmarachi Queen Olufemi-PhillipsOnyeka Chrisanctus OfodileNsisong Louis Eyo-UdoSomto Emmanuel Ewim. (2024). Conceptualizing green business practices in SMEs for sustainable development. *International Journal of Management & Entrepreneurship Research, 6*(11), 3778–3805. DOI: 10.51594/ijmer.v6i11.1719

Onu, P., & Mbohwa, C. (2021). Industry 4.0 opportunities in manufacturing SMEs: Sustainability outlook. *Materials Today: Proceedings, 44*, 1925–1930. DOI: 10.1016/j.matpr.2020.12.095

Oppong, N. B. (2022). Sustainable development goals and small and medium enterprises: A comparative study of emerging economies and sub-Saharan Africa. *Global Business Review*, 1-14, 09721509221087848.

Pham, C., Do, T., Doan, T., Nguyen Thi, X. H., & Pham, T. (2021). The impact of sustainability practices on financial performance: Empirical evidence from Sweden. *Cogent Business & Management, 8*(1), 1912526. DOI: 10.1080/23311975.2021.1912526

Reem, M., Rasoolimanesh, S. M., & Wijesinghe Sara, N. R. (2022). Sustainability indicators in hotels: A systematic literature review. *Asia-Pacific Journal of Innovation in Hospitality and Tourism, 11*(1), 149–178.

Revell, A., Stokes, D., & Chen, H. (2009). Small businesses and the environment: Turning over a new leaf? *Business Strategy and the Environment, 19*(5), 273–288. DOI: 10.1002/bse.628

Rubaj, P. (2023). Emerging markets as key drivers of the global economy. *European Research Studies, XXVI*(4), 431–445. DOI: 10.35808/ersj/3294

Sarango-Lalangui, P., Álvarez-García, J., & del Río-Rama, M. D. L. C. (2018). Sustainable practices in small and medium-sized enterprises in Ecuador. *Sustainability (Basel), 10*(6), 2105. DOI: 10.3390/su10062105

Saunders, M. Lewis, P., & Thornhill, A. (2012). Research methods for business students. Published by Pearson Education, prentice Hall New York USA.

Scott, W. (2004). Institutional carriers: Reviewing modes of transporting ideas over time and space and considering their consequences. *Industrial and Corporate Change, 12*(4), 879–894. DOI: 10.1093/icc/12.4.879

Sekarini, S., & Selvabaskar, S. (2024). AI-Powered Branding: Enhancing Consumer Experience in Emerging Markets. In *Integrating AI-Driven Technologies into Service Marketing* (pp. 19–48). IGI Global. DOI: 10.4018/979-8-3693-7122-0.ch002

Shad, M. K., Lai, F. W., Fatt, C. L., Klemeš, J. J., & Bokhari, A. (2019). Integrating sustainability reporting into enterprise risk management and its relationship with business performance: A conceptual framework. *Journal of Cleaner Production, 208*, 415–425. DOI: 10.1016/j.jclepro.2018.10.120

Sidek, S., Rosli, M. M., Khadri, N. A. M., Hasbolah, H., Manshar, M., & Abidin, N. M. F. N. Z. (2020). Fortifying small business performance sustainability in the era of Ir 4.0: E-marketing as a catalyst of competitive advantages and business performance. *Journal of Critical Reviews, 7*(13), 2143–2155.

Smith, H., Discetti, R., Bellucci, M., & Acuti, D. (2022). SMEs engagement with the Sustainable Development Goals: A power perspective. *Journal of Business Research, 149*, 112–122. DOI: 10.1016/j.jbusres.2022.05.021

Sommer, C. (2017). *Drivers and constraints for adopting sustainability standards in small and medium-sized enterprises (SMEs)* (No. 21/2017). Discussion Paper.

Sunyani Municipal Assembly. (2019). Medium-Term Development Plan 2018-2021 Sunyani, Ghana.

Sy, M. (2016). Impact of sustainability practices on the firms' performance. *Asia Pac. Bus. Econ. Perspect, 4*, 4–21.

Thorlakson, T., De Zegher, J. F., & Lambin, E. F. (2018). Companies' contribution to sustainability through global supply chains. *Proceedings of the National Academy of Sciences of the United States of America, 115*(9), 2072–2077. DOI: 10.1073/pnas.1716695115 PMID: 29440420

Tomšič, N., Bojnec, Š., & Simčič, B. (2015). Corporate sustainability and economic performance in small and medium sized enterprises. *Journal of Cleaner Production*, *108*, 603–612. DOI: 10.1016/j.jclepro.2015.08.106

Toromade, A. S., & Chiekezie, N. R.Adekunle Stephen ToromadeNjideka Rita Chiekezie. (2024). Driving sustainable business practices in SMEs: Innovative approaches for environmental and economic synergy. *Int. J. Manag. Entrep. Res*, *6*(8), 2637–2647. DOI: 10.51594/ijmer.v6i8.1411

Tyler, B. B., Lahneman, B., Cerrato, D., Cruz, A. D., Beukel, K., Spielmann, N., & Minciullo, M. (2024). Environmental practice adoption in SMEs: The effects of firm proactive orientation and regulatory pressure. *Journal of Small Business Management*, *62*(5), 2211–2246. DOI: 10.1080/00472778.2023.2218435

Van Marrewijk, M., & Werre, M. (2003). Multiple levels of corporate sustainability. *Journal of Business Ethics*, *44*(2), 107–119. DOI: 10.1023/A:1023383229086

Veleva, V. (2021). The role of entrepreneurs in advancing sustainable lifestyles: Challenges, impacts, and future opportunities. *Journal of Cleaner Production*, *283*, 124658. DOI: 10.1016/j.jclepro.2020.124658 PMID: 33078048

Wach, K. (2015). Entrepreneurial orientation and business internationalisation process: The theoretical foundations of international entrepreneurship. *Entrepreneurial Business and Economics Review*, *3*(2), 9–24. DOI: 10.15678/EBER.2015.030202

Webb, K. J., Hodge, T. G., & Thompson, J. H. (2012). Small business sustainability: What is the CPA's role? *International Journal of Business and Social Justice*, *3*(12), 1–7.

Wilhelm, M. M., Blome, C., Bhakoo, V., & Paulraj, A. (2016). Sustainability in multitier supply chains: Understanding the double agency role of the first-tier supplier. *Journal of Operations Management*, *41*(1), 42–60. DOI: 10.1016/j.jom.2015.11.001

Yakymchuk, A., & Rataj, M. (2024). Development of the green policy strategies of enterprises: A decent work. *Ecology & Environment*, *89*(2), 752–752.

Chapter 9
Foreign Direct Investment and ICT: Catalysts for a Low Carbon Economy in Africa?

Paul Adjei Kwakwa
https://orcid.org/0000-0002-6516-217X
University of Energy and Natural Resources, Ghana

ABSTRACT

African countries continue to attract FDI, and the advent of ICT has also modified daily activities on the continent as well. Scholars have presented divergent views about the effects of FDI and ICT on environmental degradation. Achieving a low-carbon economy is a global goal. However, Africa has seen a rise in its carbon emissions over the years. This chapter, therefore, examines the impacts of foreign direct investment and ICT on carbon dioxide emissions in Africa. The study used 32 African countries over the 2002-2022 period from the World Bank for analysis. Regression analyses show that FDI and ICT have a U-shaped relationship with carbon emissions. Thus, an increased level of FDI and ICT will eventually increase carbon emissions in Africa. Suggested recommendations based on the results are offered for government and policymakers' consideration

1. INTRODUCTION

For many years, technological innovations have shaped economic activities and human lives in diverse ways. Technological innovations have, for instance, eased the way of doing work and they have improved the health of mankind. Consequently, significant financial resources have been invested to drive technological innovations.

DOI: 10.4018/979-8-3373-1822-6.ch009

The World Development Indicators of the World Bank (2024) show that global expenses on research and development, which is a main driver of technological innovation, have increased from 1.93% of GDP in 2007 to 2.62% of GDP in 2021. Expenses on technological innovation increased from US$ 396 billion in 2016 to US$5,260 billion in 2024 (Statista 2025) are projected to reach $4 Trillion by 2027 (IDC 2024).

The huge financial commitment toward technological innovation is meant to improve the fortunes of economies. For instance, some studies have found that ICT promotes economic growth (Remeikiene and Vebraite 2021; Appiah-Otoo and Song 2021; Awad and Albaity 2022). Others have also reported that ICT improves health outcomes (Shao et al., 2022; Văidean and Achim 2022). ICTs have been confirmed in empirical studies to promote education (Das 2019; Eyieyien et al., 2024; Hatamleh and Hatamleh 2024), and financial sector development (Raifu et al., 2024; Dhahri et al., 2024; Kwakwa et al., 2023).

Despite these gains from technological innovations, there remains an unsettled debate about its environmental impact. When it comes to the environmental impact of ICTs, the literature offers contrasting views. Some scholars suggest that ICTs promote efficiency to reduce environmental damage (Awad 2022) and they also facilitate the development of clean innovation (Brahmia et al., 2024). On the other hand, ICTs are said to be detrimental to the environment through their production effect, transportation or distribution effect, usage effect (Appiah-Otoo et al., 2023) and electronic waste effect (Brahmia et al., 2024). Studies estimate that ICTs contribute between 1% and 4% to global carbon emissions (Ayers et al., 2024). The production of ICTs triggers environmental damage because some of the inputs used for manufacturing ICT gadgets are extracted from the natural environment. Hence, the increased production of ICT gadgets would require more resources extracted from the environment which can be detrimental. After ICT gadgets have been manufactured, they are transported to various parts of the world. This process increases the consumption of fuel which may increase carbon dioxide emissions. Also, the usage of ICT gadgets relies heavily on electricity which can increase carbon emissions. After ICT gadgets have served their purpose or when they get spoilt, they are disposed of. However, in some cases, they are disposed of indiscriminately compounding the e-waste menace. The World Health Organization (WHO, 2024) reports that e-waste is one of the fastest-growing categories of solid waste globally. The organization estimated that with a global electronic production of 62 million tons in 2022, only 22.3% is known to be collected and recycled.

Since the mid-2000s, scholars have explored the relationship between ICT and environmental degradation. Many of the results have, however, been conflicting. For instance, Atsu et al., (2021) assessed the effect of ICT, energy consumption, financial development, and environmental degradation in South Africa. Among the variables that increased carbon dioxide emission was ICT. In the long run, they reported that a 1% increase in mobile phone adoption leads to a 0.19-0.40% increase in carbon dioxide emissions. Adebayo et al., (2022) focused on the carbon emissions from the top 10 ICT nations (Denmark, Japan, Luxemburg, South Korea, Netherlands, Iceland, Norway, Sweden, Switzerland, and the United Kingdom). A regression analysis revealed that carbon emissions in Denmark, Japan, Luxemburg, Netherlands, Norway, Sweden, the United Kingdom and Switzerland are reduced by ICTs while CO2 emissions are positively affected by ICT in the Netherlands, South Korea and Iceland.

Khan et al., (2024) explored how ICT and education affect environmental degradation among Asian countries. In both the short and long run periods, ICTs were found to exert an environmental reduction effect. Chien et al., (2021) tested the effect of ICT on carbon emissions in the BRICS economy and found ICT supports the reduction of carbon dioxide emissions. Majeed (2018) investigated the linkage between ICT and environmental sustainability in developed and developing countries. A regression analysis showed that ICT supports environmental sustainability for developed countries but degrades environmental quality among developing countries.

Khan et al., (2020) analyzed the role of ICT on environmental sustainability among 91 developed and developing countries. Their estimation showed that ICT helps in reducing environmental degradation. However, this reduction was prominent for developed countries but for developing countries, it was the opposite in the sense that ICT increases environmental degradation. Albaity and Awad (2023) also performed a heterogeneous effect of ICT on countries with different levels of ecological degradation and income (50 high-income economies, 49 upper-middle-income economies, and 72 low-income economies) between 2000 and 2020. They found that ICT increases environmental deterioration in some countries but increases environmental quality in other countries. Kwakwa et al., (2023) also found South Africa's carbon dioxide emissions are increased by ICT usage.

Closely related to technologies and the environment is the issue of FDI. According to the Pollution Haven Hypothesis FDI pollutes the environment when energy-intensive firms move from a stricter-environmentally regulated countries to countries with lax regulations on the environment (Gill et al., 2018). Such firms engage in practices that do not contribute to the promotion of environmental sustainability. On the other hand, FDIs come with the transfer of efficient technology that could promote environmental quality (Gill et al., 2018). Empirically, Huang et al., (2022) analyzed the effect of FDI on G20 economies and reported a positive effect. Wang et al., (2021) found that FDI increases carbon dioxide emissions in China. For the Ghanaian economy, Kwakwa et al., (2023) reported that FDI increases carbon dioxide emissions. Assessment of the carbon emissions effect of FDI in the Belt and Road Initiative (BRI) region by Wei et al. (2023) also confirmed that FDI directly increases carbon emissions. Ullah et al., (2022) assessment also revealed that carbon emissions in Vietnam are positively affected FDI.

However, Apergis et al., (2023) examined the effect of FDI from OECD to BRICS countries. It was established that the effect of FDI on BRICS countries was subject to the country of origin. For instance, FDI moving from Denmark and the UK positively affects environmental degradation among BRICS countries while FDI from France, Germany, and Italy promotes environmental quality via lower carbon emissions. It was also found that FDI from Austria, Finland, Japan, Netherlands, Portugal and Switzerland did not exert any significant impact on BRICS carbon emissions. Osei (2024) reported that while FDI inflows increase carbon emissions in sub-Saharan Africa, FDI outflows reduce carbon emissions for the sub-region. Yue et al., (2024) found that FDI reduces carbon emissions of China.

The above shows that the effects of ICT and FDI on carbon dioxide emissions are empirically unsettled. This offers reasons for further interrogation. A look at the African terrain suggests that owing to the low pace of capacity to develop home-grown technology, countries on the continent have relied on imported technologies through FDI in particular. Also, African countries have attracted FDIs (World Bank 2024) mainly to boost economic growth and enhance financial sector development which could still have some associated technological transfer.

The continent has also witnessed some improvement in its ICT infrastructure, especially mobile phone technology. Data from the World Bank (2024) shows mobile phone adoption on the continent. This has shaped the pattern of businesses and daily lives. While leaders of the continent are determined to reduce carbon dioxide emissions there has been an increase in the level of carbon dioxide emissions from 15.9 billion tons in 1990 to 51.01 billion tons in 2022 (Global Carbon Budget 2023). Scholars have therefore shown interest in assessing the factors of carbon dioxide emissions in Africa. The findings suggest that industrialization, economic growth, trade openness, population pressure, FDI, ICT, natural resources and financial development are among the key forces behind carbon dioxide emissions (Appiah-Otoo et al., 2023; Marfo et al., 2023; Kwakwa 2023). However, the effect of ICT and FDI on carbon emissions in Africa has received little comparative attention. Considering the pace of carbon emissions on the continent (Global Carbon Budget 2023), the increasing trend of FDI inflows (World Bank 2024a), huge investment towards ICT development on the continent (International Trade Administration, 2022), and the determination by African leaders to lower carbon economies (African Union 2015) it is important to assess the carbon emissions effect of FDI and ICT in Africa.

This chapter therefore adds to the literature first of all by providing further evidence on the carbon emission effect of FDI and ICT from Africa where the issue has been under-researched. Secondly, the chapter performs a non-linear assessment of the carbon emission effect of FDI and ICT. This analysis departs from the previous studies which have mainly assessed the linear relationship to provide further insights into the nexus between ICT and FDI on one side and carbon emissions on the other hand.

2. METHODOLOGY

2.1 Theoretical and Empirical Modelling

The study adopts the Stochastic Impacts by Regression on Population, Affluence, & Technology (STIRPAT) Model to assess the carbon emissions effect of ICT and FDI in Africa. The STIRPAT model is one of the frameworks used by scholars to assess the effect of economic and non-economic factors on environmental degradation. The model posits that impact on the environment (I) emanates from the population (P), affluence (A) and technology (T) which mathematically is expressed as:

$$I = a.P^{\beta_0}A^{\beta_1}T^{\beta_2}e \quad (1)$$

The variables a, β_0, β_1, β_2 *and* e are parameters to be estimated. Equation 1 can be linearized as:

$$\ln I = a + \beta_0 \ln P + \beta_1 \ln A + \beta_2 \ln T + e \quad (2)$$

Based on the objective of the study, carbon dioxide emissions (CO2E) which are a global environmental challenge represent environmental impact (I). Africa countries are becoming more urbanized in recent years (World Bank 2024). This trend puts significant pressure on environmental quality due to population growth (Kwakwa 2023). Hence urbanization (U) shall replace P. Economic growth (E), which denotes the level of economic activities, is used to present the affluence of a country. Technology in the model denotes all the other factors that affect environmental degradation, aside from population and affluence. For this reason, ICT and FDI which can affect the level of technological advancement are used to replace technology (T). From the above arguments equation (2) therefore is converted to become:

$$\ln CO2E_{it} = a + \beta_0 \ln U_{it} + \beta_1 \ln E_{it} + \beta_2 \ln ICT_{it} + \beta_3 \ln FDI_{it} + e_{it} \quad (3)$$

As it has been argued already, the effect of FDI on carbon emissions may be non-linear. Hence a square term of FDI is introduced to the model to get:

$$lnCO2E_{it} = a + \beta_0 lnU_{it} + \beta_1 lnE_{it} + \beta_2 lnICT_{it} + \beta_3 lnFDI_{it} + \beta_4 lnFDI^2_{it} + e_{it} \quad (4)$$

Also to test for the possible non-linear effect of ICT, a square term of ICT is introduced in the model:

$$lnCO2E_{it} = a + \beta_0 lnU_{it} + \beta_1 lnE_{it} + \beta_2 lnICT_{it} + \beta_3 lnFDI_{it} + \beta_4 lnICT_{it}^{2} + e_{it} \quad (5)$$

If the coefficients of ICT and ICT^2 are positive, it means that lower levels and upper levels of ICT exert a positive effect on carbon emissions. If the coefficient of ICT is positive but ICT^2 has a negative coefficient ^{it} indicates that a lower level of ICT development harms the environment by increasing carbon emissions while an advanced level of ICT development reduces carbon emissions. The opposite of the above holds. Also, the same interpretation applies to FDI and its square terms.

2.2 Data and Estimation Techniques

Data for the study were taken from the World Bank (2024a). Owing to data availability the study period was limited to 2002-2023. This period is also significant for this study considering the events related to FDI and ICT that have taken place in Africa within this period. In all 32 African countries listed in Table 1 were chosen for the study. This selection was based on data availability. In Table 2 the measurement of the variables is presented.

Table 1. List of countries used for analysis

	Countries	
Angola	Congo. DR	Algeria
Nigeria	Mauritania	Ghana
Benin	Tunisia	Namibia
Guinea. Bissau	Cote D'Ivoire	Guinea
Senegal	Mozambique	Congo Rep
Cameroon	Togo	Liberia
Kenya	Egypt	Sierra-Leone
Uganda	Mauritius	Comoros
Rwanda	Gambia	Madagascar
Botswana	Morocco	South Africa
	Burkina Faso	Tanzania

Table 2. Data measurement adopted

Variable	Measurement	Notation	Previous studies with similar measurement
Carbon dioxide emissions	CO2 emissions (metric tons per capita)	LCO2E	Somoye et al., (2023), Kwakwa (2023)
Income	GDP Per capita	LYPC	Xing et al., (2023)
Population pressure	Urbanization	LUB	Somoye et al., (2023)
Foreign Direct Investment	Foreign direct investment, net inflows (BoP, current US$)	LFDI	Chirilu and Costea (2023); Terzi and Pata (2020)
ICT	Mobile cellular subscriptions	LICT	Powanga and Kwakwa (2023)
Education	School enrollment, primary (% gross)	LEDU	Rani et al., (2022)

2.3 Estimation Approach

Panel data are associated with unit root which makes them non-stationary. This situation hampers the attainment of reliable results from such data. As a result, econometrics procedure requires data for analysis to be free from unit root. Unit root tests particularly, the Levin, Lin and Chu (LLC) and the Im, Pesaran and Shin (IPS) tests are used to determine the stationarity property of the variables. When variables are not stationary at levels the difference is taken to attain stationarity.

A cointegration assessment is also done to investigate the existence of long run relationship between carbon dioxide emissions on one hand and income, FDI, urbanization, ICT and education. The Pedroni cointegration test is adopted for this investigation. Once a long-run relationship exists between carbon emissions and the explanatory variables a cointegration regression technique is chosen to assess their long-run effects on carbon emissions.

3. RESULTS

3.1 Unit Root and Cointegration Results

The results in Table 3 show that the LLC test indicates majority of the variables are stationary at levels with few particularly, urbanization and mobile technology to be stationary at first difference. When it comes to the IPS results, they reveal that only income is stationary at first difference. The overall outcome is that the variables are stationary at levels or at the first difference and that the chosen variables for analysis can produce reliable results.

The Pedroni Cointegration results in Table 4 indicate there is a cointegration between carbon dioxide emission, and ICT, income, FDI, population pressure and education. In Table 5 it is observed also that there is confirmation for the model with the square term of FDI, and in Table 6 when the square term of ICT is added cointegration is found to be present. This conclusion is based on the fact that there are at least three of the statistics that reject the null hypothesis of no cointegration in each of the models. This denotes that in the long run, Africa's carbon dioxide emissions can be determined by ICT, income, FDI, population pressure and education. From these results, the impact exerted by ICT, income, FDI, population pressure and education on carbon emission is estimated through regression analysis.

Table 3. Unit root test results

	LLC Unit root test			**IPS Unit root test**
Series	At levels	First difference	At levels	First difference
LCO2E	-4.118***		--1.519*	
LUB	2.15	-14.167***	-1.457*	
LYPC	-5.681***		2.38 -	2.28**
LICT	-0.717	-3.47***	-2.313**	
LFDI	-11.443***		-5.513***	
LEDU	-5.265***		-1.376*	

*** and * denote 1% and 10% level of significance respectively

Table 4. Panel cointegration results

Alternative hypothesis: common AR coefs. (within-dimension)				
			Weighted	
	Statistic	Prob.	Statistic	Prob.
Panel v-Statistic	-1.549082	0.9393	-7.751361	1.0000
Panel rho-Statistic	2.799827	0.9974	-2.242141	0.0125
Panel PP-Statistic	-1.905947	0.0283	-9.593157	0.0000
Panel ADF-Statistic	5.931791	1.0000	3.602896	0.9998
Alternative hypothesis: individual AR coefs. (between-dimension)				
	Statistic	Prob.		
Group rho-Statistic	4.396526	1.0000		
Group PP-Statistic	-7.145189	0.0000		
Group ADF-Statistic	-68.95863	0.0000		

Table 5. Pedroni cointegration results for LCO2E LYPC LUB LFDI LFDI² LICT LED

Alternative hypothesis: common AR coefs. (within-dimension)				
			Weighted	
	Statistic	Prob.	Statistic	Prob.
Panel v-Statistic	-2.645157	0.9959	-6.010170	1.0000
Panel rho-Statistic	5.814683	1.0000	6.274707	1.0000
Panel PP-Statistic	-5.879865	0.0000	-16.41418	0.0000
Panel ADF-Statistic	-0.016646	0.4934	-2.965874	0.0015
Alternative hypothesis: individual AR coefs. (between-dimension)				
	Statistic	Prob.		
Group rho-Statistic	7.644773	1.0000		
Group PP-Statistic	-24.89982	0.0000		
Group ADF-Statistic	-2.140807	0.0161		

Table 6. Pedroni cointegration results for LCO2E LYPC LUB LFDI LICT LICT² LEDU

Alternative hypothesis: common AR coefs. (within-dimension)					
				Weighted	
		Statistic	Prob.	Statistic	Prob.
Panel v-Statistic		-2.308275	0.9895	-8.067903	1.0000
Panel rho-Statistic		5.583583	1.0000	6.105262	1.0000
Panel PP-Statistic		-8.230083	0.0000	-17.15478	0.0000
Panel ADF-Statistic		3.077386	0.9990	-2.495716	0.0063
Alternative hypothesis: individual AR coefs. (between-dimension)					
		Statistic	Prob.		
Group rho-Statistic		7.841480	1.0000		
Group PP-Statistic		-22.55901	0.0000		
Group ADF-Statistic		0.132502	0.5527		

3.2 Regression Results

The regression results in Table 7 focuses on a linear-relationship analysis for ICT and FDI and their impacts on carbon emissions. It is observed that income and urbanization have positive effects on carbon emissions. However, ICT, FDI and education have negative effects on carbon emissions. When attention is focused on the nonlinear analyses, the results in Tables 8 and 9 show that the effect of income and urbanization remains positive and education maintains its negative effect. However, while lower levels of ICT and FDI maintain their negative effects on carbon emissions, higher levels of ICT and FDI exert positive impacts. This suggests that higher ICT and FDI increase carbon emissions and may impede low carbon economy efforts in Africa. The above results are discussed below.

Income exerts a positive effect on carbon emissions as shown in Tables 7-9. The positive effect means that growth in income levels causes an increase in carbon dioxide emissions. This result confirms previous works including Krishnendu and Patra (2025), Yu et al., (2023) and Somoye et al., (2023). The intuition behind this is that an increase in income is associated with increased production and consumption that involve energy consumption which exerts more carbon dioxide. It also suggests that the quest for African countries to obtain higher economic growth rates has led to the growth of energy-intensive sectors thereby increasing carbon emissions.

Urbanization also has a positive effect on carbon emissions. This result is not surprising when compared with observation in Africa's towns and cities. Urban areas in Africa struggle with efficient transportation which increases fossil fuel consumption by vehicles. There is a high incidence of urban poverty in Africa which might have caused many urban dwellers to engage in behaviors that contribute to environmental degradation. Past studies like Krishnendu and Patra (2025) and Lv et al., (2023) found similar results.

Education has a negative effect on carbon emissions. This is an indication that as more people receive formal education, they become enlightened about environmental challenges which may influence them to engage in practices that promote. The results reveal that the improvement in the educational sector on the continent has promoted a reduction in environmental degradation via carbon dioxide emissions. This supports the findings reported by Huang et al. (2021) and Yuan et al. (2024).

Foreign direct investment and ICT have a negative effect on carbon dioxide emissions. This suggests that increasing FDI inflows to Africa helps reduce carbon dioxide emissions. This implies that the Pollution Halo effect holds for Africa. However, the nonlinear assessment results in Table 8 show that this negative effect changes to positive as FDI keeps increasing. This gives rise to a U-shaped relationship between FDI and carbon emissions in Africa. This is an indication that for African countries, although the Pollution Halo effect holds, it prevails at lower levels of FDI since at higher levels of FDI Pollution Haven effect is confirmed. A justification for this outcome is that institutional bodies that regulate the operations of foreign firms may appear stricter at the initial stages of the activities. However, with time such bodies become weak in enforcing the laws that regulate the operations of firms that should protect the environment. The reported U-shaped effect of FDI on carbon emissions, which confirms the Pollution Haven effects, aligns with evidence from Apergis et al. (2023), Huang et al. (2022), and Wang et al. (2021).

The results show that at lower levels the adoption of ICT helps in reducing carbon dioxide emissions. This may be because it helps in promoting efficient usage of resources (Awad 2022). Although this may imply that ICT proves to be a panacea for environmental degradation by lowering carbon dioxide emissions, the non-linear assessment results in Table 9 show that increased ICT adoption harms the environment by increasing carbon dioxide emissions. This denotes there is a U-shaped relationship between ICT and carbon dioxide emissions. One can argue that this holds because at lower levels of ICT usage, energy consumption may be minimal, contributing to lower carbon emissions. However, as more ICT are adopted energy usage increases thereby increasing carbon emissions. Higher levels of ICT usage may be associated with e-waste which could compound the environmental degradation. The outcome is in line with Sun et al. (2023), Naseem et al. (2023)

and Powanga and Kwakwa (2024) who reported a positive effect of ICT on carbon dioxide emissions.

Table 7. Linear regression results

Variable	DOLS			FMOLSs		
	Coefficient	Std. Error	Prob.	Coefficient	Std. Error	Prob.
LYPC	2.966467	0.019723	0.0042	1.189370	0.002475	0.0000
LUB	17.35835	0.301082	0.0110	0.512237	0.005371	0.0000
LFDI	-0.080232	0.002991	0.0237	-0.014360	0.001720	0.0000
LICT	-1.086523	0.013802	0.0081	-0.006197	0.001696	0.0000
LEDU	-15.94358	0.246540	0.0098	-2.322966	0.005160	0.0000
AdjR-squared	0.99			0.83		

Table 8. Nonlinear regression results for FDI

Variable	DOLS			FMOLS		
	Coefficient	Std. Error	Prob.	Coefficient	Std. Error	Prob.
LYPC	1.222723	0.028575	0.0000	1.229001	0.001407	0.0000
LUB	0.724613	0.046482	0.0000	0.820092	0.003161	0.0000
LFDI	-1.057153	0.054743	0.0000	-1.233863	0.003511	0.0000
$LFDI^2$	0.029072	0.001452	0.0000	0.032372	8.97E-05	0.0000
LICT	-0.020553	0.012769	0.0000	-0.041221	0.000966	0.0000
LEDU	-0.539763	0.098135	0.0000	-0.073736	0.006877	0.0000
Adj R-squared	0.89			0.91		

Table 9. Nonlinear regression results for ICT

	DOLS			FMOLS		
Variable	Coefficient	Std. Error	Prob.	Coefficient	Std. Error	Prob.
LYPC	1.220215	0.003731	0.0000	1.225028	0.042843	0.0000
LUB	0.821105	0.008404	0.0000	0.677756	0.098035	0.0000
LFDI	-0.024115	0.002588	0.0000	-0.004104	0.026706	0.0000
LICT	-1.587239	0.012054	0.0000	-1.276957	0.126949	0.0000
$LICT^2$	0.052165	0.000388	0.0000	0.043110	0.004139	0.0000

	DOLS			FMOLS		
LEDU	-0.032102	0.018740	0.0876	-0.582814	0.190466	0.0024
Adj R-squared	0.91			0.86		

4. CONCLUSION

Developing countries including African countries have targeted FDI inflows as a growth and development mechanism over the years. Ongoing development suggests that African countries will continue to rely on FDI to realize their growth and development potential. Moreover, African countries have taken steps to leverage ICT in support of economic activities. The African Agenda 2065 seeks to enhance low carbon economy on the continent. Yet, the continent has witnessed a significant jump in the level of carbon dioxide emissions which has serious implications for climate change. The effects of FDI and ICT on carbon dioxide emissions have become a concern. While some scholars suggest these factors can promote a low-carbon economy, others disagree.

Consequently, this study examines the effect of FDI and ICT on carbon dioxide emissions. The assessment was based on panel data from 32 African countries over the 2002-2022 period. Based on the conflicting results in previous studies, this work analyzed the non-linear effect of ICT and FDI on carbon emissions. Regression analyses were employed to estimate their effect on carbon emissions. The study found that both FDI and ICT have non-linear carbon emission effects in Africa. Specifically, lower levels of ICT and FDI reduce emissions while higher levels increase carbon emissions. Thus, ICT and FDI offer a short-term solution to the carbon dioxide emissions challenge in Africa; however, in the long run, they become a detriment. This underscores the idea that FDI and ICT have a limited role in propelling the continent to attain its low carbon economy status.

Other findings are that income growth and urban growth increase carbon dioxide emissions while education reduces carbon dioxide emissions in Africa. Until drastic actions are taken the touted carbon reduction effect of ICT and FDI cannot be experienced in Africa and also, urbanization and economic growth shall continue to impede sustainable environment. Consequently, a number of policy recommendations are made for government, scholars and policymakers to consider.

There is a need for African countries to review existing policies that shape the activities of foreign firms on the continent. Where policies are inadequate to support a low-carbon economy, revisions to safeguard environmental sustainability will be necessary. Where the existing laws and regulations are good enough to promote low carbon economy, attention should be given to the enforcement agencies. State institutions that should be implementing established policies and laws are usually under-resourced and as such governments and policymakers should consider resourcing them to effectively function.

Recycling ICT waste products can help reduce their long-term harmful environmental degradation effect. Incorporating renewable energy in production, distribution and usage of ICT tools can be helpful toward the attainment of a low carbon economy in Africa.

Improving educational outcomes on the continent can support the fight against increased carbon dioxide emissions. It is important for authorities to therefore minimize the numerous challenges that education confronts in Africa. Specifically, by increasing enrollment, retention, and the quality of teaching carbon emissions can be lowered in Africa. School curriculums, starting at the basic level, should include environmental degradation and sustainability.

The positive effect of income on emissions calls for the promotion of cleaner and more efficient methods for economic activities. Pro-environmentalism should be inculcated in the citizens as the economy grows to help reduce carbon dioxide emissions. To avert the positive effect urban growth has on carbon emissions, governments in Africa need to work towards sustainable urbanization. This will enhance efficient transportation, buildings and resource usage.

This study is limited to African countries and analyzes total carbon dioxide emissions, mobile technology, and net FDI. Future studies may explore how institutional quality may affect the relationship established between ICT, FDI, and carbon dioxide emissions. Future studies can explore how ICT and FDI affect other forms of environmental degradation in Africa.

REFERENCES

Adebayo, T. S., Agyekum, E. B., Altuntaş, M., Khudoyqulov, S., Zawbaa, H. M., & Kamel, S. (2022). Does information and communication technology impede environmental degradation? fresh insights from non-parametric approaches. *Heliyon*, *8*(3), e09108. DOI: 10.1016/j.heliyon.2022.e09108 PMID: 35313485

Albaity, M., & Awad, A. (2023). The heterogeneous effect of ICT on countries with different levels of ecological degradation and income: A panel quantile approach. *Journal of Open Innovation*, *9*(2), 100055. DOI: 10.1016/j.joitmc.2023.100055

Apergis, N., Pinar, M., & Unlu, E. (2023). How do foreign direct investment flows affect carbon emissions in BRICS countries? Revisiting the pollution haven hypothesis using bilateral FDI flows from OECD to BRICS countries. *Environmental Science and Pollution Research International*, *30*(6), 14680–14692. DOI: 10.1007/s11356-022-23185-4 PMID: 36163571

Appiah-Otoo, I., Acheampong, A. O., Song, N., & Chen, X. (2023). The impact of information and communication technology (ICT) on carbon dioxide emissions: Evidence from heterogeneous ICT countries. *Energy & Environment*, *34*(8), 3080–3102. DOI: 10.1177/0958305X221118877

Appiah-Otoo, I., & Song, N. (2021). The impact of ICT on economic growth-Comparing rich and poor countries. *Telecommunications Policy*, *45*(2), 102082. DOI: 10.1016/j.telpol.2020.102082

Atsu, F., Adams, S., & Adjei, J. (2021). ICT, energy consumption, financial development, and environmental degradation in South Africa. *Heliyon*, *7*(7), e07328. DOI: 10.1016/j.heliyon.2021.e07328 PMID: 34258464

Awad, A. (2022). Is there any impact from ICT on environmental quality in Africa? Evidence from second-generation panel techniques. *Environmental Challenges*, *7*, 100520. DOI: 10.1016/j.envc.2022.100520

Awad, A., & Albaity, M. (2022). ICT and economic growth in Sub-Saharan Africa: Transmission channels and effects. *Telecommunications Policy*, *46*(8), 102381. DOI: 10.1016/j.telpol.2022.102381

Brahmia, S. Y., & Mannai, S. (2024). Environmental Degradation in Gulf Cooperation Council: Role of ICT Development, Trade, FDI, and Energy Use. *Sustainability (Basel)*, *17*(1), 54. DOI: 10.3390/su17010054

Chien, F., Anwar, A., Hsu, C. C., Sharif, A., Razzaq, A., & Sinha, A. (2021). The role of information and communication technology in encountering environmental degradation: Proposing an SDG framework for the BRICS countries. *Technology in Society*, *65*, 101587. DOI: 10.1016/j.techsoc.2021.101587

Chirilu , A., & Costea, A. (2023). The effect of FDI on environmental degradation in Romania: Testing the pollution Haven hypothesis. *Sustainability (Basel)*, *15*(13), 10733. DOI: 10.3390/su151310733

Das, K. (2019). The role and impact of ICT in improving the quality of education: An overview. *International Journal of Innovative Studies in Sociology and Humanities*, *4*(6), 97–103.

Dhahri, S., Omri, A., & Mirza, N. (2024). Information technology and financial development for achieving sustainable development goals. *Research in International Business and Finance*, *67*, 102156. DOI: 10.1016/j.ribaf.2023.102156

Eyieyien, O. G., Idemudia, C., Paul, P. O., & Ijomah, T. I. (2024). The Impact of ICT Projects on Community Development and Promoting Social Inclusion.

Gill, F. L., Viswanathan, K. K., & Karim, M. Z. A. (2018). The critical review of the pollution haven hypothesis. *International Journal of Energy Economics and Policy*, *8*(1), 167–174.

Global Carbon Budget. (2022), "Per capita CO 2 emissions, 1750-2022", available at: https:// ourworldindata.org/co2-and-greenhouse-gas-emissions

Hatamleh, A. I., & Hatamleh, I. M. (2024, January). The Importance of Using ICT in Improving the Quality of Education. In *2024 ASU International Conference in Emerging Technologies for Sustainability and Intelligent Systems (ICETSIS)* (pp. 95-99). IEEE. DOI: 10.1109/ICETSIS61505.2024.10459655

Huang, J., Li, X., Wang, Y., & Lei, H. (2021). The effect of energy patents on China's carbon emissions: Evidence from the STIRPAT model. *Technological Forecasting and Social Change*, *173*, 121110. DOI: 10.1016/j.techfore.2021.121110

Huang, Y., Chen, F., Wei, H., Xiang, J., Xu, Z., & Akram, R. (2022). The impacts of FDI inflows on carbon emissions: Economic development and regulatory quality as moderators. *Frontiers in Energy Research*, *9*, 820596. DOI: 10.3389/fenrg.2021.820596

IDC. (2024). Worldwide Spending on Digital Transformation. Accessed from https:// www.idc.com/getdoc.jsp?containerId=prUS52305724

Khan, F., Zahir, S., Rahman, H. U., Raza, A., & Noor, S. (2024). Exploring the role of ICT and education in reducing environmental degradation among Asian countries. In *Natural Resources Forum*. Oxford, UK: Blackwell Publishing Ltd. DOI: 10.1111/1477-8947.12548

Khan, F. N., Sana, A., & Arif, U. (2020). Information and communication technology (ICT) and environmental sustainability: A panel data analysis. *Environmental Science and Pollution Research International*, *27*(29), 36718–36731. DOI: 10.1007/s11356-020-09704-1 PMID: 32566985

Krishnendu, J., & Patra, B. (2025). Environmental impacts of economic growth: A STIRPAT analysis using machine learning algorithms. *Sustainable Futures : An Applied Journal of Technology, Environment and Society*, *9*, 100404. DOI: 10.1016/j.sftr.2024.100404

Kwakwa, P. A. (2023). Climate change mitigation role of renewable energy consumption: Does institutional quality matter in the case of reducing Africa's carbon dioxide emissions? *Journal of Environmental Management*, *342*, 118234. DOI: 10.1016/j.jenvman.2023.118234 PMID: 37327731

Kwakwa, P. A., Adjei-Mantey, K., & Adusah-Poku, F. (2023). The effect of transport services and ICTs on carbon dioxide emissions in South Africa. *Environmental Science and Pollution Research International*, *30*(4), 10457–10468. DOI: 10.1007/s11356-022-22863-7 PMID: 36071365

Kwakwa, P. A., Adzawla, W., Alhassan, H., & Oteng-Abayie, E. F. (2023). The effects of urbanization, ICT, fertilizer usage, and foreign direct investment on carbon dioxide emissions in Ghana. *Environmental Science and Pollution Research International*, *30*(9), 23982–23996. DOI: 10.1007/s11356-022-23765-4 PMID: 36331739

Kwakwa, P. A., Dankwah, J. B., Adu Boahen, E., & Hammond, P. (2023). Financial development in South Africa: The role of natural resources, IT infrastructure, and government size. *Cogent Economics & Finance*, *11*(2), 2281844. DOI: 10.1080/23322039.2023.2281844

Lv, T., Hu, H., Xie, H., Zhang, X., Wang, L., & Shen, X. (2023). An empirical relationship between urbanization and carbon emissions in an ecological civilization demonstration area of China based on the STIRPAT model. *Environment, Development and Sustainability*, *25*(3), 2465–2486. DOI: 10.1007/s10668-022-02144-6

Majeed, M. T. (2018). Information and communication technology (ICT) and environmental sustainability in developed and developing countries. *Pakistan Journal of Commerce and Social Sciences*, *12*(3), 758–783.

Naseem, S., Hu, X., Shi, J., Mohsin, M., & Jamil, K. (2023). Exploring the optical impact of information communication technology and economic growth on CO2 emission in BRICS countries. *Optik (Stuttgart)*, *273*, 170339. DOI: 10.1016/j.ijleo.2022.170339

Osei, A. (2024). Investing for a greener future: Unpacking the dual impact of FDI on carbon emissions and the transformative role of renewable energy. *SN Business & Economics*, *5*(1), 14. DOI: 10.1007/s43546-024-00779-3

Raifu, I. A., Okunoye, I. A., & Aminu, A. (2024). The effect of ICT on financial sector development in Africa: Does regulatory quality matter? *Information Technology for Development*, *30*(3), 424–451. DOI: 10.1080/02681102.2023.2233458

Rani, T., Amjad, M. A., Asghar, N., & Rehman, H. U. (2022). Revisiting the environmental impact of financial development on economic growth and carbon emissions: Evidence from South Asian economies. *Clean Technologies and Environmental Policy*, *24*(9), 2957–2965. DOI: 10.1007/s10098-022-02360-8 PMID: 35874941

Remeikiene, R., & Vebraite, V. (2021). The role of ICT development in boosting economic growth in transition economies. *Journal of International Students*, *14*(4).

Shao, M., Fan, J., Huang, Z., & Chen, M. (2022). The Impact of Information and Communication Technologies (ICTs) on Health Outcomes: A Mediating Effect Analysis Based on Cross-National Panel Data. *Journal of Environmental and Public Health*, *2022*(1), 2225723. DOI: 10.1155/2022/2225723 PMID: 35990542

Somoye, O. A., Ozdeser, H., Seraj, M., & Turuc, F. (2023). The determinants of CO2 emissions in Brazil: The application of the STIRPAT model. *Energy Sources. Part A, Recovery, Utilization, and Environmental Effects*, *45*(4), 10843–10854. DOI: 10.1080/15567036.2023.2251921

Statista (2024). Information technology (IT) worldwide spending from 2005 to 2024. Accessed from https://www.statista.com/statistics/203935/overall-it-spending-worldwide/

Sun, Y. (2023). Income inequality, carbon emissions, and green development efficiency. *Environmental Science and Pollution Research International*, *30*(8), 21081–21091. DOI: 10.1007/s11356-022-23583-8 PMID: 36264465

Terzi, H., & Pata, U. (2020). Is the pollution haven hypothesis (PHH) valid for Turkey? *Panoeconomicus*, *67*(1), 93–109. DOI: 10.2298/PAN161229016T

Ullah, S., Nadeem, M., Ali, K., & Abbas, Q. (2022). Fossil fuel, industrial growth and inward FDI impact on CO2 emissions in Vietnam: Testing the EKC hypothesis. *Management of Environmental Quality*, *33*(2), 222–240. DOI: 10.1108/MEQ-03-2021-0051

Văidean, V. L., & Achim, M. V. (2022). When more is less: Do information and communication technologies (ICTs) improve health outcomes? An empirical investigation in a non-linear framework. *Socio-Economic Planning Sciences*, *80*, 101218. DOI: 10.1016/j.seps.2021.101218

Wang, Y., Liao, M., Xu, L., & Malik, A. (2021). The impact of foreign direct investment on China's carbon emissions through energy intensity and emissions trading system. *Energy Economics*, *97*, 105212. DOI: 10.1016/j.eneco.2021.105212

Wei, G., Bi, M., Liu, X., Zhang, Z., & He, B. J. (2023). Investigating the impact of multi-dimensional urbanization and FDI on carbon emissions in the belt and road initiative region: Direct and spillover effects. *Journal of Cleaner Production*, *384*, 135608. DOI: 10.1016/j.jclepro.2022.135608

WHO. (2024). Electronic waste (e-waste). Accessed from Information technology (IT) worldwide spending from 2005 to 2024. Accessed from https://www.who.int/news-room/fact-sheets/detail/electronic-waste-(e-waste)

World Bank. (2024). *Measuring the Emissions and Energy Footprint of the ICT Sector: Implications for Climate Action.* Washington, D.C. Accessed https://documents.worldbank.org/curated/en/099121223165540890/P17859712a98880541a4b71d57876048abb

World Bank. (2024a). World Development Indicators. Accessed from https://databank.worldbank.org/source/world-development-indicators

Xing, L., Khan, Y. A., Arshed, N., & Iqbal, M. (2023). Investigating the impact of economic growth on environment degradation in developing economies through STIRPAT model approach. *Renewable & Sustainable Energy Reviews*, *182*, 113365. DOI: 10.1016/j.rser.2023.113365

Yu, S., Zhang, Q., Hao, J. L., Ma, W., Sun, Y., Wang, X., & Song, Y. (2023). Development of an extended STIRPAT model to assess the driving factors of household carbon dioxide emissions in China. *Journal of Environmental Management*, *325*, 116502. DOI: 10.1016/j.jenvman.2022.116502 PMID: 36274310

Yuan, Q., Wang, R., Tang, H., Ma, X., & Zeng, X. (2024). A study on the potential of higher education in reducing carbon intensity. *PLoS One*, *19*(11), e0309546. DOI: 10.1371/journal.pone.0309546 PMID: 39591416

Yue, L., Yan, H., & Han, L. (2024). The carbon emission reduction effect of two-way foreign investment from the perspective of government environmental attention. *Environment, Development and Sustainability*, •••, 1–26. DOI: 10.1007/s10668-024-05679-y

Chapter 10
Optimizing Technology Innovation to Enhance Environmental Bonds for Improving Environmental Quality.

Keshav Anand
https://orcid.org/0009-0001-6285-4306
Christ University, India

Babita Chaudhary
https://orcid.org/0009-0003-2854-6670
Christ University, India

ABSTRACT

This new incorporation of such high technologies transform the efficiency and effectiveness of environmental bonds as a tool for sustainable finance. This is because most of the pitfalls that may characterize greenwashing, lack of transparency, and weak stakeholder engagement are taken care of by these technologies while undermining the impact of environmental bonds. For instance, accountability, trust and measurable environmental outcomes are ascertained through timely monitoring, and enhanced reporting standards. Blockchain blocks the alteration of data, while IoT promotes continuous environmental data monitoring; AI links to environmental impact assessment through predictive analytics. Implementation of such technologies will require careful ethical consideration in terms of privacy of data and access. Technology adoption should be responsible to achieve its full potential in improving environmental quality and reaching the SDGs. The approach aligns

DOI: 10.4018/979-8-3373-1822-6.ch010

corporate innovation with public sustainability objectives and facilitates real and social development.

INTRODUCTION

Green bonds are among the most innovative financial instruments so far that are used in order to invest in projects which may foresee results for achieving favorable impacts on the environment and contribute towards the meeting of Sustainable Development Goals. In contrast to the traditional regulatory mechanism of imposing Pigouvian taxes to internalize environmental externalities, green bonds instead resort to the mechanism of market. The firms are rewarded for undertaking environmentally good projects with monetized returns.

However, despite their future prospects, the widespread usage and efficiency of green bonds are seriously challenged. The first critical issue is definitely greenwashing: projects announced as green but not providing benefits in terms of ecology. Such a misrepresentation may dissolve trust from investors, hence reducing the credibility of such green financing initiatives.

The skepticism regarding the real environmental outcome of financed projects is caused by lack of transparency about how funds are allocated and used for projects. This usually leads to very low stakeholder confidence and is another barrier to the growth and success of the green bond market.

The promising solutions to these challenges come through digital technologies. Blockchain technology can bring in an immutable ledger system to increase the transparency of recording transactions in a safe and transparent record. This allows money spent to be channeled exactly for the purpose intended at use, creating trust between the investors and the stakeholders. The direct connection of IoT devices gives real-time monitoring of the environment, hence possible assessment in terms of its ecological impacts. Artificial Intelligence complements these technologies further by adding predictive analytics and impact assessments in such a manner that business objectives are aligned with measurable environmental outcomes.

All of these-implementation of blockchain, IoT, and AI in an intertwined manner-bring large-scale benefits from the aspects of accountability to those of reporting. Companies may, then apply them to have verifiable data over the environmental impacts of the project and therefore bring stakeholder confidence thereby. Investors would be attracted. However, they spur innovation at the private level toward attainment of public sustainability goals.

However, these need to be fitted into the framework of green bond with careful concern for ethical values such as keeping data private and equitable access by all to technological facilities and by algorithmic justice to avoid exclusions of vulnerable people and for integrity of solution. The good adoption of those technologies will win trust and thereby facilitate long-lasting collaboration among stakeholders to generate the maximum results from green bonds in environmental transformations.

This study examines how blockchain, AI, and IoT unlock the full potential of environmental bonds as sustainable finance instruments. In this regard, the study makes cases through various interviews with the different stakeholders, as well as corporate sustainability reports, as a way technology closes the gaps within these structures to provide measurable environmental results. Generally speaking, the best approach would be one that combines technological innovation and ethics to maximize the applicability of environmental impacts toward achieving scalar and meaningful outcomes.

METHODOLOGY

This is a qualitative study based on secondary sources of information from journals, industry reports, credible websites, and leading authors. This research study deals with a theme in the environmental finance; hence the subject matter here is which of these subjects has relationships with technological innovations as it primarily focuses on topics related to environmental bonds. Previous literature is reviewed closely to critically investigate the efficiency of technologies in developing bonds for environmental protection purposes. The best practices and potential areas for improving applications of AI, blockchain, and Big Data in environmental finance are identified during critical analysis. Additional theoretical frameworks' development is integrated for critical review of case studies using information created from corporate sustainability reports and other regulatory documents for the purpose of understanding practical adaptation and application in real-world systems of environmental finance.

Key points will then be ethical issues in terms of data privacy and transparency as well as equitable distribution of financial and technological means. The debate has to balance these challenges and opportunities in presenting the discussion regarding the advanced technologies as well as the roles of environmental finance that ought to offer clearer insights into how effective and more credible technological innovation in improving the environmental bonds actually is.

LITERATURE REVIEW

The current environmental finance trajectory has been very well set off by sophisticated technologies like blockchain, artificial intelligence, and the Internet of Things.All of them promise higher transparency, accountability, and measurability for environmental finance, which very well aligns with the pursuit of Sustainable Development Goals. In this regard, environmental bonds get a review of the contribution that these technologies are making towards them. In this respect, pros and cons along with ethical considerations have also been taken care of (Yadav et al., 2023).

Blockchain Technology in Environmental Bonds

Blockchain is one of the new environmental finance technologies which is perhaps opening degrees of transparency and immutability that have never been felt before in the environmental bond market. Blockchain refers to a situation where all transactions are seen by stakeholders; decentralization is sure to not sacrifice sustainability claims (Chen & Wang, 2022).

The benefits of using blockchain for environmental bonds include immutable ledger technology. Such an application makes each transaction traceable and auditable. This will further reduce cases of greenwashing because stakeholders can believe that environmental projects are indeed being undertaken as expected (Naderi & Tian, 2022). For instance, a reforestation finance program sponsored by an environmental bond will put in the blockchain all donations. It tracks the amount of money and how efficiently it is being used in doing the project (Chen & Wang, 2022).

Blockchain technology has also reformed sustainability reporting because it made tracking data automation and hence reliable and effective as well. Consequently, it thus eliminates fraud ability because this technique gives an indelible record of every published report (Bhandari et al., 2025).

As the blockchain technology is implemented, it ensures that reporting will go live in real-time. Consequently, the carbon offset measures of biodiversity-related data will be right for the concerned parties (Chen et al. 2022).

Artificial Intelligence (AI) in Environmental Bonds

Predictive analytics, real-time monitoring of data, and proper decision-making shall be the significant functions of AI for the optimization of environmental bond projects.

Kinds of AI: Machine learning and deep learning are used in order to assess the environmental risks so that it may be likely to succeed or not; however, critical information needs to be provided for optimising environmental factors. The biggest application of AI in environmental finance is predictive analytics. AI measures humongous amounts of data produced by environmental projects such as carbon sequestration rates, amount of energy saved, or biodiversity improvement (Chen & Wang, 2022).

For example, AI-based systems can predict how much carbon dioxide is absorbed by forest regeneration, hence allowing investment decisions based on more quantifiable environmental outcomes (Alotaibi & Nassif, 2024).

The AI-based NLP models are very easy to relate with the stakeholders and conform to environmental policies. AI can, in real terms, be set to automatically adapt to such a level that cumbersome regulations will never have any human interaction to enable the fact that an environmental bond will never default on its project through illegal standards (He et al., 2025). This is the same case when AI captures the stakeholder sentiment by making inferences on how the investors, community, or even other regulatory departments think. Then it will construct opinions which bring this to contribute to the deciding policy (Alotaibi & Nassif, 2024).

Internet of Things (IoT) in Environmental Bonds

Internet of Things (IoT) basically refers to the connected devices that, in real time, can capture and send data about the environmental conditions for further study. For these reasons, technologies become sources of information on conditions at a previously unseen level of detail, whereby the progress of projects financed with environmental bonds could be monitored at all times. Such inventions will make the IoT-based devices such as smart cameras and smart sensors forever collect data on the complete level of parameters that include air quality, water level, and other similar parameters pertaining to environmental issues. All this collected data is extremely important in the real-time business cycle in order to conduct the performance analytics related to the sustainability-based projects so that one can remain within track to fulfill the desired value of environmental factors (Ambilwade et al., 2024).

For example, sensors installed during river restoration can monitor the water quality almost in real-time so that feedback is immediate on the health status of the ecosystem. It has been reported that these devices have already successfully demonstrated the generation of data in tremendous volume transmitted to a central server for analysis and reporting (Chen & Wang, 2022).

IoT also reduces human interference, which therefore makes it easier and efficient in the collection and reporting process. The automated systems will take information relating to the progress of the project directly to the environmental bond without necessarily involving manual input (Ambilwade et al., 2024).

According to the research, smart water management systems have enhanced operational efficiency and the correctness of data compared to traditional reporting and monitoring (Naderi & Tian, 2022).

ETHICAL CHALLENGES AND GOVERNANCE IN TECHNOLOGICAL ADOPTION

While all such promising applications of blockchain, AI, and IoT in environmental finance entail numerous ethical issues concerning data privacy and bias from algorithms, and access. All these are to be handled with extreme caution so that everyone involved in the deal is able to win fair and square (Adeyelu et al., 2024).

Data Privacy and Security Issues

The more the environmental data stored in blockchain and IoT devices, the more issues of data privacy and security. That blockchain permits more openness is setting up trust in many applications. However, these are not always appreciated when it comes to access to information that needs to be kept out of the limelight (Adeyelu et al., 2024). For instance, information about local communities or private landowners can be published without their knowledge. Protection through encryption has to be of very high complexity in this case (Yadav et al., 2023).

Algorithmic Bias and Fairness

Although AI algorithms are efficient, biases can propagate when they are not controlled. Biased predictions in environmental finance change the project's assessment; thus, resources might be distributed unevenly, or some stakeholders get unfair treatment (Adeyelu et al., 2024). Biases can be avoided only if AI models are designed and tested for fairness. This means monitoring the algorithm in real time, and it's fine-tuning with these considerations for results justifiable to stakeholders and communities (Tan et al. 2021).

Ethical Technology Governance Framework

There is a need to establish an all-encompassing governance framework in order to address ethical concerns involving blockchain, AI, and IoT when using the same in the context of environmental bonds. This architecture should be based on privacy protection, algorithmic fairness in its predictions, and equal access to technological resources (Adeyelu et al., 2024). The government will also be expected to foster sustained stakeholder engagement in ensuring that the above technologies are put in a way that works for everybody. Suitable regulatory frameworks will be required to ensure the responsible development of the said technologies (He et al., 2025).

EMERGING TRENDS IN TECHNOLOGY: DIGITAL TWINS AND DEFI

Digital twins and DeFi, which stands for decentralized finance, are the two trends that are going to emerge from this technology space and will transform the structuring and financing of environmental bonds. This is where accessibility is heightened, and complexity in environmental finance is decreased (Naderi & Tian, 2022).

Digitised Twins for Environmental Simulation

In a nutshell, the digital twin refers to virtual modeling that simulates the behavior and performance of the physical object or system. Environmental bonds can also use digital twins to simulate the environmental impact of the project so that stakeholders may be able to see possible outcomes and risks (Chen & Wang, 2022). For example, for a marine conservation project, its digital twin might simulate the effects of any biodiversity conservation strategies such that more rational decisions can be made over which strategies should be funded (He et al., 2025).

Decentralized Finance, or DeFi, is Peer-to-Peer Lending

The new peer-to-peer lending and investment options in DeFi platforms are a likely source for financing environmental bonds. The systems operate on a blockchain that bypasses third-party intermediaries such as banks, creating freer and cheaper access to the platforms (Alotaibi & Nassif, 2024).

DeFi may be used in small-sized environment projects that might not access traditional capital sources to acquire a much larger investor base while democratizing access to capital resources for environmental sustainability causes (Naderi & Tian, 2022).

TECHNOLOGY-BASED PERSONALIZATION IN GREEN BONDS

In the emerging world of green finance, smart technologies such as Artificial Intelligence and the Internet of Things have altered the face of environmental bonds substantially. Personalization of the environmental bond then means the adjustment in terms of terms to fit those environmental and finance goals of involved parties. This more specific approach would pay much more attention toward more responsive, dynamic funding and therefore better address environmental projects within more aligned objectives from investors to local communities while avoiding government controls.

The use of dynamic interest rates and performance-based rewards would make the green bonds even more attractive while further increasing the level of commitment towards achieving impacts in the environmental aspects. Green bond personalization comes from adjusting the financial terms, contractual arrangements, and those aligned with achievable and measurable green targets. Personalized green bonds shall depend on the following: dynamic interest rates, milestone-driven incentive schemes, and penal schemes for default performance.

This focused approach links financial returns to ecological outcomes, making investors fund projects that are clearly producing environmental value. Conventional environmental bonds, on the other hand, are generally general-purpose financial instruments funding general environmental initiatives, such as renewable energy, reforestation, and carbon abatement. Such tools are effective but lack the specificity and flexibility required to adapt to unique requirements for a project. Personalized bonds mitigate this by introducing flexibility along with performance-based incentives toward more favorable environmental impact and investor engagement.

But the rise of AI and IoT brings along with it the ability to tackle environmental issues in the local community much more precisely. For instance, such a renewable energy initiative could be promoted, and quantified reduction levels for emissions; the bonds for restorations can also be designed such that the trees' growth and healthy soil increases can be traced. This makes the bond, in its form of personalization, align environmental and financial ends in one line with each other (Moursi et al., 2025).

AI plays a crucial role in the personalization of green bonds because it goes through huge volumes of data and identifies patterns that lead to actionable insights. With the help of AI processing, a large-scale environment scheme can be assessed to study its feasibility and predict success. For example, in renewable energy project, AI analysis of historical data on wind speed, solar radiation, and other meteorological factors can make an assessment based on those inputs on whether the output of the project meets both environmental and financial targets. This analysis corrects the terms of the green bond where aspects like interest rates, repayment time frames, and risk ratings are defined. In doing so, AI balances bond terms against project viability, reduces the risks to investors, which, in turn lowers the interest rates as the financing is made more efficient and attractive.

However, AI is a strong driver to improve the rate of the bond in which high risks have made such high-investment more attractive in a high amount so that more amount is raised as a result that pays for its high risk (Katten, 2021).

One of the best advantages is that AI can provide a dynamic interest rate for customizing green bonds. In traditional bonds, fixed interest rates are there for the tenure of the bonds, but with AI systems, it can vary rates dynamically in accordance with real-time performance from the project. For instance, where such a project, renewable energy that is possibly funded through a green bond, has performed better than anticipated in terms of emission reduction or production of energy, the rate needs to be reduced in advance. In this phase, the individuals developing the project are motivated directly in maintaining focus for attaining goals by green-linking financial benefits with environmental performance directly. Therefore, it converts the sense of responsibility to the environment into measurable progress and actionable results towards sustainability.

In the same manner, in the case of IoT, this technology also aids real-time, personal feedback along with sophisticated monitoring devices like environment sensors and camera-laden drones. The machines track important parameters of the environment over time related to air quality, health of soil, temperature, and biodiversity. In this regard, all the stakeholders receive these data in real-time for immediate insights regarding the impact made by the project on the environment. AI and IoT bring joint accuracy, efficiency, and accountability to green bond investments. Adjustments in green bond terms are made in light of such real-time feedback on environmental effects and that this fiscal funding system becomes output-driven. This is to ensure that capital invested will be associated more closely with ecological end results, ensuring transparency through sustainability in green finance.

With Internet of Things (IoT)-based environmental monitoring, real time intelligence about activities is provided to diverse stakeholders. In a reforestation scheme, the IoT sensors would monitor growth and soil conditions in real time and report actual data on the levels of carbon being sequestered. For instance, in case of sharp rise in new growths and the offsetting benefit exceeded expectations, then such bonds issued would be re-collated at a premium on an interest differential basis. Such an approach makes the environmental output equivalent to the financial returns that are generated through payouts from investment as well as pushes more investment in sustainability-related activity.

The second use of IoT monitoring is linked to social and economic indicators that will determine the implementation of projects. IoT networks will keep track of the participation of the community in forest restoration activities or even employment statistics from renewable energy projects, such as wind power, or solar. It becomes part of how green bonds can become more accountable and inclusive and show the social good for the economy in general. This information comes from the social and economic consequences of the project on a larger scale, allowing green bonds to evolve as a platform so that they can more effectively address these **"spin-off"** impacts.

AI and the Internet of Things allow us to gather real time data or to make on the spot decisions whether a project is finished on time or not. This information will enable the argument for retrofitting or modification of a bond's terms (e.g., interest rate or incentives) to depend on project viability. This flexibility allows bondholders to time their action to see how the environmental goal can be achieved while the system can be built through trusted interactions. Investors now have an easy on-site means of seeing that their investments have a tangible impact in the real world, thereby creating a strong basis for long-term, sustainable finance.

This comes on the grounds of transparency while calling more people to the aspect of sustainable finances since one would now clearly observe how finances tally with the resulting impact on the environment that would be ascribed to those very finances. The scope that is offered by the integration of AI and IoT with green bonds is endless, but challenges that integration brings have to be faced so that the innovations will be sustainable for a long time.

The main areas are data security, privacy, and ethical aspects. There have been growing apprehensions in many quarters that such information, as might be garnered with real-time environmental data via IoT devices, may raise severe data-security and privacy issues, as it gathers much sensitive information that pertains to land use, environmental health, and local ecosystems, and the collected data should also ensure security storage and transmission.

Another importance is in the form of protecting it against cyber attacks, along with unapproved access, so that no disturbance is brought to the genuineness of the green bond system. Privacy is yet another issue for which this information might be pertaining to private owners and communities; hence, rigorous data protection processes such as encryption and anonymization, etc accompanied by access control need to be implemented.

Optimizing the Engagement and Retention of Stakeholders in Environmental Bonds Through Technology

This technology, for maximum deployment and sustainability over time, has to engage and keep the stakeholders concerned in environmental bonds. This will be due to increased demands among the issuers, investors, and implementing parties for a greater level of transparency, accountability, and the actual measurable effects of the implementation on the environment. Examples are advanced artificial intelligence, the Internet of Things, and blockchain. Environmental bonds with state-of-the-art technologies of AI, IoT and blockchain will bring a varied aspect as several moves are still at hand in line to bring down the expectation line of stakeholders considering environmental aspects well aligned with. With a view to achieving up-to-date real-time updated data; crafting actionable insight under customized policy frameworks of communications to engage stakeholders but any stakeholder endures in an engagement activity to this type of bond for pretty long times ahead. This is because environmental bonds' success hinges on how successful the interaction would be between all parties; therefore, such an interaction's formation would be key to ensuring that there are continual investments in and support for as well as promoting the same through mouth-of-mouth effect. Afterwards, with a traditional approach which is annual reporting and a static unidirectional environmental reporting will not be beyond expectations in addressing the current time information needed as well as the current level of dynamic information needed.

Technology-mediated communication is a technological/operational promise of speed, fidelity and auditing, which in turn leads to greater transparency/translucency and hence dialogue in the running to better value-deliver to corporate stakeholders, over–and towards–the world they care about.

Real-time capability and reporting environmental performance are areas in which technology is strong. In the field of reforestation, for instance, IoT sensors can track tree biomass, soil characteristics, and carbon sequestration. An AI then learns from the data to derive an estimate of the environmental long-term effect of the project. These tools guarantee that every one who wants to contribute to the green bond design is in any case immediately aware of it. Enough of such data is also able to allow issuers to act promptly, leading to trust and subsequently to raise further capital [low to moderate quality of evidence].

Blockchain technology is also a powerful instrument for encouraging the stakeholders' participation. Because of its security, transparency, and immutability, environmental performance and fund usage record is tamperproof. Blockchain allows the enforcement of smart contracts, which can, autonomously, modify the terms of a bond when a set of sustainability targets is specified. To cite an example, achieving targets (e.g., hitting the targets of carbon emission reduction or biodiversity rise) may lead to a drop of interest rates or further disbursements. Through such a process, there will be ownership and a long-term personal involvement, as the individuals will be able to control the project and see the physical result of their investment.

In addition, the public ledger that is a core characteristic of blockchain technology will ensure that environmental bond funds will be used in the way they are intended and therefore help to almost eliminate the risk of diversion for misuse. At this stage of disclosure, it gives assurance to the stakeholders and thus, leads to better retention and motivation to which the stakeholders are subjected to in the projects over a period of time.

AI extends this ecosystem in the fact that it allows timely updates and reports to be specific to the user's decision. Through the application of AI algorithms to segment stakeholders by their investment goals, risk appetite, and to selectively deliver this information to the right place, stakeholders can obtain only the information they can possibly need, i.e. This aseptic communication fosters the relationships with stakeholders and guarantees that they are informed and kept in the loop in the path to project success.

For instance, institutional investors will obtain more information about the project's financial performance as well as the carbon credit performance, and community stakeholders will be informed about local environmental impacts such as improvement in air quality or biodiversity. Therefore, targeted communication leaves the stakeholders valued and knowledgeable; thus, the stakeholders have a strong bond with the project, and most probably will continue to invest. It may not only predict the disengaged or at-risk stakeholders before they change their minds and retract their support but also their changing minds. It then shares with the issuers so that they can obtain right incentives and participate in engagement forums to improve issues and rejuvenate stakeholder engagement. For example, perhaps he would require a request for an invitation to attend the stakeholder as such webinars, at an appropriate time so that it might offer him all the advantages for getting detail reports up to the interested project, but in such a situation, where from time to time it would come out to form any interest would seem like happening anything latest to do with a specific project.

Through technology, stakeholder motivation can be sustained overtime by facilitating consistent support and engagement till the end of the project. It has massive implications on how we interact with stakeholders, yet bringing this technology into society is a double-edged sword. When technology integrates itself into environmental bonds, stakeholders want to know that their personal information is safe and AI-driven systems which form a hyper-personalised experience for them are ethical. It is paramount to set up regulative frameworks that justify the use of data and who is the algorithm's decision. This way stakeholders get the process explained in layman language and also have little trust that this time there is no party favoritism. Green Bonds Issuers must specify the type of data accessed, possessed and used and provide stakeholders with an "opt out" or limited applicability.

True transparency and stakeholder collaboration builds trust, something that will be required imminently if we want broader stakeholder support for the continued longevity and commitment.

Also, by nature technology massively elevates stakeholder seats at the table on decision-making. Through the use of AI and data-driven insights, issuers are able to engage stakeholders in real ways which will ensure that their voices are heard. This also underpins retention and commitment in turn forming a strong basis upon which to build a successful environmental bond.

It may be possible to allow the stakeholders to be included in the decision-making process as per their opinion about the project, which could comment on views regarding the project at large and give their feedback on important decisions, for example, online forums or even online surveys that may be found in getting views from community-based stakeholders about the environmental priority that the project will address.

The era of environmental bonds as one-way (nonliving) investments is becoming a collaborative ecosystem where various stakeholders get more active in bringing value to the project well beyond passive investing. The next frontier on this front is advanced technologies, including AI, IoT sensors and blockchain which are changing the dynamics of personalization, transparency, and adaptive engagement. This innovation is the bond to unlock stakeholder insights, which leads to fulfilling trust, satisfaction, and long-term engagement in environmental projects from bond issuers. In the engine of the bond market ecosystem, these technologies make processes for environmental quality goals to be implemented more efficient and effective. The right amount of creativity and responsible tech use strikes the balance between all, so that environmental bonds can bring trust, inclusiveness and openness in fulfillment of global environmental goals.

Undoubtedly, continued success of environmental bonds will depend on the skills of the issuers to unlock and maximize the very best that technology can offer in a context, of course, where stakeholder interests are at the forefront when choices are made. In these respects, then, environmental bonds will progressively be that much more essential for projects trying to shape up a more friendly and greener world.

Ethical Implications of the Implementation of Technology Innovation to Optimize on Environmental Bond

Technological innovation brought into environmental bonds to improve them has created a big number of ethical issues.

Since these connections introduce a new situation in sustainable finance, the coming together of such technologies as AI, IoT, and blockchain results in very serious questions regarding privacy, equity, and long-run sustainability, among others. These technologies are at the forefront of the efficiency, transparency, and measurable change in environmental impact, which serves as the basis for building trust and helping environmental projects sustain themselves.

Most importantly, the most critical issue of ethics related to technology concerning environmental bonds is the use of private information. Rocket speed with which sensors, satellites, and AI systems mushroom, collecting tremendous amounts of data related to environments increases the possibility of leakages concerning the sensitive information in ecosystems or in the use of lands, a local community might have. Such data, therefore, needs to be handled responsibly so that it does not leak into the private affairs of any individual or group. Data collection and usage ethics mean that personal information about local population groups or landowners should never be identifiable so that no one is allowed unauthorized surveillance or exploitation. Mechanism of sharing of data and interaction among parties is another element with regard to which privacy is concerned about it. This would ensure that sensitive information winds up with the wrong people and is used for purposes other than that intended to better facilitate action for the achievement of environmental sustainability. Companies whose firms are likely to hold an environmental bond should therefore provide efficient means of security other than adequate guidance towards privacy on data.

The second dimension is much more than just data privacy-the ethics of algorithm transparency of use in AI, and decision criteria. AI system adoption has increased to optimize the environmental bonds using project outcome forecast and sustainability evaluation. The issue of opaqueness about AI algorithms has increased some ethics dimensions to include lack of accountability and a lack of fair play. It is at this point that it fails to give any clue of how an AI model arrived at the decision most vulnerable to break the trust and it would not do any harm for a dash of manipulation too. Environmental bonds projects for firms have to use transparent and explainable AI systems. It will then build confidence for investors, regulators, and local communities. This will keep environmental bonds projects from biased or unbalanced predictions by firms. Transparency and understanding of the AI systems would then lead to trust in the business houses. And when applied responsibly and ethically, technological developments will find their way through society.

There is another more significant ethical consideration which arises due to the advent of new environmental bonds. This should bring about important gains for some corporations, such as if the technologies favor wealthier companies or advanced economies. Equitable distribution of the benefits brought by these innovations to all parties is of utmost importance, especially that progress made through environmental bonds will benefit even the most marginal or least developed regions.

This is something of concern about accessible and adaptable technologies that are scalable towards the needs of different communities regardless of their economic statuses.

Environmental bonds should be structured to spur technological innovations for the good of all regions, particularly the neediest. The technologies adopted by the wealthier regions to monitor and track their performance regarding the environment must be made accessible to developing countries and disadvantaged communities. Scaling up these technologies, and adapting them to local contexts, can be an effort to address particular community needs while helping fill the gap on global sustainability goals.

The third major ethical concern that is relevant to environmental bonds is bias in AI models. Predictive models of AI and decision-making on large datasets will reflect the inherent bias in those datasets. Datasets that do not represent diversity are prone to unconscious biases, which will continue to be transmitted by biased decision-making and the distribution of resources. For instance, if the data sets on which the AI models are trained contain inadequate knowledge about certain environmental factors or other marginalized communities, then predictions would likely become biased.

Such problems can be prevented if the data sets used by AI machines are diversified, inclusive, and representative of global environmental contexts. The use of AI models should also be audited periodically so that the existing biases can be identified and rectified before such models are deployed. Fairness and equity encouraged within AI systems can provide fairer and effective outcomes within environmental bonds.

The correct application of this approach will ensure that the technological products applied in the environmental bond deals will improve environmental justice in terms of making sure the dividends on the sustainable financing provided are just and fair. It should also pay attention to long-term impacts of the technology applied, avoiding its degradation.

AI, IoT, and blockchain can enhance environmental bonds in quite a good manner, but then again, so does their cost to the environment. The power consumption of operating AI systems, blockchain networks, and deep learning models usually tends to be too high for the good that is done to the environment. Indeed, blockchain networks are very energy intensive, particularly those operating with proof-of-work consensus mechanisms. Firms should instead adopt technologies that are environmentally friendly, such as the green data center, renewable sources of energy, or sustainable blockchain systems which lower energy consumption.

Ethical leadership is essential to guarantee that technology-based innovations in environmental bonds are undertaken responsibly and sustainably.

Leadership engagement with the process of organizing, deployment, and reviewing the technology to be used in environmental bonds will include data privacy, transparency, and fairness guidelines among others and have them audited periodically to ascertain that the technologies applied are consistent with sustainability objectives.

This means that there is a culture of accountability in the establishment, which makes decisions regarding the use of technology with regard to the best interests of all the stakeholders involved in the activities and is clearly transparent. It is, therefore, by placing the ethical considerations above the development and implementation of the technology that ensures the environmental bonds are not only viable financially but socially and environmentally as well. An increase in the transparency of funds is also made by application of blockchain and that is because for example how blockchain can first securely account openly the places for money and are used by spending money. It ensures that it meets the environment bonds raised funds and applies its set responsibilities regarding application.

Most importantly, blockchain-based tracking systems will let stakeholders have real-time awareness about the impact that is being created with the funds. It makes the whole movement transparent so that people may believe in it, and it will allow investors to have sure fruit of their work, contribution of local communities to be satisfactory, and regulatory bodies be sure that they actually put their money into good use for the environmental projects.

This can be interpreted as a reality that technology innovation for optimizing environmental bonds carries multilateral ethical dimensions. Stronger ethical constructs such that any such innovation be an enabler that supports environment and the welfare good of social necessity will keep firms capable of increasingly creating the kind of transparent as well as accountability driven environmental bonds market that is genuinely value giving towards stakeholders who enable realization for attaining globalization with sustainability aspects, which means that the technology-based environmental bonds will be successful only if they encourage the firms to innovate responsibly, so that the fruits of such technologies are appropriately and equitably dispensed among all the levels. This approach harmonizes technological advances with the two-fold aim of achieving long-term environmental sustainability and social equity. With responsibility and inclusivity as commendatory elements, the innovations by the firms can be achieved not just for environmental goals but for generating a positive meaningful impact on society, too.

Future Trends in Environmental Bonds: Driving Innovation Through Technology

This new approach to sustainability has now emerged as a completely integrated strategy all around the world toward achieving environmental goals and objectives with the help of technology along with environmental solutions. The digital technologies, AI, blockchain, and IoT have opened wide scopes toward the most efficient utilization of environmental resources. The technology will eventually become the bedrock through which the environmental bonds become effective, finally converting them into mightier tools in the struggle for environmental sustainability.

Some of the technologies that are currently shaping the future of environmental bonds are blockchain, AI, and Big Data analytics reform how they apply and impact toward solving some of the burning global issues with climate change, biodiversity loss, and environmental pollution. Innovations that let environmental bonds become more than just simple financial tools; dynamic and transparent solutions contributing directly to sustainability and environmental progress.

Artificial Intelligence and Big Data: The Enabler of Real-Time Impact Analysis

The future does look bright after all, especially since Environmental bonds locked up in AI, Big Data, and blockchain open new channels for increased transparency, efficiency, and accountability. In real-time, AI analyzes data, inputs, and reports fundamental sustainability metrics such as carbon emissions, energy efficiency, and biodiversity impact. Machine learning also predicts trends of environmental performance, thereby providing a guideline to investors in making decisions on resource allocation and funding strategies. Big Data further makes a project feasible because it allows the analysis of socio-economic conditions, regulatory factors, and environmental influences. This improves the decision-making process and increases stakeholder trust through authenticity in project data, eliminating misinformation, thus increasing investor confidence.

The other ways through which blockchain technology can increase transparency and accountability in the environmental bond market is that it can create an immutable and tamper-proof record of transactions.

Blockchain, therefore, allows stakeholders to trace how the funds are being used in real time and the outcome of the projects through a tamper-proof ledger, offering higher oversight. Using smart contracts, this disbursed money is only upon satisfactory meeting the prescribed pre-agreed conditions such performance as in its application may secure meeting set ecological goals preceding distribution to discourage or avert misapplication fraud as well as an increase in an impact such ecological bonds made cheaper and of superior impact. Further, automated compliance checks and secure data sharing advanced technologies instill confidence among investors, governments, and communities that the funds are actually spent for real environmental progress.

IoT and Remote Monitoring: Maintaining the Accuracy of Data

The Internet of Things, therefore represents a kind of linking between the physical world of devices and sensors to the virtual world. With IoT, actual data gathering would be in real-time, therefore easier to monitor and accurately assess those environmental bonds. It opens up avenues for tracking project implementation status based on no report and site visit, improved decisions because of enhanced fund allocation relative to maximum environmental facility usage.

IoT minimizes the dependence on manual monitoring and provides accurate, transparent outcomes of the projects. It is integrable with AI and blockchain that may allow predictive analytics and safe data sharing. That gives a holistic view of how environmental bonds will perform therefore optimizing future investments and encouraging stakeholders to engage in the process. This means there is more accountability, and everyone's voice has a say in the decision-making. However, it also brings problems like risk of data security and privacy issues. Huge volumes of sensitive data need to be encrypted and accessed in a limited manner to prevent misuse. Huge deployment costs prevent its adoption by small-scale projects.

Despite all these challenges, IoT revolutionizes environmental bond monitoring through real-time tracking and data-driven decisions. With time, this technology will reduce in price, and environmental finance will be very transparent, efficient, and sustainable.

Personalization and Customization of Environmental Bond Investments

The introduction of AI, Big Data, and blockchain technology in the environmental bond market changes personal investment. Investor financial objectives will be aligned with sustainability objectives to create tailored investments in reforestation, water conservation, or clean energy. The traditional bonds cannot be applied as the use of AI and Big Data analyze the environmental performance and preference of investors for an outcome-based investment.

The investment return fits in sustainability since it bases the portfolio on location or risk or environmental focus so that funding can go precisely to where it most urgently needs to be and traceability and trust increase. Blockchain promotes transparency; AI-driven automation reduces administrative costs but increases responsibility. It helps the investor to modulate his risk appetite; from high-risk innovative projects to stable low-risk investments. Challenges abound though, such as how to establish credibility and consistency in the data of a project, for instance, and how to handle the myriad number of personalized investments. IoT and AI enhance transparency and engagement because the data is real-time.

The other problem is scalability since big projects make up the majority of investment inflows. However, technology advancement will make the bonds personalized and increase investor participation and effectiveness in sustainability outcomes. Innovation would then allow the environmental bonds to provide impact, precision, and transparency in environmental solutions while giving the investors a feeling of ownership and responsibility.

Smart Regulation: Technologies-Enabled Governance for Environmental Bonds

Technology-driven regulation transforms environmental standards from passive monitoring to active enforcement and, therefore, makes environmental bonds much more effective sources of funding sustainability projects. Governance, accountability, and efficiency increase through real-time tracking, data-driven decision making, and the automated enforcement of compliance using AI, blockchain, IoT, and Big Data. The IoT sensors will be monitoring the air, water, and emissions. The AI will go through humongous data sets to determine environmental trends so that the project's targets can be met. Transparency: The allocations remain immutable through blockchain technology, therefore, proper funds allocation. It dispenses automatically through smart contracts based on their performance, and therefore, minimizing the administrative cost. Big data lets the regulator monitor in real time and then intervene accordingly. These innovations give more strength to investor confidence, thereby making it effective in project implementation.

Despite these advantages, there are some disadvantages: data standardization, regulatory gaps in expertise, and large data management. Training investment, and coordination between the government and technology providers will be required to successfully utilize intelligent regulation.

As technology advances, the environmental bonds are going to develop the core part of smart regulation. The approach adds the use of automation, predictive analytics, and decentralized ledgers to strengthen transparency, accountability, and sustainability. Intelligent regulation assures that environmental finance stays effective in achieving global climate goals with trust with stakeholders.

Sustainability in Technology: Addressing the Carbon Footprint of Innovation

Environmental bonds are the future of sustainability, and yet, AI, blockchain, and IoT pose a lot of environmental challenges as these consume a lot of energy. AI models, blockchain proof of work, and IoT infrastructure consume lots of electricity, and carbon footprints get threatened. Sustainability would only come from energy-efficient computing, optimized algorithms, and green-powered data centers only (solar, wind, hydro). Circular economy must be applied for e-waste management and hardware design. Green technologies need low-energy AI, effective blockchain, and low-power IoT support; sustainable development is real. Balance between their innovative steps in harmony with what will be eco-friendly must come from governments together with technological providers who are liaised with the groups of environmentalists to not lose out on climate goals as the environment bonds begin on the way toward actual and sustained sustainability.

CONCLUSION

All this-AI, blockchain, IoT, and data analytics-make the environmental bonds more transparent, efficient, and real-time followed. Hence, the above innovation technologies can improve the quality of decision making and ensure trustworthiness with ethical adoption to drive the goals of sustainability (SDG 13). Further, the progress of innovations will support public-private cooperation followed by stronger policy support and accountability. Ethical deployment prevents mal adaptations and benefits the investors as well as the environment. The greater the demand for environmental bonds, the greater is the importance of technology to ensure that they create lasting quality to the environment. The merging of environmental finance with technology represents an important opportunity to research and invest in a much more sustainable future.

REFERENCES

Adeyelu, A., Ugochukwu, C., & Shonibare, P. (2024). Ethical implications of AI in financial decision-making: A review with real-world applications. *International Journal of Applied Research in Social Sciences*, *6*(4), 608–630.

Alotaibi, E., & Nassif, N. (2024). Artificial intelligence in environmental monitoring: In-depth analysis. *Discover Artificial Intelligence*, *4*, 84.

Ambilwade, R. P., Kumar, R., Narayana, K. S. S., Srinivas, P., Yadav, B., & Rusia, S. (2024). AI-powered environmental monitoring: Machine learning approaches for air and water quality assessment. *Nanotechnology Perceptions*, *20*(1), 45–58.

Bhandari, M., Tiwari, G., & Dhakal, M. (2025). Enhancing transparency and accountability in sustainable finance through blockchain technology: A systematic review of the literature. *Journal of Intelligent Management Decision*, *4*(1), 23–43.

Chen, Y., & Wang, L. (2022). Scaling up sustainable investment through blockchain-based project bonds. *Development Policy Review*, *40*(6), e12582.

He, J., Zhao, Y., Zhang, B., Chen, L., & Ma, X. (2025). The impact of blockchain technology on green investment decisions for a sustainable supply chain with an overconfident manufacturer. *Sustainability*, *17*(1), 284.

Katten, G. (2021). Issuing green bonds on the Algorand blockchain. *arXiv* preprint arXiv:2108.10344.

Moursi, M., Wehn, N., & Hammoud, B. (2025). Smart environmental monitoring of marine pollution using edge AI. *arXiv* preprint arXiv:2504.21759.

Naderi, N., & Tian, Y. (2022). Leveraging blockchain technology and tokenizing green assets to fill the green finance gap. *Energy Research Letters*, *3*(3), 1–9.

Yadav, M. P., Kumar, S., & Mukherjee, D. (2023). Do green bonds offer a diversification opportunity during COVID-19? Empirical evidence from energy, crypto, and carbon markets. *Environmental Science and Pollution Research International*, *30*(1), 7625–7639.

Chapter 11
Policies of Strategic Communication for Green Tourism Sustainability Using Artificial Intelligence

Ghada Seif
https://orcid.org/0000-0001-9127-2607
Abu Dhabi University, UAE & Minia University, Egypt

Puja Mahesh
Symbiosis International University, UAE & Symbiosis International University, India

ABSTRACT

The authors in this study review the importance of applying (AI) technologies in achieving sustainability in green tourism. The UAE directs sustainable investments in the tourism sector towards green transformations. The study explores the importance of applying AI technologies in green tourism sustainability and highlights the UAE case, The study adopts descriptive and qualitative thematic analysis approaches. The results provide insights into strategic communication strategies for a sustainable future of green tourism within the framework of AI. The findings showed that digitalization of tourism and using different ways can be applied to use AI effectively in the tourism industry such as chatbots and social media platforms. This will enhance through increase visitor happiness, and improve operational efficiencies. Virtual tourism is provided by AR and VR tools that promote sustainable services and a high-quality customer experience. Adopting AI faces several challenges, including

DOI: 10.4018/979-8-3373-1822-6.ch011

the lack of standardized datasets, the digital divide, and gaps in policy frameworks.

INTRODUCTION

The role of artificial intelligence (AI) technologies in tourism is changing the paradigm of sustainability practice development, implementation, and communication. Given the fact that international tourism is under growing demands to comply with sustainable development goals (SDGs), practitioners within the industry are resorting to empowering technologies like artificial intelligence to increase efficiencies, decrease adverse impacts on the environment, and create sustainability. These activities are beneficial in green tourism, where AI-applied advanced features, including smart mobility systems, IoT, virtual tourism, and energy management tools, are key in carbon emission reduction and environmental sustainability (Singh & Hassan, 2024; Raihan et al., 2024).

The United Arab Emirates (UAE), one of the top world tourism centers, issued a smart tourism strategy encouraging AI technology integration within tourism. Concepts like Smart Dubai combine tourism activities with a national sustainability agenda like Vision 2021 Strategy or Green Growth Plans. These concepts seek to embed AI and IOT along with strategic communicative approaches into tourism services to improve the tourist experience and lessen the negative environmental impact (Sudipa Majumdar, 2022; Smart Dubai, 2022). Through public-private partnerships and the use of AI, the UAE has given a boost to sustainable tourism practices.

AI is not only limited to operations but is central in value communication on green tourism initiatives. AI-centric applications like chatbots and advanced marketing platforms elevate public discourse and understanding around green tourism initiatives (De Andrade & Tumelero, 2022). However, their effective utilization is hampered by technical difficulties, privacy considerations, and lack of access that must be overcome to capitalize on AI's potential in tourism. These barriers emphasize the need for supporting measures to engage and integrate different stakeholder groups into collaboration (Radjabov & Serrano, 2024).

The key focus of this study is on the embedding of AI in green tourism practices, particularly in the context of the UAE tourism industry. This research seeks to provide insights into (i) the strategic frameworks that guide the use of AI technologies in tourism activities, (ii) the communication strategies used in promoting sustainability, and (iii) the barriers and benefits brought about by AI strategies. Explaining these elements, this paper aims at equipping stakeholders who wish to operate within the bounds of sustainable tourism development with the necessary activities to take.

The role of AI in sustaining green tourism practices encompasses inter-stakeholder strategic communication, which is invaluable in strengthening eco-friendly practices. The resource optimization and operational efficiencies that stem from applying Artificial Intelligence technologies in tourism activities are utilized in eco-smart cities, which rely on smart technologies for urban sustainability (Asrifan et al., 2024). AI also improves accessibility and provides depth of information about attractions, which are essential for the development of local economies (Kumari et al., 2024). On the other hand, Artificial Intelligence, more so in generative forms, raises some ethical questions that should be adhered to ensure there are no adverse effects on tourism stakeholders (Aron et al., 2024). Such AI-enabled applications meet sustainable tourism objectives and reinforce the entire tourism industry.

There are AI systems for enhancing customer service. In this way artificial intelligence allows a customer to be assisted with several questions and transactions about tourism events. In addition, artificial intelligence allows tourism events to be performed more accurately and efficiently by having technology generate responses as feedback to the questions being asked (Simon et al., 2020).

It is also critical to develop cleaner and more sustainable automated systems for tourism to integrate AI with other forms of intelligence. In this regard, scholars encourage an interdisciplinary approach that joins AI automation with speculative thinking as well as other forms of intelligence.

Sustainability is commonly acknowledged to have three dimensions: environmental, social, and economic (Jose, et al., 2023)· Strategic communication is crucial in bridging the gap between complex AI applications and practical implementation, ensuring transparency, and fostering public trust (Mishra, 2024). Moreover, green marketing communication can effectively communicate sustainability efforts, highlighting both environmental and economic benefits, thus avoiding greenwashing (Pizzichini, 2024).

LITERATURE REVIEW

AI applications and tools can promote sustainable tourism practices; in their study (Singh & Hassan, 2024) investigated the effects of AI on sustainable tourism. They aim to guide industry stakeholders for betterment through AI. The results consist of (Dalipi, F., Kastrati, Z. & Öberg, T. 2023) results regarding AI significantly impacting tourism operations and sustainability. The study identified challenges including dataset standardization and evaluation metrics. In the same context, (Li, D., Du, P., & He, H. 2022) revealed the role of AI in promoting Sustainable tourism development in natural heritage sites. High Average of sustainable tourism practices as a part of ecological sustainability.

According to Rane, N., Choudhary, S., & Rane, J. (2023), IoT devices enhance visitor safety and satisfaction. Augmented Reality (AR) and Virtual Reality (VR) technologies offer immersive experiences that redefine how tourists interact with destinations. AR enriches the cultural and historical context of attractions, fostering a deeper understanding and appreciation of local heritage. Meanwhile, VR enables virtual tourism experiences, reducing the need for physical travel and consequently mitigating carbon footprints. The findings underscore the potential of a technology and AI-driven approach to ensure long-term environmental, social, and economic sustainability. This consists of Guo, Z. (2023) results, which indicated the economic side of AI. The study revealed that tourists increased by 7.58% after AI management and operating costs decreased by 19.19%. The AI innovations in energy, transportation, biodiversity, and water management are new tools that can be employed to sustain the resources in the environment and attract more tourists (Raihan, et al., 2024).

The findings of (Louati, et al., 2024) (Kashem, et al., 2023) and Chakraborty, P. P. (2024). emphasize how technology enhances sustainable tourism practices. Innovations like blockchain, IoT, and AI address sustainability challenges. Machine learning and AI can contribute to sustainable tourism practices, focusing on resource management, economic resilience, and environmental stewardship. This integration aims to support stakeholders in developing resilient strategies for the tourism sector. Integrating AI and blockchain into tourism can transform the sector, making it more resilient against economic downturns and geopolitical instability while enhancing socio-economic conditions. This is agreed with (Majid, 2024) regarding The integration of AI chatbots which can play a significant role in fostering sustainable travel practices among tourists. And (Sinha, et al., 2021) regarding the strategic positions of a hotel in terms of sustainability, AI and IoT technology are important components that Hotels will consider for the strategic intention of adopting AI and IoT for environmental sustainability. Moreover using the following green transportation policies: According to the 2000 European Conference of Ministers of Transport (ECMT) report on Sustainable transport policy (ECMT 2000), which identifies a set of recommendations, the principal ones are in relation to sustainability(Jose, et al., 2023): (a) Reducing greenhouse gas emissions, (b) Improving air quality/reducing noise, (c)Improving transport safety, (d) Improving access, (e) Reducing congestion and (f) Supporting the economy.

These findings of Tiago, et al., (2020) suggest that consumers consider eco-labels as warranties that firms are acting on sustainability principles. The same was found to be partially true in the tourism sector. The results indicate that (1) eco-labeling promotion by local official entities does not have a significant impact on the rates of local accommodation certification but (2) smaller firms value eco-labeling when promoting their offers online.

Researchers highlight the potential for AI to improve or replace visitor interactions, potentially transforming tourist destinations (Kumari, et al., 2022).

Additionally, AI can boost efficiency and productivity by lowering service costs, improving safety and security, assisting decision-making processes, and assisting visitors in navigating through smart places (Wang et al.,2020).

In the future, it's expected that the use of intelligent automation in travel and tourism will grow, because of the development of advances in artificial intelligence (AI) and associated technologies (Tussyadiah, 2020a).

Two studies by Tussyadiah, et al. (2020) b on autonomous robots and beverage-serving robots showed that customer trust in intelligent robots, and that customer trust in intelligent robots designed for social interactions in travel environments is influenced by negative attitudes toward technology and a tendency to trust technology.

Artificial intelligence in tourism facilitates portability, social integration, and territorial marketing by implying that smart cities are expanding the tourist and hospitality business (Dias et al., 2021).

Strategic communication enhances AI-driven sustainability initiatives and stakeholder engagement. The study highlighted challenges including technical complexity, data privacy, and cultural sensitivities (Mazza, 2023).

The literature on AI and strategic communication indicates developing marketing plans and the effective use of AI-based solutions, such as the use of chatbots and voice-activated AI support, for companies providing services including public relations, marketing communications, relational marketing, and political communication (De Andrade & Tumelero, 2022). Adopting AI in strategic communications could enhance public relations and marketing efforts, thereby enhancing the overall effectiveness of sustainability messages.

Although AI offers great potential to enhance sustainability, The research on strategic communication for AI-driven sustainability emphasizes challenges such as technical complexity, data privacy, and cultural sensitivities, (Radjabov & Serrano,2024) and the digital divide may hinder the implementation potential to enhance sustainability in green tourism initiatives. advocating for comprehensive approaches with stakeholder analysis and adaptive communication strategies to address these issues effectively.

The current literature lacks a comprehensive understanding of the role of technology applications in sustainable tourism development (Kumari et al., 2024). Advancement of the latest technology application in tourism industry is recommended for sustainable tourism development) Hossain, et al., 2022). The referenced paper investigates economic asset administration in tourism, centering on optimizing asset utilization and minimizing natural impacts. It emphasizes the significance of circular economy models and life cycle appraisals in tourism, highlighting the part of approach systems and imaginative advances in accomplishing supportability objectives. The consideration concludes by distinguishing and investigating crevices related to asset proficiency and the requirement for multi-stakeholder collaborations in sustainable tourism.

Green tourism is similar in meaning to ecotourism, both focus on environmental factors (Al-Sakkaf et al, 2024). Green tourism encourages the use of alternative (renewable) energies, such as solar, wind, and other energies, to reduce climate change and its repercussions on the environment, in line with the United Nations' efforts under the Kyoto Protocol in 2005, especially to reduce gas emissions from industrialized countries(Alawsi et al, 2024).UNEP and UNWTO defined Sustainable tourism as fully considering current and future economic, social, and environmental impacts and addressing the needs of visitors, industry, the environment, and host communities. It is not a special form of tourism; rather, all forms of tourism may strive to be more sustainable. The diversity of stakeholders in sustainable strategic communication in tourism emphasizes the importance of the right marketing communication strategy.

It is preferable to rely on communication strategies that include a multi-channel marketing approach, as a combination of marketing strategies that are delivered through multiple communication channels.

RESEARCH METHODOLOGY

The papers were reviewed, and the findings were compiled using the following research approach: Peer-reviewed journal articles, reports from databases like MDPI, and government policy documents about AI, sustainable tourism, and strategic communication frameworks were used to find pertinent research publications. The search was conducted using keywords like "AI in tourism," "sustainable tourism practices," "strategic communication," and "green tourism initiatives. A purposive sample of only recent research on AI, sustainable practices, or strategic tourism frameworks published between 2020 and 2024 was chosen. Studies that addressed the operational and environmental concerns of integrating AI in tourism were given priority. Articles that don't discuss AI-driven frameworks or green tourism, or that have outdated methodology weren't considered.

The current study has adapted the thematic analysis approach. The literature was categorized using thematic coding, including resource management, public-private partnerships, AI-enabled tourism operations, IoT and AR/VR technologies, and sustainability policies (Singh & Hassan, 2024; Rane et al., 2023).

By concentrating on important findings about AI's influence on strategic communication and sustainable tourism practices, manual coding assisted in identifying trends and connections among research.

To identify patterns, inconsistencies, and new trends, a comparison study of several studies was carried out. Studies like Guo (2023) and Raihan et al. (2024), for instance, indicated AI increased productivity and decreased operating costs, but others highlighted issues such as data standardization and privacy issues (Mazza, 2023; Radjabov & Serrano, 2024).

Studies were grouped by their focus areas to better align with the research questions and the objective of the study.

FINDINGS AND CONCLUSION

The study results demonstrate that using artificial intelligence (AI) technologies greatly enhances the sustainability of tourism-related operations. AI apps improve travel experiences and encourage environmental stewardship and green tourism by providing individualized services, environmentally friendly suggestions, and operational efficiencies. From the evaluated studies the important conclusions drawn are:

a. *Effect on Efficiency and Operations*

Through improved service delivery, AI integration has increased operational efficiency, resulting in a 19.19% decrease in operating costs and a 7.58% rise in tourism inflows (Guo, 2023).

In Middle Eastern countries, mobile devices are used in every aspect of planning a trip starting from booking trips online, buying tickets, and looking up hotels. This gives vast opportunities to hospitality firms, airlines, and tour planners who utilize the devices to connect with their customers and plan their business ideas by customers' feedback. As the generations are becoming more tech-savvy, there has been a noticeable shift in travel patterns. It was reported that the tourism and hospitality market in the UAE is driven by a rise in visitors (Hossain et al, 2022).

b. *Sustainable Smart Technologies*

IoT, AR, and VR applications in tourism promote visitor safety and engagement. VR lessens the need for actual travel, reducing the carbon footprint, while AR improves visitors' engagement with cultural heritage sites. (Rane et al., 2023). AI techniques ensure sustainable use of natural resources by optimizing energy, water, transportation, and biodiversity management (Raihan et al., 2024).

c. *Consumer and Stakeholder Behavior*

Chatbots and other strategic AI-powered communication tools increase client happiness and encourage environmentally friendly travel practices. While eco-labeling has been shown to have little impact on larger tourism companies, AI also helps businesses align with sustainability standards (Tiago et al., 2020; Majid, 2024).

d. AI Governance and Public-Private Partnerships

One example of how public-private partnerships spur innovation in sustainable tourism is the United Arab Emirates' Smart Dubai initiative. AI tools are utilized to manage tourism flows, while government policies encourage alignment with SDGs (Smart Dubai, 2022; Sudipa Majumdar, 2022). The impact of AI applications on tourism in the UAE is evident in hotels in three main areas: improving market share, service effectiveness, and service efficiency, according to Al-Shami et al., (2021). The efficiency of sustainable tourism services in the UAE as AI technology facilitates the effectiveness of trip management and provides information about tourist sites, including popular places, food, and cultures. AI technologies have helped improve the quality and efficiency of services: Hotel robots are used with other enabling technologies such as facial recognition, automated payment, and self-driving cars. These reduce the time and cost to complete hotel tasks (CSRE, 2024).

Discussion

AI technology integration is revolutionizing green tourism sustainability in terms of operations and policy alignment. The findings are summed up in four main areas in this section:

AI-driven Strategic Communication and Public Engagement: Using chatbots and automated assistance systems, AI-driven communication strategies assist travel agencies in better interacting with travelers (De Andrade & Tumelero, 2022). However, research indicates that for these technologies to reach their full potential, issues including data protection, technological difficulties, and cultural sensitivity must be resolved (Mazza,2023). Social media platforms and digitalization of tourism a great enabler of using AI in green tourism.

Regarding Operational Impact and Efficiency, AI solutions save expenses, increase visitor happiness, and improve operational efficiencies. These results are consistent with those of Wang et al. (2020), who pointed out that AI can also increase tourist productivity and safety.

The Economic and Environmental Advantages, AI technologies are essential for reducing carbon emissions and resource utilization.

For example, virtual tourism options are provided by AR and VR tools, which promote sustainable practices without sacrificing visitor experiences (Kumari et al.,2024). Longterm environmental sustainability is also promoted by AI's capacity to optimize energy, water, and transport -ation management.

Challenges, despite its advantages the adoption of Artificial Intelligence is hampered by Challenges with dataset standardization and the digital divide, policy frameworks, and flexible communication tactics that consider the interests of many stakeholders necessary to address these issues (Radjabov & Serrano, 2024).

Limitations & Recommendations

This literature review underscores the critical role of AI technologies in sustainable tourism development, with positive impacts on visitor management, cost reduction, and environmental stewardship. However, several challenges must be addressed, including technical limitations, data governance, and cultural sensitivity.

According to the study's findings, AI technologies promote the growth of sustainable tourism by enhancing productivity, managing the environment, and facilitating strategic communication. However, only by tackling issues like the digital divide and data privacy concerns can these technologies reach their full potential.

Further research is needed to explore strategic communication frameworks that can better integrate AI tools and foster sustainable tourism practices, especially in the travel industry in the United Arab Emirates and globally.

REFERENCES

Al-sakkaf, M. A., Al-Attas, W. A., Khalufi, N., & Murshid, M. A. (2024), "Defining Green and Responsible Tourism Destinations", Alnoor, A., Bayram, G.E., XinYing, C. and Shah, S.H.A. (Ed.) *The Role of Artificial Intelligence in Regenerative Tourism and Green Destinations* (*New Perspectives in Tourism and Hospitality Management*), Emerald Publishing Limited, Leeds, pp. 51-69. DOI: 10.1108/978-1-83753-746-420241004

Alawsi, H., Emran, A. Q., Ebrahim, A. M., Ibrahim, F. M., & Mohammed, M. N. (2025). Remarks on Sustainable Tourism. In AlDhaen, E., Braganza, A., Hamdan, A., & Chen, W. (Eds.), *Business Sustainability with Artificial Intelligence (AI): Challenges and Opportunities. Studies in Systems, Decision and Control* (Vol. 566). Springer., DOI: 10.1007/978-3-031-71318-7_14

Chakraborty, P. P. (2024). The role of technology in enhancing sustainable tourism practices. In *Advances in Hospitality, tourism and the Services Industry (AHTSI) book series* (pp. 195–230). DOI: 10.4018/979-8-3693-5903-7.ch011

CSRE,. (2024). Research project the UAE As A Sustainable Tourism Destination, Stage1, The Center for Sustainability Through Research and Education- CSRE.

Dalipi, F., Kastrati, Z., & Öberg, T. (2023). "The Impact of Artificial Intelligence on Tourism Sustainability: A Systematic Mapping Review," *International Conference on Computational Intelligence and Knowledge Economy (ICCIKE)*, Dubai, United Arab Emirates, 2023, pp. 119-125, DOI: 10.1109/ICCIKE58312.2023.10131818

De Andrade, I. M., & Tumelero, C. (2022). Increasing customer service efficiency through artificial intelligence chatbot. *Revista De Gestão*, *29*(3), 238–251. DOI: 10.1108/REGE-07-2021-0120

Dias, A., Santinha, G., Rodrigues, M., Queirós, A., Rodrigues, C., & Rocha, N. P. (2021). Smart cities and accessible tourism: A systematic review. In Eusébio, C., Teixeira, L., & Carneiro, M. J. (Eds.), *ICT tools and applications for accessible tourism* (pp. 96–114). IGI Global. DOI: 10.4018/978-1-7998-6428-8.ch005

Hossain, S. F. A., Ahsan, F. T., Nadi, A. H., Ahmed, M., & Neyamah, H. (2022). Exploring the Role of Technology Application in Tourism Events, Festivals and Fairs in the United Arab Emirates: Strategies in the Post-Pandemic Period. In Hassan, A. (Ed.), *Technology Application in Tourism Fairs, Festivals and Events in Asia*. Springer., DOI: 10.1007/978-981-16-8070-0_19

Hossain, S. F. A., Ahsan, F. T., Nadi, A. H., Ahmed, M., & Neyamah, H. (2022). Exploring the Role of Technology Application in Tourism Events, Festivals and Fairs in the United Arab Emirates: Strategies in the Post-Pandemic Period. In Hassan, A. (Ed.), *Technology Application in Tourism Fairs, Festivals and Events in Asia*. Springer., DOI: 10.1007/978-981-16-8070-0_19

Jose, S. V., Sandhya, H., & Varghese, B. (2023). Artificial intelligence towards smart green transportation. In *Advances in marketing, customer relationship management, and e-services book series* (pp. 102–124). DOI: 10.4018/978-1-6684-8140-0.ch005

Jose, S. V., Sandhya, H., & Varghese, B. (2023). Artificial intelligence towards smart green transportation. In *Advances in marketing, customer relationship management, and e-services book series* (pp. 102–124). DOI: 10.4018/978-1-6684-8140-0.ch005

Kashem, M. A., Shamsuddoha, M., Nasir, T., & Chowdhury, A. A. (2023). The role of artificial intelligence and blockchain technologies in sustainable tourism in the Middle East. *Worldwide Hospitality and Tourism Themes, 15*(2), 178–191. DOI: 10.1108/WHATT-10-2022-0116

Kumari, M., Guleria, S. and Kumar, S. (2024). Sustainability in tourism and hospitality: Artificial intelligence role in eco-friendly practices in Indian hotels. *Journal of tourism theory and research*, 10(2):73-83. DOI: 10.24288/jttr.1523976

Kumari, M., Guleria, S. and Kumar, S. (2024). Sustainability in tourism and hospitality: Artificial intelligence role in eco-friendly practices in Indian hotels. *Journal of tourism theory and research*, 10(2):73-83. DOI: 10.24288/jttr.1523976

Li, D., Du, P., & He, H. (2022) Artificial Intelligence-Based Sustainable Development of Smart Heritage Tourism, Wireless Communications and Mobile Computing 2022, 13 p. DOI: DOI: 10.1155/2022/5441170

Louati, A., Louati, H., Alharbi, M., Kariri, E., Khawaji, T., Almubaddil, Y., & Aldwsary, S. (2024). Machine learning and Artificial intelligence for a Sustainable tourism: A case study on Saudi Arabia. *Information (Basel), 15*(9), 516. DOI: 10.3390/info15090516

Majid, G. M., Tussyadiah, I., & Kim, Y. R. (2024). Exploring the potential of chatbots in extending tourists' sustainable travel practices. *Journal of Travel Research*, 00472875241247316. Advance online publication. DOI: 10.1177/00472875241247316

Mazza, B. (2023). A Theoretical model of Strategic Communication for the Sustainable development of sport tourism. *Sustainability (Basel), 15*(9), 7039. DOI: 10.3390/su15097039

Mishra, D. (2024). Strategic Communication for AI-Driven Sustainability Initiatives: Bridging Technology, Policy, and Public Engagement in Achieving SDGs. In Riswandi, B., Singh, B., Kaunert, C., & Vig, K. (Eds.), *AI Applications for Clean Energy and Sustainability* (pp. 187–211). IGI Global., DOI: 10.4018/979-8-3693-6567-0.ch010

Pizzichini, L. (2024). Green tourism. In *Edward Elgar Publishing eBooks* (pp. 220–242). DOI: 10.4337/9781789904833.00023

Radjabov, O. O., & Serrano, M. D. T. (2024). Analysis of tourism Development indicators and their Relationship with communication at Cultural Heritage Sites. HEDIT 2024 - International Congress for Heritage Digital Technologies and Tourism Management.

Raihan, A., Paul, N. A., Rahman, M. S., Islam, S., Paul, P., & Karmakar, S. (2024). Artificial Intelligence (AI) for Environmental Sustainability: A concise review of technology innovations in energy, transportation, biodiversity, and water management. *Journal of Technology Innovations and Energy*, *3*(2), 64–73. DOI: 10.56556/jtie.v3i2.953

Rane, N., Choudhary, S., & Rane, J. (2023). Sustainable Tourism Development Using Leading-edge Artificial Intelligence (AI), Blockchain, Internet of Things (IoT), Augmented Reality (AR) and Virtual Reality (VR) Technologies (October 31, 2023). Available at *SSRN*: https://ssrn.com/abstract=4642605 or DOI: 10.2139/ssrn.4642605

Sinha, M., Fukey, L. N., & Sinha, A. (2021). Artificial Intelligence and Internet of Things readiness: inclination for hotels to support a sustainable environment. In *Elsevier eBooks* (pp. 327–353). DOI: 10.1016/B978-0-323-85769-7.00015-X

Tiago, F., Gil, A., Stemberger, S., & Borges-Tiago, T. (2020). Digital sustainability communication in tourism. *Journal of Innovation & Knowledge*, *6*(1), 27–34. DOI: 10.1016/j.jik.2019.12.002

Tussyadiah, I. (2020). a. A review of research into automation in tourism: Launching the annals of tourism research curated collection on artificial intelligence and robotics in tourism. *Annals of Tourism Research*, *81*, 102883. DOI: 10.1016/j.annals.2020.102883

Tussyadiah, I. P., Zach, F. J., & Wang, J. (2020). b. Do travelers trust intelligent service robots? *Annals of Tourism Research*, *81*, 102886. DOI: 10.1016/j.annals.2020.102886

Wang, J., Xie, C., Huang, Q., & Morrison, A. M. (2020). Smart tourism destination experiences: The mediating impact of arousal levels. *Tourism Management Perspectives*, *35*, 100707. DOI: 10.1016/j.tmp.2020.100707

Chapter 12
The Challenge of Artificial Intelligence in Environmental Sustainability: Opportunities, Risks, and Policy Implications

Muhammad Usman Tariq
https://orcid.org/0000-0002-7605-3040
Abu Dhabi University, UAE & University College Cork, Ireland

ABSTRACT

This chapter sustainably examines the dual role of artificial intelligence (AI) in the environment, highlighting its potential opportunities and associated risks. AI offers significant opportunities to improve environmental protection through improved resource management, biological diversity conservation, and climate change reduction. Sector applications such as agriculture, energy, water management, and waste reduction demonstrate the ability of AI to promote sustainable practices. However, AI poses challenges, particularly concerning the ecological footprint, ethical concerns, data, and governance questions. The energy consumption and potential inequality of large-scale AI models when using AI solutions must be carefully considered. This chapter also highlights the role of political decision-makers, researchers, and industry leaders in ensuring the responsible development of AI and the need for ethical guidelines, transparency, and inclusion in AI applications.

DOI: 10.4018/979-8-3373-1822-6.ch012

INTRODUCTION

The role of AI in environmental compatibility has grown rapidly. It has become essential to dealing with the most urgent environmental issues. From fighting climate change to efficient management of natural resources, artificial intelligence is seen as a transformational tool that can promote innovation and improve environmental outcomes. The growing influence in this area shows how technological advancements can complement sustainability efforts. Algorithms for machine learning can process significant data records, including satellite images and weather data, to provide more accurate predictions of climate factors. These forecasts can impact climate policy, governments, and organizations, allowing you to make data-controlled decisions about where you can focus your resources and efforts. AI can also be used to optimize the production of renewable energy. For example, machine learning algorithms can predict weather patterns to manage better integrating solar and wind energy into networks and improve efficiency and reliability. Transitioning to a more sustainable energy infrastructure is essential, as renewable energy sources such as wind and solar are intermittent and unpredictable.

AI can provide knowledge that helps authorities take immediate action to reduce pollution with its ability to analyze air and water quality data in real-time. Sensors and drones operated by AI can monitor air pollution, recognize harmful emissions from factories, and even pursue illegal waste. AI systems can help enforce environmental laws and improve public health by providing more accurate and timely data. AI can monitor river, lake, and ocean water quality (Shuford, 2024). It means that contamination and potential risks are identified more efficiently than through traditional methods. It helps ecosystem health and ensures that local governments have access to clean water, an essential resource for human survival. Traditional resource extraction and use methods often lead to inefficiency, waste, and environmental degradation. AI can optimize these processes and lead to more sustainable practices. Agriculture uses AI-controlled precision cultivation to reduce the use of water, pesticides, and fertilizers, ensuring that plants are grown with minimal environmental impact. Machine learning algorithms can analyze data from on-site sensors to determine the ideal amount of water, nutrients, and pesticides needed for a particular system, leading to healthier plants and reduced resource consumption.

Furthermore, AI can improve waste disposal systems by predicting waste-produced materials, optimizing recycling processes, and identifying opportunities for circular economy practices. These skills are essential to manage increased pressure on natural resources and reduce human activities' ecological footprint. One of the most critical questions is how AI can improve the environment's quality in the long term. Researchers will explore how AI models can be integrated into environments within their environments to support sustainability efforts. For example, how can AI be used to predict the long-term impact of climate policy on ecosystems and human society? How can an algorithm train to identify environmental risks before they become a serious problem? These questions are important as they highlight the need for AI to address immediate problems and contribute to long-term solutions that prevent further environmental losses (Ueda et al., 2024). One of the main concerns is the possibility that AI can exacerbate existing inequality. For example, AI solutions developed in wealthy countries may not be easily accessible to developing countries most susceptible to climate change and environmental degradation.

Furthermore, AI models can be biased as they rely on data that can reflect historical inappropriate or inaccuracy. For example, suppose an AI system is trained with data that underestimates a particular community or ecosystem. In that case, its predictions and recommendations for these areas may not be applied or effective. It can alienate population groups that require protection and undermine efforts to achieve global sustainability goals.

Another ethical issue revolves around the transparency and accountability of AI systems. In many cases, AI algorithms act as "black boxes." In other words, the decision-making process is challenging to understand and explain. This lack of clearness can make it tough to assess the fairness and accuracy of AI-controlled environmental decisions (Camilleri, 2024). It is vital that these processes are transparent and that stakeholders can retain the system for the outcomes, allocate resources, enforce environmental regulations, and predict climate change's impact using AI systems. Without this transparency, there is a risk that AI systems could be operated or abused to impair environmental compatibility.

The environmental impact of AI itself is another problem that cannot be overlooked. AI has the potential to mitigate environmental issues. However, the development and use of AI technology have unique environmental issues. Training large-scale AI models require considerable computing power and consume significant energy. The CO2 footprint associated with data centers and cloud computing infrastructure is a growing problem as demand for AI services increases. However, compensation for the benefits of AI with environmental costs remains an ongoing challenge. For example, AI systems that optimize resource usage can incorrectly promote the consumption or exhaustion of specific resources if not managed carefully. Similarly, AI applications in agriculture that promote precision reproduction can lead to excessive reliance on monoculture or specific technologies, reducing biological diversity and ecosystem resilience. These risks highlight the requirement for vigilant consideration and regulation of AI applications regarding environmental capabilities, ensuring they are genuinely advantageous and do not inadvertently pose new problems.

Despite these challenges, AI's possibility to contribute to ecological sustainability is immeasurable. The ability to process large amounts of data, identify patterns, and provide real-time knowledge is a powerful tool to solve the complex, interconnected problems faced in environmental protection. However, addressing the ethical, social and ecological risks associated with its use is essential to exploit AI's possibilities fully. It will allow AI systems to be transparent, accountable, justified, and developed with sustainability and equity in mind. Investigating the environmental impact of AI itself must continue to ensure that the benefits of technology cannot be resolved. However, the integration of AI in sustainability practices is not without challenges. Ethical considerations must be treated with caution, including the risk of worsening inequality, the need for transparency in AI itself, and the risk of environmental costs. Only by navigating these issues can AI be a positive force for environmental change and ensure efforts to generate more sustainable and equitable world support for future generations.

BACKGROUND

Artificial intelligence (AI) has been developed as a transformative device in various sectors, particularly climate change, sustainable energy management, and environmental protection. The possibility that AI could revolutionize ecological sustainability by improving climate forecasts to optimize energy systems is immeasurable (Zavrazhnyi et al., 2024). This broad discussion explores how AI plays a key role in climate change, sustainable energy management, and pollution and waste management management. It is one of the key areas where AI influences climate forecasting and disaster risk assessment. Machines - Learning models, especially deep learning techniques, can process large amounts of data from various sources, such as satellites, sensors, and weather stations. These models can analyze historical climate patterns, predict future trends, and simulate potential climate tapes accurately. In this way, they support scientists and political decisions (Anwar & Sakti, 2024). Carbon dioxide emissions (CO2) are primarily due to burning fossil fuels, the most common cause of global warming. AI control systems are being developed to monitor and optimize CO2 recording technologies that capture CO2 emissions from power plants and industrial procedures before they are released into the atmosphere. Algorithms for machine learning can improve the efficiency of CO2 recordings by determining optimal operating conditions and detecting adapted leaks and processes in real-time.

Additionally, AI can help develop new materials and methods for carbon covering, storage, and use that could potentially change the landscape of emission reduction technologies (Regona et al., 2024). One of the most promising applications of AI in energy management is the creation of smart grids. Smart Grid is an unconventional power system that uses AI and machine learning to optimize power distribution and consumption. AI systems can predict energy requirements, recognize inefficiencies, and ensure seamless network behavior by integrating real-time sensor data, weather forecasts, and usage patterns. It leads to more reliable, cheaper, and environmentally friendly energy management.

Additionally, AI can compensate for energy loads, prevent power outages, and help incorporate renewable energy sources such as wind and solar into the network. In other words, starting performance fluctuates depending on weather conditions. AI algorithms are developed to predict these variations and optimize the process of renewable energy systems. For example, AI can predict solar radiation or wind speeds, allowing operators to plan better energy generation and storage (Rolando & Mulyono, 2024). AI also improves the efficiency of energy storage systems such as batteries by optimizing energy requirements and load and emission cycles.

Air and water quality monitoring is one of the region's primary uses in AI. Traditional methods for monitoring environmental pollution often include time-consuming and expensive manual sampling and laboratory analysis. However, AI-powered sensors can provide real-time air and water quality data. Combined with machine learning algorithms, these sensors identify a combination of contamination sources, predict pollution trends, and suggest corrective actions (Mohammed et al., 2024). For example, AI particle (PM) levels, nitrogen dioxide (NO2) concentrations, and other contaminants can be recognized.

Furthermore, AI improves your ability to pursue and manage water quality. Industrial emissions, agricultural wastewater, and water pollution from urban waste are growing issues in many parts of the world. AI algorithms can analyze water samples for contamination and predict possible water supply risks. By monitoring parameters such as PH values, dissolved oxygen, and temperature, AI helps ensure water resources' safety and quality, reduce the risk of water diseases, and promote sustainable water consumption practices (Ajirotutu et al., 2024). A circular economy is an economic system that minimizes waste and uses the resources available to the best extent by reusing, restoring, renovating, and recycling materials. AI is used to optimize waste collection, sorting, and recycling processes. For example, an AI-attached robot can identify and separate sources of different types of plastic, metal, and waste with computer vision to restore and reuse valuable resources. AI can optimize the operation of waste energy systems by predicting waste supply, adjusting treatment parameters, and improving energy recovery.

By integrating AI into the data in waste management systems, operators can maximize the energy generated from waste while at the same time minimizing environmental impact. Using AI to analyze the product lifecycle, companies can determine opportunities to reduce material use, minimize waste, and extend product life. AI algorithms can also support the design of products that contribute to reducing e-waste and other forms of waste. A change in the direction of cyclical approaches for production and consumption is essential to achieve long-term sustainability goals and reduce the ecological footprint of industrial activities. Machine learning models improve climate forecasting and disaster risk assessments, allowing for better preparation for extreme weather events. AI will help optimize strategies to reduce carbon capture and emission reductions and contribute to efforts to slow the pace of global warming. In the energy sector, AI will help improve the efficiency of smart grids, renewable energy systems, and energy storage, helping to accelerate the transition to a low-carbon energy future.

Furthermore, AI will change pollution and waste management controls by enabling real-time air and water quality monitoring, improving waste sorting and recycling processes, and improving support models for circular economic models. Using AI solutions requires significant investment in infrastructure, research, and development (Bibri et al., 2024). Additionally, ethical considerations such as data protection and algorithm distortion must be addressed to ensure that AI technology is used responsibly and fairly. Political decisions - Manufacturers, scientists, and industry leaders must work together to develop framework conditions and regulations that maximize AI's benefits while minimizing risk. Its applications in reducing climate change, sustainable energy management, and pollution control are promising and will develop as technology matures. Using the power of AI can create more intelligent, more efficient, and more sustainable systems that will help defend the planet for future generations.

ETHICAL AND POLICY CHALLENGES IN AI FOR SUSTAINABILITY

Artificial intelligence (AI) is becoming essential in environmental science, from forecasting climate mutants to assessing biological diversity. However, like all tools, AI has flaws (Salman et al., 2024). One of the main concerns of AI applications in environmental modeling is the risk of distorted data that leads to inaccurate conclusions. AI models, including those used in climate and environmental assessments, are strongly based on data input. Suppose the data used to train these models is incomplete, not representative, or skewed.

In that case, the results reflect these distortions and can produce misleading knowledge or predictions. Suppose these historical data points are concentrated in a particular geographical region or period. In that case, the model may not represent global or future conditions correctly. For example, suppose AI is primarily trained on urban temperature and precipitation data (Nguyen et al., 2024). In that case, this may not be able to predict rural and remote conditions. Similarly, AI-dependent environmental reviews may not ignore specific ecosystems or predict future environmental conditions in a way that benefits all involved.

Environmental risks like pollution can affect disproportionately weak or marginalized communities. Suppose AI models are not correctly designed for these differences. In that case, they may overlook essential risks with serious social consequences. It could lead to decisions or guidelines that exacerbate existing inequality instead of mitigating it. Many of the data inputs used to form AI systems are based on previous practices, laws, and assumptions regarding the sustainability of ecological sustainability, which are no longer relevant or suboptimal. For example, suppose AI models are trained with data derived from previous industrial activities that are now understood as environmental compatibility.

In that case, we can maintain that certain practices are sustainable. It may hinder more effective and fair environmental guidelines. Scientists and political decision-makers should work together to ensure that the data used in AI-controlled environmental assessments are comprehensive and reflect attention to the future environment. Furthermore, AI models must be subjected to rigorous verification to ensure that predictions and recommendations are reliable and that no distortions can impair their effectiveness. Environmental issues. These technologies provide unprecedented insight into everything from air quality to biological diversity, helping scientists pursue and respond to environmental changes. However, in addition to this progress, there is considerable concern about data protection and safety, mainly as environmental surveillance is increasingly based on surveillance-based technologies. For example, sensors embedded in urban and rural areas can gather information about human activity, from transport patterns to energy consumption. This data is invaluable for pursuing environmental impacts and developing sustainable solutions. However, it also includes important questions about individual privacy.

This data can often be used to pursue a particular person or group and can violate your data protection rights. In some regions, environmental sensors or AI-enabled cameras can be used without public awareness or agreement, which raises concerns about transparency and accountability. The increased data for these technologies increases the risk of abuse or unauthorized access. Suppose the data collected by AI control systems is not properly protected. In that case, malicious actors can use it for purposes unrelated to environmental protection, such as marketing or government surveillance. These systems often work on cloud-based platforms that store, process, and analyze data. Suppose these platforms are not properly protected against cyberattacks or data injuries. In that case, sensitive environmental data cannot be exposed to authorized people or organizations (Tariq, 2025).

Additionally, there is the risk of manipulating or hacking AI models, which can lead to false environmental predictions and misinformed political decisions. Governments and organizations must ensure that data collected through environmental surveillance technology is used responsibly and that data protection rights are respected. Transparency related to data collection practices and the intended use of data is essential to developing public trust (Olawade et al., 2024). AI exists to optimize political decisions by analyzing large amounts of data and forecasts of future trends, there is concern that excessive reliance on automated systems can undermine human judgment and surveillance, especially in complex, context-sensitive questions such as environmental protection.

AI-controlled systems can process and analyze data at speeds and scales far exceeding human capabilities. For example, AI can assess industrial activities' environmental impact, predict political interventions' outcomes, and optimize resource allocation for sustainability efforts. These AI systems often provide more accurate, data-controlled recommendations than traditional methods. It could lead to more effective and efficient environmental guidelines. One of the main concerns is that even if they are well-designed, AI systems can either be inflexible or explain the broader social, cultural, and ethical implications of political decisions. Environmental issues often include complex networks of stakeholders with a variety of interests. They may not be equipped with AI models to navigate these complexities. For example, AI control decision systems for carbon reduction can optimize the economic impact on communities that rely on industries contributing to carbon emissions. If the data input is incomplete or skewed, recommendations generated by the AI model can lead to undesired results. For example, suppose AI models prioritize efficiency and cost savings compared to stocks and social justice. In that case, guidelines that disproportionately harm populations that require protection can be recommended.

Therefore, human oversight must remain essential to AI-controlled decision-making in environmental policy. For example, political decision-makers are more equipped to consider sustainability policies' long-term social and cultural consequences and integrate input from affected communities. Additionally, human decision-makers can submit critical judgments when compromises must be made between AI predictions or competing priorities. AI must be used to support and improve human decision-making and cannot be replaced entirely. Political decisions - Manufacturers need to use the power of AI to inform and guide decisions while also ensuring that human values, ethical considerations, and social needs remain at the forefront of sustainability policy. It requires continued cooperation between engineers, environmental experts, and social scientists, creating AI systems that complement human judgment and contribute to the broader goals of sustainable development (Tariq, 2025).

One of the main concerns is the high energy consumption associated with the training and operation of AI models. Computing capabilities are required to develop and deliver advanced AI systems, particularly in climate modeling, data analysis, and predictive environment review. It can have an essential environmental footprint, contributing to energy consumption and greenhouse gas emissions. This process uses special hardware, such as a GPU (Graphics Processing Unit), to perform the complex mathematical calculations required to improve the model's accuracy. The energy required to run these models can be significant, especially when the model is trained on a solid dataset or used for continuous actual time analysis. With increasing demand for more powerful AI models, energy consumption is also increasing. In many parts of the world, servers used for AI processing are based on fossil fuels that contribute to carbon emissions and further exacerbate the impact of climate change. While there are efforts to switch to renewable energy sources in data centers, the growing energy consumption associated with AI continues to raise concerns about general environmental capabilities. The production and disposal of hardware used in AI systems also contribute to environmental degradation. The production of GPUs, processors, and other components requires raw materials, energy, and water, and the disposal of outdated or outdated devices can lead to e-waste. Considering the environmental impact of AI technology throughout the lifecycle, you need to ensure that your benefits are not at the expense of the environment. Researchers are working on generating AI models with less computing power and reducing energy consumption. The transition from data centers to renewable energy sources and the elevation of sustainable hardware production and disposal practices are essential steps to reducing the AI ecological footprint. From climate model bias to data protection and security concerns, AI's use in environmental solutions must be carefully considered and monitored.

POLICY FRAMEWORKS AND GOVERNANCE FOR AI IN SUSTAINABILITY

Artificial intelligence (AI) has become a transformative device for managing global environmental challenges. From predicting climate change patterns to optimizing energy consumption and reducing waste, AI offers a variety of solutions to create a more sustainable future. However, the AI technology process offers a complex challenge that requires effective governments to ensure that they are used responsibly and ethically in the context of ecological sustainability. Global guidelines and regulations play an important role in AI's development, use, and management for environmental purposes, balancing the need for innovation with national wells and environmental protection. Various organizations and governments have set ambitious climate and environmental targets at the international level. These international frameworks form the basis for how AI can be used in the fight against climate change and environmental degradation. However, the rapid growth of AI technologies requires developing corresponding global guidelines that ensure effective contributions to sustainability goals, using rules on how these technologies are used. It requires a comprehensive approach to political design, which includes not only the environmental impact of AI technology itself but also the broader impact of its provisions. For example, AI systems that support energy efficiency or reduce emissions are advantageous. However, the environmental costs of the production and operation of these AI systems, such as energy consumption, hardware production, and e-waste disposal, also need to be considered. Effective guidelines promote the use of renewable energy for the performance of AI systems, promote the design of energy-efficient algorithms, and provide incentives for businesses to develop and deliver AI that minimizes environmental damage. Environmental issues are global, and climate change loss means that environmental pollution and biological diversity are cross-border. It can be challenging as countries have different technological developments, resources, and priorities. However, international organizations such as the United Nations (UN), the World Economic Forum (WEF), and the International Telecommunication Union (ITU) are working to determine guidelines and framework conditions for the responsible use of AI in environmental management. These efforts aim to ensure that AI technology is used in a way that is accessible to all countries, including developing countries, and that its benefits are very, very distributed. In 2021, the EU proposed the Artificial Intelligence Act. This first legal framework deals with AI and its ethical impacts, among other things. The law's main focus is generally on AI systems but includes environmental compatibility provisions. The EU approach emphasizes the transparency, accountability, and fairness essential to using AI-controlled environmental solutions responsibly (Hong & Xiao, 2024). AI systems develop quickly, and regulations can often remain behind technological advances. Therefore, a dynamic regulatory framework is required and can be adapted to new AI applications. Governments and international organizations should work closely with AI researchers and industry leaders to monitor trends, ensure current regulations, and reflect the latest scientific and technological developments. It requires the promotion of continuous dialogue between political decision-makers, the private sector, and civil society. AI models are based on large data records to create predictions and recommendations. However, the resulting model can lead to false conclusions if the data is biased or incomplete (Ahmad et al., 2024). It is essential for environmental reviews where biased model results are important. Regulations must ensure that AI application data is collected transparently and ethically to address data protection and confidentiality concerns. Data exchange protocols must be set up to promote global cooperation and, at the same time, protect individual rights. While AI has great potential to support sustainability, energy consumption associated with developing and delivering AI models is important. Data centers operating AI systems consume much electricity (Radanliev, 2024). However, most of them still come from non-renewable sources. To mitigate the environmental impact of AI, regulations should encourage the use of energy-efficient hardware and renewable energy sources, as well as the development of AI algorithms with low computing capabilities. Political decision-makers should also promote the recycling and responsible disposal of e-waste created by AI hardware. Developing and using AI technologies for environmental compatibility requires considerable research, infrastructure, and innovation investment. As a result, cooperation between governments, private companies, academic institutions, and non-governmental organizations (NGOs) is essential for developing AI solutions, scaling, and implementing effectively, fairly, and sustainably.

In AI-controlled sustainability initiatives, public-private partnerships can take various forms, ranging from joint research and development projects to fundraising programs and political interest groups. Governments can provide financial support, incentives, and regulatory framework conditions encouraging private companies to invest in sustainable AI technology. Public funding, for example, can help develop AI applications that address environmental issues such as renewable energy, water management, biodiversity conservation, and waste reduction. Second, private companies bring technical know-how, resources, and innovation to the table, helping to accelerate the development and deployment of AI solutions. Public-private partnerships often contribute to creating the necessary infrastructure and guidelines to promote the use of AI in the energy sector. For example, governments can fund the installation of intelligent meters and sensors. At the same time, private companies can develop AI algorithms that process data to optimize and reduce waste. These partnerships are crucial for scaling renewable energy technologies and achieving global climate targets (Tariq, 2025).

AI can be similarly used in agriculture to optimize water, fertilizers, and pesticides, improve sustainability, and reduce environmental harm. Public-private partnerships in this field, agricultural experts, AI researchers, and technology companies can rally together to create AI control solutions that help farmers reduce their environmental impact and, at the same time, increase productivity. However, private companies can help bring these innovations to the market. AI can monitor ecosystems, pursue wildlife populations, and predict environmental threats like habitat drafting and destruction. While governments and NGOs often have specialized knowledge and mandate to protect biological diversity, private companies bring technical know-how to develop AI-powered tools. Through collaboration, these sectors can ensure that AI improves maintenance efforts and protects endangered species and ecosystems. Governments can determine AI's regulatory and political framework and ensure it is used in ways that contribute to global sustainability goals (Ukoba et al., 2024). The private sector can work with governments to focus on AI technology being responsible and used and minimizing negative ecological and social impacts. These partnerships can promote knowledge exchange and capacity structure, particularly in developing countries with limited access to AI technologies and capabilities.

However, public-private partnerships are not without challenges. The key topic is to ensure that these partnerships are fair and transparent and have an appropriate expression for all involved. Governments must ensure that public funds and resources are used effectively and that private companies are responsible for AI solutions' ecological and social impacts (Gikunda, 2024). Transparency in the decision-making process and data and knowledge exchange is essential to building trust and enabling public-private partnerships to achieve their intended outcomes. In some cases, private companies can prioritize profits before sustainability, leading to decisions not in the environment's or society's best interests. Governments and other stakeholders must work together to determine clear guidelines and standards to ensure that AI-controlled sustainability initiatives will help the well-being of the masses and not only be driven by commercial interests (Tariq, 2024). Through collaboration, these sectors can accelerate the development and use of AI technologies to address the global environment's challenges while ensuring that these technologies are used responsibly and ethically.

Integrating artificial intelligence (AI) into the persecution of the Sustainable Development Goals (SDGs) has attracted considerable attention due to its potential to pose some of the world's most urgent challenges. Although AI is under development, it is essential to determine ethical guidelines to ensure its use will positively contribute to sustainable development. These guidelines must ensure that AI technologies are responsible, comprehensive, and transparently used while aligning with the UN SDG's broader goals. Ethical considerations regarding AI are complicated as they affect questions such as fairness, the obligation to explain, transparency, and the likelihood of unintended social, economic, and ecological outcomes. Developing such ethical guidelines is necessary to ensure that AI-controlled solutions contribute, rather than exacerbate, environmental damage or contribute to social justice, economic inclusion, and environmental compatibility (Ogbeibu et al., 2024). These principles must lead the development of AI technologies that apply to initiatives for sustainable development. For example, AI systems to improve agricultural practices and energy efficiency should ensure that the benefits of different segments of society are equally distributed, including population groups of marginalized or endangered species. In AI , fairness involves minimizing data and algorithm distortions that can lead to discriminatory outcomes, particularly in companies already borne by inequality. To ensure that AI systems are designed as comprehensive, you need to carefully consider interactions with various stakeholders, including those from disadvantaged communities. Transparency is another important principle. AI models, particularly those used in guidelines and resource management, should be understood by experts and non-experts so that the discussions behind the decision are clear and checked. It is vital in the SDGS context, as decisions affected by AI can have serious social, economic, and ecological consequences. It is because people with access to advanced technology disproportionately benefit from AI innovation, and people in undeveloped areas are left behind. This fight against explanation requires a global approach to ethical AI, knowledge exchange, capacity structure, and a fair distribution of resources and technology (Tariq, 2025).

Furthermore, ecological sustainability must be at the forefront of AI ethics to ensure that AI does not contribute to environmental degradation. These environmental costs must be considered in developing AI solutions for sustainability (Lodhi et al., 2024). Ethical AI guidelines should promote the development of energy-efficient algorithms and work to use renewable energy in AI-powered systems. It not only supports the performance of SDG 7 (Affordable Clean Energy) but also helps reduce the CO2 footprint of AI technology and aligns with the climate target set at SDG 13. One of the strengths of AI is its ability to analyze and interpret significant data records from various sources. However, the context in which these AI systems are used is essential, especially when dealing with global challenges. For example, AI models that work well in a region or country may not necessarily apply to other parts of the world, as local culture, infrastructure, or economic situations are differentiated. Ethical guidelines should encourage the development of flexible, context-sensitive AI solutions that can be adapted to the specific needs and situations of different regions, communities, and sectors.

Governments and international organizations need to work together to create regulatory framework conditions that ensure that AI technology contributes to sustainable development and, at the same time, prevents the possibility of damage (Leal Filho et al., 2024). These regulations must promote transparency, equity, accountability, and human rights protection while stimulating innovation and SDGS-compliant AI technology. Global standards of AI governance, such as those proposed by organizations such as the European Union and the OECD, can serve as valuable benchmarks for the responsible development of AI (Rohde et al., 2024). However, these frameworks need to be updated continuously to meet technological advancements and emerging challenges so that they are genuinely effective.

It is essential to ensure that human supervision is centrally important in using AI systems when AI is more deeply integrated with decision-making processes, particularly SDGs. Ethical AI guidelines should support and improve human decision-making and promote the use of AI rather than completely replacing it. AI can provide valuable knowledge and data with controlled recommendations. However, human judgment and ethical considerations should guide the final decision. It is imperative to address complex and diverse topics, such as climate change, poverty management, and social inequality, where human values and ethics play an essential role in designing results (Rauf et al., 2024). Therefore, developing SDG ethical AI guidelines must emphasize balancing automation and human supervision. Many AI systems are designed to optimize for immediate results. B. Cost reduction or efficiency improvements. However, sustainable development requires a long-term perspective considering long-term AI solutions' social, economic, and ecological impacts. Ethical guidelines should encourage the development of AI models optimized for short-term efficiency and consider future generations' wider impact. It corresponds to the principles of SDG 12 (Responsible Consumption and Production). It requires a transition to sustainable production and consumption patterns. ESG reports are essential for assessing how businesses and organizations do things regarding sustainability and social responsibility. Accurate and comprehensive reporting is essential for stakeholders, including investors, supervisors, and consumers, to make appropriate decisions. AI can revolutionize ESG reports by automating data collection, improving accuracy, and providing deeper insights into environmental performance (Voigtlaender et al., 2024). This transformation could lead to more transparent, reliable, and implementable ESG reports that will help promote sustainability in the private sector. Traditional ESG reports often involve collecting large amounts of data from various sources, including corporate information, government reports, and third-party databases. This data should be processed, analyzed, and presented in an essential way to the stakeholders. For example, AI companies can analyze disclosures, sustainability reports, and regulatory registrations to extract relevant information about environmental output, such as carbon emissions, water consumption, and waste disposal practices (Ribeiro, 2024). example, AI can analyze satellite imagery, sensor data, and climate models to give insights into environmental factors such as air quality, deforestation, and water scarcity. These findings can be used to generate more detailed environmental performance indicators to help businesses better assess their environmental impact and identify areas of improvement. Additionally, AI can create predictive models that predict future environmental trends, helping businesses predict and mitigate potential risks associated with climate change and resource praise (Wang et al., 2024). By including these findings in ESG reports, businesses can help stakeholders with more comprehensive and future information and promote long-term sustainability.

Additionally, AI helps standardize ESG reporting by providing a consistent frame for measurements and reports for environmental performance. One of the challenges of ESG reporting is the lack of uniform standards and metrics in industries and regions. Companies can use different methods to measure environmental impact, making it difficult for investors and regulators to compare performance between organizations.

Companies often report Traditional ESG reports, which can lead to considerations regarding the greenwashing or operation of environmental services data. AI can help to resolve these concerns by providing an independent review and analysis of ESG claims. For example, AI can be used to view data using third-party provider sources such as satellite images, sensor data, and state databases to verify the accuracy of environmental claims. It helps to ensure that ESG reports are reliable and reliable, providing the information needed for interest groups to make well-discovered decisions. Additionally, AI can pursue and monitor the corporate environmental output over time, providing a more dynamic, real-time image of sustainability.

Traditionally, ESG focused primarily on environmental factors such as carbon emissions and resource consumption. However, there is growing recognition that social and governance factors such as labor practices, human rights, and corporate management are extremely important for sustainable development. AI helps to integrate these factors into ESG reports by analyzing data on social and governance issues and environmental performance indicators. For example, AI social media data, news articles, and corporate applications can be analyzed to assess the company's work practices, human rights laws, and corporate governance structure. This approach to ESG reports provides a more comprehensive overview of the company's sustainability services. It can help investors and stakeholders make more informed decisions. Collecting data, improving reporting accuracy and depth, standardizing metrics, and improving transparency and accountability can change how businesses and organizations report on environmental, social, and governance performance (Wörsdörfer, 2024). However, AI applications must be directed towards ESG reporting of ethical principles that prioritize fairness, transparency, and accountability, ensuring that AI is used responsibly to support the transition to a more sustainable and better future.

CASE STUDIES AND FUTURE TRENDS IN AI FOR ENVIRONMENTAL QUALITY

Artificial intelligence (AI) is the foundation of environmental protection and sustainability efforts worldwide. The ability to process large amounts of data and provide predictive knowledge has proven AI invaluable for addressing environmental challenges, improving resource management, and promoting sustainable practices (Al-Raeei, 2025). While we progress, AI's use in environmental quality control continues to expand and develop, with emerging trends showing even greater possibilities in the interface between AI and green technology. By examining lessons from successful AI operations, climate protection initiatives, and emerging trends, we can better understand how AI shapes the future of environmental management and sustainability. AI-powered systems are increasingly used to monitor ecosystems, pursue endangered species, and recognize illegal activities such as poaching and deforestation. In Africa, for example, AI is used with satellite imaging and drone technology to monitor wildlife populations and protect national parks from poaching (Durai et al., 2024). AI algorithms can analyze satellite data to identify changes in the landscape, such as deforestation and illegal human activities, send warnings to Parken-Ranger, and take timely measures.

Thus, not only does AI help protect wildlife, but it also serves the broader goal of maintaining biological diversity. It is critical to maintaining ecosystem balance and global health. In some cases, AI is also used to predict the likelihood of a forest fire based on environmental data such as temperature, moisture, and wind patterns. This predictive ability helps efficiently allocate resources before escalating and responding to the threat of bushfires. These AI control solutions improve environmental efforts and help us achieve long-term sustainability goals by promoting responsible land management (Alzoubi & Mishra, 2024). AI algorithms are used to process intelligent networks and help to compensate for energy demand and supply by predicting consumption patterns and optimizing distribution. By analyzing data from intelligent measurement devices, weather forecasts, and historical usage trends, AI can make real-time adjustments to ensure energy is used efficiently, minimizing waste and CO2 footprints.

In areas where water is limited resources, AI algorithms can help optimize irrigation systems to ensure that water is being used efficiently and waste is reduced. By analyzing data from sensors placed on soil and plants, AI systems can determine the optimal amount of water needed for irrigation, considering soil moisture, weather conditions, and harvesting factors. It helps save water, increases agricultural productivity, and is an essential tool for sustainable agriculture.

It allows you to analyze data from water sensors to identify contaminants and potential contamination sources. It will allow faster responses to water quality issues and protect water resources for future generations (Olawade et al., 2024). The importance of data quality and accuracy is a huge success. AI systems are based on a large amount of data to make predictions and inform decisions. The effectiveness of these systems is directly connected to the quality of the data they process.

In many cases, the success of AI applications in environmental protection depends on the availability of high-quality real-time data from sensors, satellites, and other monitoring tools. It highlights the need for improved data infrastructure and standardization of environmental data so that AI systems can work effectively and provide wise knowledge. The success of AI deployment in environmental protection often includes partnerships between these various sectors, with expert knowledge and resources contributing to the success. For example, when monitoring wildlife, AI technology companies work with conservation organizations and local governments to provide the technology infrastructure needed for effective monitoring. Similarly, AI companies are working with suppliers and governments in the energy sector to integrate AI into existing network systems (Satornino et al., 2024). These collaborations are essential to the fact that AI solutions are implemented in a way that meets local needs and priorities and contributes to long-term sustainability goals.

Additionally, AI-driven climate management initiatives must be designed to consider justice and inclusiveness. AI has the possiblity to accelerate progress in the direction of sustainability. However, its benefits in society must be distributed quite a bit. It means that AI solutions should be accessible to both developed and developing countries, and they need to consider the protection needs of smallholder farmers, marginalized communities, and people living in areas most susceptible to climate change. In this regard, AI can play an essential role in supporting climate adaptation and strength, helping communities better understand and respond to climate change's impacts.

Blockchain, a decentralized, immutable ledger technology, needs to revolutionize the possibilities of tracking and sharing environmental data. Integrating AI into the blockchain allows you to create more transparent and secure systems to monitor your environment. For example, blockchain can pursue a product's CO2 footprint throughout the lifecycle from production to consumption, ensuring that companies are responsible for their environmental impact. Similarly, blockchain can be used to create transparent, continuous systems for renewable energy financing so that consumers and businesses can verify the reliability of their green energy requirements (Martini et al., 2024). While businesses can use AI to pursue the environmental impact of their supply chains, blockchains can help ensure that data is transparent, secure, and manipulated. This combination of technologies helps businesses identify supply chain inefficiencies, reduce waste, and ensure their products are sustainable. By providing greater transparency into the environmental impact of products, these technologies allow consumers to make more informed decisions and encourage businesses to use more sustainable practices. With the ability to collect real-time data from the environment, IoT devices can provide valuable insights into air and water quality, energy consumption, and waste management. In combination with AI, this data can be used to optimize resource use, predict environmental trends, and inform political decisions. Deploying 5G networks will improve IoT devices' capabilities, allowing faster and more reliable data transmission. It is essential for real-time monitoring and decision-making (Ghobakhloo et al., 2024). For example, AI can optimize urban planning, reduce traffic congestion and air pollution, and improve disaster efforts to improve community resilience to extreme weather events. Ethical considerations such as fairness, transparency, and inclusiveness must be paramount when designing and implementing AI systems for environmental compatibility. It includes the need for AI solutions to be at risk for all population groups, especially at risk, and to minimize negative social and ecological impacts. When AI is integrated through decision-making processes, it is essential to maintain human supervision and ensure that AI improves human judgment. From improving resource management to promoting maintenance efforts and renewable energy, AI plays a critical role in addressing some of our time's most urgent environmental challenges. As new technologies such as blockchain, IoT, and 5G are developed, AI will become even more critical to environmental governments and green technology. However, the future of AI in ecological sustainability relies solely on responsible development, local access, and cooperation. You can make the most of AI's potential and follow ethical guidelines to accelerate progress in a more sustainable, resistant, and fairer future direction for everyone.

CONCLUSION

AI offers unique challenges, particularly concerning environmental impacts, ethical concerns, and the potential to exacerbate existing social inequality. Understanding the dual role of AI is extremely important for political decision-making - researchers, researchers, and industry leaders are taking advantage of their advantages and simultaneously covering risks. AI enables real-time monitoring and data analysis to enable better decision-making in agriculture, energy, water management, and waste reduction. For example, AI algorithms can help optimize irrigation systems and reduce water waste in areas with water shortages.

For example, the AI-operated system was used to pursue wildlife populations, monitor illegal activities such as poaching, and recognize design patterns for satellite imagery. In forestry, AI can predict the likelihood of a forest fire based on environmental factors, allowing authorities to allocate resources and respond quickly to prevent or mitigate damage. AI is increasingly being used to monitor ecosystem health by analyzing data from sensors, satellites, and drones and providing more detailed insight into environmental conditions than traditional methods. AI applications optimize production processes, reduce material waste, and increase energy efficiency in manufacturing. Similarly, AI can help businesses minimize environmental impact by identifying supply chain patterns, explaining raw materials more sustainably, and reducing unnecessary consumption. Despite these promising applications, AI's role also offers several environmental care challenges. One of the most critical concerns is the environmental impact of AI itself, particularly large models of deep learning and machine learning. The energy consumption of AI systems has a significant CO2 footprint, especially when trained on large data sets. The ecological footprint of AI is often hidden by its potential benefits. However, it is essential to recognize and address the sustainability issues inherent in AI development. Research into more energy-efficient AI models and using renewable energy sources Powering AI infrastructure is intended to reduce the CO2 footprint of AI.

AI systems are as good as the data they are trained. Suppose the dataset used to form the AI model is incomplete, biased, or not representative of the ecosystem you want to analyze. In that case, it can distort the results, leading to unfair or ineffective solutions. For example, biased AI models in agriculture can lead to recommendations that do not meet marginalized communities and smallholder farmers' unique needs, reinforcing existing inequalities in accessing resources and sustainable practices. Similarly, pre-tensioned data in environmental monitoring can lead to inaccuracies in environmental health assessments that may overlook the areas where intervention is most urgently needed. For example, the increased use of AI in environmental monitoring based on surveillance monitoring increases data protection and security concerns.

In some cases, AI operation systems collect large amounts of data about individuals, e.g... These concerns must be considered by clear regulations and ethical guidelines that compensate for the benefits of AI with the protection of individual freedoms and rights. When AI systems are more autonomous and integrated into environmental decisions, ensuring transparency, explanation, and accountability is becoming increasingly difficult. Manufacturing processes are often opaque, and it may not always be clear how or why a particular decision is made. In the context of environmental compatibility, this lack of transparency can undermine trust in AI systems and their consequences. Effective governance frameworks are needed to ensure that AI systems are transparent and accountable, especially concerning decisions with significant ecological and social effects. Political decisions - Manufacturers need to create a supportive regulatory environment that promotes the responsible development and use of AI for environmental protection while at the same time reducing its potential risks. It includes creating guidelines to promote transparency and equity in AI applications ensuring that AI solutions align with sustainability and social justice goals. Political decision-makers should also invest in AI-controlled environmental surveillance and develop data infrastructures that support decisions to ensure high-quality, accurate, current data is available in AI models.

Regarding data management, it is essential to ensure that all stakeholders are open, transparent, and accessible to record and share environmental data. The development of standardized ambient data formats and open-source AI tools can help promote regional cooperation and exchange of knowledge and resources to manage global environmental challenges. Additionally, in political decisions, manufacturers must address ethical and privacy issues in AI-related environmental surveillance to ensure that AI uses do not violate communities that disproportionately affect individuals or communities needing protection. Investigating CO2 footprint reductions through hardware and software innovations from AI is not a potential benefit to ensure the environmental impact of AI. Researchers should also address the biases available in AI models, ensuring they are trained on various representative data that reflect the complexities of environmental systems and human society.

By incorporating environmental compatibility at the core of your business strategy, businesses can advance the adoption of AI solutions that reduce resource consumption, emissions, and cycle support. In this way, we must ensure that AI applications are designed to minimize their ecological footprint, particularly concerning energy consumption and waste production. From AI with other emerging technologies such as blockchain, Internet of Things (IoT), and 5G. Continuous improvements in AI algorithms related to sensor network growth and real-time data analysis will improve the ability to monitor and manage environmental systems at a global level.

Furthermore, integrating AI and blockchain could improve transparency and accountability for environmental management, enable secure persecution of environmental data, and ensure the integrity of sustainability claims. promises of AI as a means of environmental protection and sustainability are enormous, but they must be used carefully with its potential risks and unintended consequences. To further develop it, it is essential to ensure that AI promotes trends, equity, and environmental justice and cast the UN goals of the Sustainable Development Goals (SDG). By addressing challenges related to AI, such as environmental impacts, data distortions, and ethical concerns, we can ensure that it contributes to a more sustainable and fairer future. The future trajectory of AI-controlled environmental solutions is shaped by continuous innovation, responsible governance, and a shared commitment to addressing global environmental challenges. Through careful and thoughtful implementation, AI can play a transformative role in achieving a sustainable, resistant, and prosperous future for everyone.

REFERENCES

Ahmad, S., Priyadharshini, S. L., Hosen, M. S., Ng, A., Islam, S., & Manik, J. A. (2024). The Impact Of Artificial Intelligence On Business & Social Values: Benefits, Challenges, And Future Directions. *Educational Administration: Theory and Practice*, *30*(6), 3174–3180.

Ajirotutu, R. O., Adeyemi, A. B., Ifechukwu, G. O., Ohakawa, T. C., Iwuanyanwu, O., & Garba, B. M. P. (2024). Exploring the intersection of Building Information Modeling (BIM) and artificial intelligence in modern infrastructure projects. *Journal of Advanced Infrastructure Studies*.

Al-Raeei, M. (2025). The smart future for sustainable development: Artificial intelligence solutions for sustainable urbanization. *Sustainable Development (Bradford)*, *33*(1), 508–517. DOI: 10.1002/sd.3131

Alzoubi, Y. I., & Mishra, A. (2024). Green artificial intelligence initiatives: Potentials and challenges. *Journal of Cleaner Production*, *468*, 143090. DOI: 10.1016/j.jclepro.2024.143090

Anwar, M. R., & Sakti, L. D. (2024). Integrating artificial intelligence and environmental science for sustainable urban planning. [ITSDI]. *IAIC Transactions on Sustainable Digital Innovation*, *5*(2), 179–191. DOI: 10.34306/itsdi.v5i2.666

Bibri, S. E., Huang, J., Jagatheesaperumal, S. K., & Krogstie, J. (2024). The synergistic interplay of artificial intelligence and digital twin in environmentally planning sustainable smart cities: A comprehensive systematic review. *Environmental Science and Ecotechnology*, *20*, 100433. DOI: 10.1016/j.ese.2024.100433 PMID: 38831974

Camilleri, M. A. (2024). Artificial intelligence governance: Ethical considerations and implications for social responsibility. *Expert Systems: International Journal of Knowledge Engineering and Neural Networks*, *41*(7), e13406. DOI: 10.1111/exsy.13406

Durai, S., Manoharan, G., & Ashtikar, S. P. (2024). Harnessing artificial intelligence: Pioneering sustainable solutions for a greener future. In *Social and Ethical Implications of AI in Finance for Sustainability* (pp. 89–117). IGI Global. DOI: 10.4018/979-8-3693-2881-1.ch003

Ghobakhloo, M., Fathi, M., Iranmanesh, M., Vilkas, M., Grybauskas, A., & Amran, A. (2024). Generative artificial intelligence in manufacturing: Opportunities for actualizing Industry 5.0 sustainability goals. *Journal of Manufacturing Technology Management*, *35*(9), 94–121. DOI: 10.1108/JMTM-12-2023-0530

Gikunda, K. (2024). Harnessing artificial intelligence for sustainable agricultural development in Africa: opportunities, challenges, and impact. *arXiv preprint arXiv:2401.06171.*

Hong, Z., & Xiao, K. (2024). Digital economy structuring for sustainable development: The role of blockchain and artificial intelligence in improving supply chain and reducing negative environmental impacts. *Scientific Reports, 14*(1), 3912. DOI: 10.1038/s41598-024-53760-3 PMID: 38365912

Leal Filho, W., Mbah, M. F., Dinis, M. A. P., Trevisan, L. V., de Lange, D., Mishra, A., Rebelatto, B., Ben Hassen, T., & Aina, Y. A. (2024). The role of artificial intelligence in the implementation of the UN Sustainable Development Goal 11: Fostering sustainable cities and communities. *Cities (London, England), 150*, 105021. DOI: 10.1016/j.cities.2024.105021

Lodhi, S. K., Gill, A. Y., & Hussain, H. K. (2024). Green innovations: Artificial intelligence and sustainable materials in production. *BULLET: Jurnal Multidisiplin Ilmu, 3*(4), 492–507.

Martini, B., Bellisario, D., & Coletti, P. (2024). Human-centered and sustainable artificial intelligence in industry 5.0: Challenges and perspectives. *Sustainability (Basel), 16*(13), 5448. DOI: 10.3390/su16135448

Mohammed, K. S., Serret, V., Jabeur, S. B., & Nobanee, H. (2024). The role of artificial intelligence and fintech in promoting eco-friendly investments and non-greenwashing practices in the US market. *Journal of Environmental Management, 359*, 120977. DOI: 10.1016/j.jenvman.2024.120977 PMID: 38678903

Nguyen, V. N., Tarełko, W., Sharma, P., El-Shafay, A. S., Chen, W. H., Nguyen, P. Q. P., Nguyen, X. P., & Hoang, A. T. (2024). Potential of explainable artificial intelligence in advancing renewable energy: Challenges and prospects. *Energy & Fuels, 38*(3), 1692–1712. DOI: 10.1021/acs.energyfuels.3c04343

Ogbeibu, S., Emelifeonwu, J., Pereira, V., Oseghale, R., Gaskin, J., Sivarajah, U., & Gunasekaran, A. (2024). Demystifying the roles of organisational smart technology, artificial intelligence, robotics and algorithms capability: A strategy for green human resource management and environmental sustainability. *Business Strategy and the Environment, 33*(2), 369–388. DOI: 10.1002/bse.3495

Olawade, D. B., Fapohunda, O., Wada, O. Z., Usman, S. O., Ige, A. O., Ajisafe, O., & Oladapo, B. I. (2024). *Smart waste management: A paradigm shift enabled by artificial intelligence.* Waste Management Bulletin.

Olawade, D. B., Wada, O. Z., David-Olawade, A. C., Fapohunda, O., Ige, A. O., & Ling, J. (2024). Artificial intelligence potential for net zero sustainability: Current evidence and prospects. *Next sustainability, 4*, 100041.

Radanliev, P. (2024). Cyber diplomacy: defining the opportunities for cybersecurity and risks from Artificial Intelligence, IoT, Blockchains, and Quantum Computing. *Journal of Cyber Security Technology*, 1-51.

Rauf, M., Kumar, L., Zulkifli, S. A., & Jamil, A. (2024). Aspects of artificial intelligence in future electric vehicle technology for sustainable environmental impact. *Environmental Challenges*, *14*, 100854. DOI: 10.1016/j.envc.2024.100854

Regona, M., Yigitcanlar, T., Hon, C., & Teo, M. (2024). Artificial intelligence and sustainable development goals: Systematic literature review of the construction industry. *Sustainable Cities and Society*, *108*, 105499. DOI: 10.1016/j.scs.2024.105499

Ribeiro, J. P. S. (2024). *The role of artificial intelligence in achieving environmental sustainability for the health and technological industry* (Master's thesis).

Rohde, F., Wagner, J., Meyer, A., Reinhard, P., Voss, M., Petschow, U., & Mollen, A. (2024). Broadening the perspective for sustainable artificial intelligence: Sustainability criteria and indicators for Artificial Intelligence systems. *Current Opinion in Environmental Sustainability*, *66*, 101411. DOI: 10.1016/j.cosust.2023.101411

Rolando, B., & Mulyono, H. (2024). Managing Risks In Fintech: Applications And Challenges Of Artificial Intelligence-Based Risk Management. [ECBIS]. *Economics and Business Journal*, *2*(3), 249–268. DOI: 10.47353/ecbis.v2i3.127

Salman, M., Wang, G., Qin, L., & He, X. (2024). G20 roadmap for carbon neutrality: The role of Paris agreement, artificial intelligence, and energy transition in changing geopolitical landscape. *Journal of Environmental Management*, *367*, 122080. DOI: 10.1016/j.jenvman.2024.122080 PMID: 39111003

Satornino, C. B., Du, S., & Grewal, D. (2024). Using artificial intelligence to advance sustainable development in industrial markets: A complex adaptive systems perspective. *Industrial Marketing Management*, *116*, 145–157. DOI: 10.1016/j.indmarman.2023.11.011

Shuford, J. (2024). Interdisciplinary perspectives: fusing artificial intelligence with environmental science for sustainable solutions. *Journal of Artificial Intelligence General science (JAIGS) ISSN: 3006-4023, 1*(1), 106-123.

Tariq, M. U. (2024). Leveraging artificial intelligence for a sustainable and climate-neutral economy in Asia. In *Strengthening sustainable digitalization of Asian economy and society* (pp. 1–21). IGI Global. DOI: 10.4018/979-8-3693-1942-0.ch001

Tariq, M. U. (2025). Empowering Learning Through Networked and Connected Education: Case Studies in Digital Engagement. In Gierhart, A. (Ed.), *Cases on Enhancing P-16 Student Engagement With Digital Technologies* (pp. 169–198). IGI Global Scientific Publishing., DOI: 10.4018/979-8-3693-5633-3.ch007

Tariq, M. U. (2025). Sustainability of Quality Processes in Higher Education: Strategies for Continuous Improvement. In Saeed, S., Rana, O., & Dhanaraj, R. (Eds.), *Higher Education and Quality Assurance Practices* (pp. 305–334). IGI Global Scientific Publishing., DOI: 10.4018/979-8-3693-6765-0.ch011

Tariq, M. U. (2025). Harnessing Emotional Intelligence for Strategic Leadership: A Contemporary Approach. In Sayed, Z. (Ed.), *Emotionally Intelligent Methods for Meaningful Leadership* (pp. 197–222). IGI Global Scientific Publishing., DOI: 10.4018/979-8-3693-7372-9.ch009

Tariq, M. U., & Sergio, R. P. (2025). Nurturing Digital Citizenship in Society 5.0 Through AI and Computational Intelligence Education. In Pandey, R., Srivastava, N., Prasad, R., Prasad, J., & Garcia, M. (Eds.), *Open AI and Computational Intelligence for Society 5.0* (pp. 59–84). IGI Global Scientific Publishing., DOI: 10.4018/979-8-3693-4326-5.ch003

Ueda, D., Walston, S. L., Fujita, S., Fushimi, Y., Tsuboyama, T., Kamagata, K., Yamada, A., Yanagawa, M., Ito, R., Fujima, N., Kawamura, M., Nakaura, T., Matsui, Y., Tatsugami, F., Fujioka, T., Nozaki, T., Hirata, K., & Naganawa, S. (2024). Climate change and artificial intelligence in healthcare: Review and recommendations towards a sustainable future. *Diagnostic and Interventional Imaging*, *105*(11), 453–459. DOI: 10.1016/j.diii.2024.06.002 PMID: 38918123

Ukoba, K., Olatunji, K. O., Adeoye, E., Jen, T. C., & Madyira, D. M. (2024). Optimizing renewable energy systems through artificial intelligence: Review and future prospects. *Energy & Environment*, *35*(7), 3833–3879. DOI: 10.1177/0958305X241256293

Voigtlaender, S., Pawelczyk, J., Geiger, M., Vaios, E. J., Karschnia, P., Cudkowicz, M., Dietrich, J., Haraldsen, I. R. J. H., Feigin, V., Owolabi, M., White, T. L., Świeboda, P., Farahany, N., Natarajan, V., & Winter, S. F. (2024). Artificial intelligence in neurology: Opportunities, challenges, and policy implications. *Journal of Neurology*, *271*(5), 2258–2273. DOI: 10.1007/s00415-024-12220-8 PMID: 38367046

Wang, Q., Li, Y., & Li, R. (2024). Ecological footprints, carbon emissions, and energy transitions: The impact of artificial intelligence (AI). *Humanities & Social Sciences Communications*, *11*(1), 1–18. DOI: 10.1057/s41599-024-03520-5

Wörsdörfer, M. (2024). Mitigating the adverse effects of AI with the European Union's artificial intelligence act: Hype or hope? *Global Business and Organizational Excellence*, *43*(3), 106–126. DOI: 10.1002/joe.22238

Zavrazhnyi, K. Y., Kulyk, A. K., Voronenko, V. I., Sokolov, M. A., & Antunes de Abreu, O. (2024). Formation of strategic directions for the use of artificial intelligence in the enterprise to achieve the goals of sustainable development.

KEY TERMS AND DEFINITIONS

Artificial Intelligence (AI): Technology replicating human intelligence for decision-making procedures

Environmental Sustainability: Managing resources to meet present and future requirements

Policy Implications: Regulatory considerations guiding liable AI development and deployment

Compilation of References

Abdul Rahman, N. S. N., Abdul Hamid, N. W., & Nadarajah, K. (2021). Effects of Abiotic Stress on Soil Microbiome. *International Journal of Molecular Sciences*, *22*(16), 9036. DOI: 10.3390/ijms22169036 PMID: 34445742

Abdullahi, M. S., Jakada, B. A., & Kabir, S. (2016). Challenges affecting the performance of small and medium scale enterprises (SMEs) in Nigeria. [JHCD]. *Journal of Human Capital Development*, *9*(2), 21–46.

Abid, A., Mehmood, U., Tariq, S., & Haq, Z. U. (2022). The effect of technological innovation, FDI, and financial development on CO2 emission: Evidence from the G8 countries. *Environmental Science and Pollution Research International*, *29*(8), 11654–11662. DOI: 10.1007/s11356-021-15993-x PMID: 34545519

Abor, J., & Quartey, P. (2010). Issues in SME development in Ghana and South Africa. *International Research Journal of Finance and Economics*, *39*(39), 218–228.

Abulibdeh, A., Zaidan, E., & Abulibdeh, R. (2024). Navigating the confluence of artificial intelligence and education for sustainable development in the era of industry 4.0: Challenges, opportunities, and ethical dimensions. *Journal of Cleaner Production*, *437*, 140527. DOI: 10.1016/j.jclepro.2023.140527

Acheampong, A. O., Adams, S., & Boateng, E. (2019). Do globalization and renewable energy contribute to carbon emissions mitigation in Sub-Saharan Africa? *The Science of the Total Environment*, *677*, 436–446. DOI: 10.1016/j.scitotenv.2019.04.353 PMID: 31059886

Acquaye, A., Feng, K., Oppon, E., Salhi, S., Ibn-Mohammed, T., Genovese, A., & Hubacek, K. (2016). Measuring the environmental sustainability performance of global supply chains: A multi-regional input-output analysis for carbon, sulphur oxide and water footprints. *Journal of Environmental Management*, *187*, 571–585. DOI: 10.1016/j.jenvman.2016.10.059 PMID: 27876164

Addai, K., Serener, B., & Kirikkaleli, D. (2022). Empirical analysis of the relationship among urbanization, economic growth and ecological footprint: Evidence from Eastern Europe. *Environmental Science and Pollution Research International*, *29*(19), 27749–27760. DOI: 10.1007/s11356-021-17311-x PMID: 34981376

Adebayo, T. S., Agyekum, E. B., Altuntaş, M., Khudoyqulov, S., Zawbaa, H. M., & Kamel, S. (2022). Does information and communication technology impede environmental degradation? fresh insights from non-parametric approaches. *Heliyon*, *8*(3), e09108. DOI: 10.1016/j.heliyon.2022.e09108 PMID: 35313485

Adeyelu, A., Ugochukwu, C., & Shonibare, P. (2024). Ethical implications of AI in financial decision-making: A review with real-world applications. *International Journal of Applied Research in Social Sciences*, *6*(4), 608–630.

Adner, R. (2002). When are technologies disruptive? A demand-based view of the emergence of competition. *Strategic Management Journal*, *23*(8), 667–688. DOI: 10.1002/smj.246

Adongo, J. A., Atugeba, I. L. A., & Insah, B. (2020). An assessment of the relationship between capital investments and financial performance of selected SMEs in the Upper East Region of Ghana. *International Journal of Innovation and Applied Studies*, *29*(3), 760–770.

Afolabi, H., Ram, R., Hussainey, K., Nandy, M., & Lodh, S. (2023). Exploration of small and medium entities' actions on sustainability practices and their implications for a greener economy. *Journal of Applied Accounting Research*, *24*(4), 655–681. DOI: 10.1108/JAAR-09-2022-0252

Afridi, F. E. A., Afridi, S. A., Zahid, R. M. A., Khan, W., & Anwar, W. (2023). Embracing green banking as a mean of expressing green behavior in a developing economy: Exploring the mediating role of green culture. *Environmental Science and Pollution Research International*, *31*(34), 46137–46147. DOI: 10.1007/s11356-023-25449-z PMID: 36694066

Ağan, B. (2024). Sustainable development through green transition in EU countries: New evidence from panel quantile regression. *Journal of Environmental Management*, *365*, 121545. DOI: 10.1016/j.jenvman.2024.121545 PMID: 38917545

Agan, Y., Acar, M. F., & Borodin, A. (2013). Drivers of environmental processes and their impact on performance: A study of Turkish SMEs. *Journal of Cleaner Production*, *51*, 23–33. DOI: 10.1016/j.jclepro.2012.12.043

Aghelie, A. (2017). Exploring drivers and barriers to sustainability green business practices within small medium sized enterprises: Primary findings. *International Journal of Business & Economic Development*, *5*(1), 41–48.

Agyekum, E. O., Fortuin, K. P. J., & van der Harst, E. (2021). Assessing circularity of electronic waste management in the context of developing countries: Evidence from Ghana. *Journal of Cleaner Production*, *292*, 126032.

Ahmad, S., Priyadharshini, S. L., Hosen, M. S., Ng, A., Islam, S., & Manik, J. A. (2024). The Impact Of Artificial Intelligence On Business & Social Values: Benefits, Challenges, And Future Directions. *Educational Administration: Theory and Practice*, *30*(6), 3174–3180.

Ahmed, B., & Halil, M. (2023). Environmental Sustainability and Competitive Advantage: A Meta-Analysis. *Competitive Research Journal Archive*, *1*(01), 80–90.

Ahmed, Z., & Le, H. P. (2021). Linking Information Communication Technology, trade globalization index, and CO2 emissions: Evidence from advanced panel techniques. *Environmental Science and Pollution Research International*, *28*(7), 8770–8781. DOI: 10.1007/s11356-020-11205-0 PMID: 33070290

Ahmed, Z., Nathaniel, S., & Shahbaz, M. (2020). The criticality of information and communication technology and human capital in environmental sustainability: Evidence from Latin American and Caribbean countries. *Journal of Cleaner Production*, *286*, 125529. Advance online publication. DOI: 10.1016/j.jclepro.2020.125529

Ahmed, Z., Wang, Z., Mahmood, F., Hafeez, M., & Ali, N. (2019). Does globalization increase the ecological footprint? Empirical evidence from Malaysia. *Environmental Science and Pollution Research International*, *26*(18), 18565–18582. DOI: 10.1007/s11356-019-05224-9 PMID: 31054053

Ajirotutu, R. O., Adeyemi, A. B., Ifechukwu, G. O., Ohakawa, T. C., Iwuanyanwu, O., & Garba, B. M. P. (2024). Exploring the intersection of Building Information Modeling (BIM) and artificial intelligence in modern infrastructure projects. *Journal of Advanced Infrastructure Studies*.

Akhtar, R., Masud, M. M., Al-Mamun, A., & Saif, A. N. M. (2023). Energy consumption, CO2 emissions, foreign direct investment, and economic growth in Malaysia: An NARDL technique. *Environmental Science and Pollution Research International*, *30*(22), 63096–63108. DOI: 10.1007/s11356-023-26246-4 PMID: 36952165

Akomea, S. Y., Agyapong, A., Ampah, G., & Osei, H. V. (2023). Entrepreneurial orientation, sustainability practices and performance of small and medium enterprises: Evidence from an emerging economy. *International Journal of Productivity and Performance Management, 72*(9), 2629–2653. DOI: 10.1108/IJPPM-06-2021-0325

Alawsi, H., Emran, A. Q., Ebrahim, A. M., Ibrahim, F. M., & Mohammed, M. N. (2025). Remarks on Sustainable Tourism. In AlDhaen, E., Braganza, A., Hamdan, A., & Chen, W. (Eds.), *Business Sustainability with Artificial Intelligence (AI): Challenges and Opportunities. Studies in Systems, Decision and Control* (Vol. 566). Springer., DOI: 10.1007/978-3-031-71318-7_14

Alayón, C. L., Säfsten, K., & Johansson, G. (2022). Barriers and enablers for the adoption of sustainable manufacturing by manufacturing SMEs. *Sustainability (Basel), 14*(4), 2364. DOI: 10.3390/su14042364

Albaity, M., & Awad, A. (2023). The heterogeneous effect of ICT on countries with different levels of ecological degradation and income: A panel quantile approach. *Journal of Open Innovation, 9*(2), 100055. DOI: 10.1016/j.joitmc.2023.100055

Alharbi, S., Felemban, A., Abdelrahim, A., & Al-Dakhil, M. (2024). Agricultural and Technology-Based Strategies to Improve Water-Use Efficiency in Arid and Semiarid Areas. *Water (Basel), 16*(13), 1842. DOI: 10.3390/w16131842

Ali, Z., Ghazali, M. A. I. M., Ismail, R., Muhammad, N. N., Azlinda, Z. A. N., & Malek, N. A. (2018). Digital Board Game: Is There a Need for It in Language Learning Among Tertiary Level Students? *Matec Web of Conferences, 150*, 5026. DOI: 10.1051/matecconf/201815005026

Ali, H., Chen, T., & Hao, Y. (2021). Sustainable manufacturing practices, competitive capabilities, and sustainable performance: Moderating role of environmental regulations. *Sustainability (Basel), 13*(18), 10051. DOI: 10.3390/su131810051

Ali, N., Phoungthong, K., Khan, A., Abbas, S., Dilanchiev, A., Tariq, S., & Sadiq, M. N. (2023). Does FDI foster technological innovations? Empirical evidence from BRICS economies. *PLoS One, 18*(3), e0282498. DOI: 10.1371/journal.pone.0282498 PMID: 36893169

Al-Khayri, J. M., Rashmi, R., Surya Ulhas, R., Sudheer, W. N., Banadka, A., Nagella, P., Aldaej, M. I., Rezk, A. A. S., Shehata, W. F., & Almaghasla, M. I. (2023). The Role of Nanoparticles in Response of Plants to Abiotic Stress at Physiological, Biochemical, and Molecular Levels. *Plants, 12*(2), 292. DOI: 10.3390/plants12020292 PMID: 36679005

Almaleeh, N. (2019). Are Sustainable Firms More Profitable? Evidence from Egypt. *International Journal of Accounting and Financial Reporting*.

Alola, A. A., Bekun, F. V., & Sarkodie, S. A. (2019). Dynamic impact of trade policy, economic growth, fertility rate, renewable and non-renewable energy consumption on ecological footprint in Europe. *The Science of the Total Environment, 685*, 702–709. DOI: 10.1016/j.scitotenv.2019.05.139 PMID: 31203164

Alotaibi, E., & Nassif, N. (2024). Artificial intelligence in environmental monitoring: In-depth analysis. *Discover Artificial Intelligence*, *4*, 84.

Al-Raeei, M. (2025). The smart future for sustainable development: Artificial intelligence solutions for sustainable urbanization. *Sustainable Development (Bradford)*, *33*(1), 508–517. DOI: 10.1002/sd.3131

Al-sakkaf, M. A., Al-Attas, W. A., Khalufi, N., & Murshid, M. A. (2024), "Defining Green and Responsible Tourism Destinations", Alnoor, A., Bayram, G.E., XinYing, C. and Shah, S.H.A. (Ed.) *The Role of Artificial Intelligence in Regenerative Tourism and Green Destinations* (*New Perspectives in Tourism and Hospitality Management*), Emerald Publishing Limited, Leeds, pp. 51-69. DOI: 10.1108/978-1-83753-746-420241004

Alshammari, M. T. (2020). Evaluation of Gamification in E-Learning Systems for Elementary School Students. *Tem Journal*, 806–813. https://doi.org/DOI: 10.18421/tem92-51

Álvarez Jaramillo, J., Zartha Sossa, J. W., & Orozco Mendoza, G. L. (2019). Barriers to sustainability for small and medium enterprises in the framework of sustainable development-Literature review. *Business Strategy and the Environment*, *28*(4), 512–524. DOI: 10.1002/bse.2261

Alzoubi, Y. I., & Mishra, A. (2024). Green artificial intelligence initiatives: Potentials and challenges. *Journal of Cleaner Production*, *468*, 143090. DOI: 10.1016/j.jclepro.2024.143090

Ambilwade, R. P., Kumar, R., Narayana, K. S. S., Srinivas, P., Yadav, B., & Rusia, S. (2024). AI-powered environmental monitoring: Machine learning approaches for air and water quality assessment. *Nanotechnology Perceptions*, *20*(1), 45–58.

Amri, F., Ben Lahouel, B., & Ben Zaied, Y. (2019). ICT, total factor productivity, and carbon dioxide emissions in Tunisia. *Technological Forecasting and Social Change*, *146*, 212–217. DOI: 10.1016/j.techfore.2019.05.028

Anderson, C. (2020). Policy interventions for promoting green technologies: A global review. *Environmental Policy and Governance*, *30*(2), 65–80.

Anser, M. K., Ali, S., Umair, M., Javid, R., & Mirzaliev, S. (2024). Energy consumption, technological innovation, and economic growth in BRICS: A GMM panel VAR framework analysis. *Energy Strategy Reviews*, *56*, 101587. DOI: 10.1016/j.esr.2024.101587

Ansu-Mensah, P., & Bein, M. A. (2019). Towards sustainable consumption: Predicting the impact of social-psychological factors on energy conservation intentions in Northern Cyprus. *Natural Resources Forum*, *43*(3), 181–193. DOI: 10.1111/1477-8947.12174

Ansu-Mensah, P., & Kwakwa, P. A. (2022). Modelling electricity consumption in Ghana: The role of financial development indicators. *Green Finance*, *4*(1), 54–70. DOI: 10.3934/GF.2022003

Ansu-Mensah, P., Kwakwa, P. A., & Nyamedor, S. T. M. (2021). Determinants of the business performance of small and medium enterprises in an emerging market economy. *International Journal of Management. Knowledge and Learning*, *10*(1), 219–233.

Antwi-Boateng, C., Asumah, S., Kumi, E., & Ansu-Mensah, P. (2025). Examining the triple impact of eco-intellectual capital on the financial, social and environmental sustainability performance of SMEs in emerging economies. *Journal of Tropical Futures*, *2*(1), 10–25. DOI: 10.1177/27538931241288959

Anugraha, G. A. P., & Padmadewi, N. N.Gede Angga Pratama AnugrahaNi Nyoman Padmadewi. (2023). The Need Analysis of Technology-Based Project-Based Learning in Higher Education. *Jurnal Pendidikan Bahasa Inggris Undiksha*, *10*(3), 242–248. DOI: 10.23887/jpbi.v10i3.58466

Anwar, M. R., & Sakti, L. D. (2024). Integrating artificial intelligence and environmental science for sustainable urban planning. [ITSDI]. *IAIC Transactions on Sustainable Digital Innovation*, *5*(2), 179–191. DOI: 10.34306/itsdi.v5i2.666

Apergis, N., Pinar, M., & Unlu, E. (2023). How do foreign direct investment flows affect carbon emissions in BRICS countries? Revisiting the pollution haven hypothesis using bilateral FDI flows from OECD to BRICS countries. *Environmental Science and Pollution Research International*, *30*(6), 14680–14692. DOI: 10.1007/s11356-022-23185-4 PMID: 36163571

Appiah-Otoo, I., Acheampong, A. O., Song, N., & Chen, X. (2023). The impact of information and communication technology (ICT) on carbon dioxide emissions: Evidence from heterogeneous ICT countries. *Energy & Environment*, *34*(8), 3080–3102. DOI: 10.1177/0958305X221118877

Appiah-Otoo, I., & Song, N. (2021). The impact of ICT on economic growth-Comparing rich and poor countries. *Telecommunications Policy*, *45*(2), 102082. DOI: 10.1016/j.telpol.2020.102082

Araújo, S. O., Peres, R. S., Barata, J., Lidon, F., & Ramalho, J. C. (2021). Characterizing the Agriculture 4.0 Landscape—Emerging Trends, Challenges, and Opportunities. *Agronomy (Basel)*, *11*(4), 667. DOI: 10.3390/agronomy11040667

Arend, R. J. (2014). Entrepreneurship and dynamic capabilities: How firm age and size affect the capability enhancement–SME performance relationship. *Small Business Economics*, *42*(1), 33–57. DOI: 10.1007/s11187-012-9461-9

Arslan, H. M., Khan, I., Latif, M. I., Komal, B., & Chen, S. (2022). Understanding the dynamics of natural resources rents, environmental sustainability, and sustainable economic growth: New insights from China. *Environmental Science and Pollution Research International*, *29*(39), 58746–58761. DOI: 10.1007/s11356-022-19952-y PMID: 35368236

Atsu, F., Adams, S., & Adjei, J. (2021). ICT, energy consumption, financial development, and environmental degradation in South Africa. *Heliyon*, *7*(7), e07328. DOI: 10.1016/j.heliyon.2021.e07328 PMID: 34258464

Awad, A. (2022). Is there any impact from ICT on environmental quality in Africa? Evidence from second-generation panel techniques. *Environmental Challenges*, *7*, 100520. DOI: 10.1016/j.envc.2022.100520

Awad, A., & Albaity, M. (2022). ICT and economic growth in Sub-Saharan Africa: Transmission channels and effects. *Telecommunications Policy*, *46*(8), 102381. DOI: 10.1016/j.telpol.2022.102381

Aydin, M., & Erdem, A. (2024). Analyzing the impact of resource productivity, energy productivity, and renewable energy consumption on environmental quality in EU countries: The moderating role of productivity. *Resources Policy*, *89*, 104613. DOI: 10.1016/j.resourpol.2023.104613

Ayeleru, O. O., Dlova, S., Akinribide, O. J., Ntuli, F., Kupolati, W. K., Marina, P. F., Blencowe, A., & Olubambi, P. A. (2020). Challenges of plastic waste generation and management in sub-Saharan Africa: A review. *Waste Management (New York, N.Y.)*, *110*, 24–42. DOI: 10.1016/j.wasman.2020.04.017 PMID: 32445915

Ayeni, O. O., & Eden, C. A.Oyebola Olusola AyeniChima Abimbola Eden. (2024). Equity and Access in Higher Education: Legal Perspectives and Management Strategies. *International Journal of Science and Research Archive*, *11*(2), 199–206. DOI: 10.30574/ijsra.2024.11.2.0391

Azam, M., & Haseeb, M. (2021). Determinants of foreign direct investment in BRICS-does renewable and non- renewable energy matter? *Energy Strategy Reviews*, *35*, 100638. DOI: 10.1016/j.esr.2021.100638

Azam, R., Malik, K., Sultana, T., Arooj, F., & Kazmi, A. (2025). Plant-mediated bioinspired iron nanoparticles as an alternative to enhance crop resistance against biotic and abiotic stress; a review. *Physiological and Molecular Plant Pathology*, *136*, 102586. DOI: 10.1016/j.pmpp.2025.102586

Aziz, G., Striełkowski, W., Sarwar, S., & Tiwari, A. K. (2024). Implications of circular economy, digitalization and technological innovation to achieve sustainable environmental goal: Pre and post-vision 2030. *Heliyon*, *10*(10), e30978. Advance online publication. DOI: 10.1016/j.heliyon.2024.e30978 PMID: 38770279

Babu, S., Madhusudanan, S., Sathiyanarayanan, M., Mortka, M. Z., Szymański, J., & Rahul, R. (2024). Soil mapping for farming productivity: Internet of things (IoT) based sustainable agriculture. *Microsystem Technologies*. Advance online publication. DOI: 10.1007/s00542-024-05608-z

Bahadır, Ş. İ. N., & Kadıoğlu, İ. (2019). Detection of Weeds Using Unmanned Aerial Vehicle (UAV) and Image Processing Techniques. *Turkish Journal of Weed Science*, *22*(2), 211–217.

Bailey-Serres, J., Parker, J. E., Ainsworth, E. A., Oldroyd, G. E. D., & Schroeder, J. I. (2019). Genetic strategies for improving crop yields. *Nature*, *575*(7781), 109–118. DOI: 10.1038/s41586-019-1679-0 PMID: 31695205

Baker, M. J. (2000). Selecting a research methodology. *The Marketing Review*, *1*(3), 373–397. DOI: 10.1362/1469347002530736

Balcı, B. (2022). *How Industry 4.0 Changes the Nature of Labor: A Qualitative Study in Enterprises Implementing Industry 4.0 Technologies in Turkey* (Unpublished Doctoral Dissertation). Kırklareli University, Institute of Social Sciences, Department of Labor Economics and Industrial Relations, Kırklareli, Turkey.

Bamal, A., Uddin, M. G., & Olbert, A. I. (2024). Harnessing machine learning for assessing climate change influences on groundwater resources: A comprehensive review. *Heliyon*, *10*(17), e37073. DOI: 10.1016/j.heliyon.2024.e37073 PMID: 39286200

Barney, J. (1991). Firm resources and sustained competitive advantage. *Journal of Management*, *17*(1), 99–120. DOI: 10.1177/014920639101700108

Barrett, H., & Rose, D. C. (2022). Perceptions of the Fourth Agricultural Revolution: What's In, What's Out, and What Consequences Are Anticipated? *Sociologia Ruralis*, *62*(2), 162–189. DOI: 10.1111/soru.12324

Barton, J., Gavalda, R., & Kim, H. (2020). Synthetic biology for environmental sustainability. *Environmental Science & Technology*, *54*(13), 7998–8009.

Bazaliy, R. (2021). Development of Ecocentric-Ecological Consciousness of Students of a Technical University. *E3s Web of Conferences, 273*, 12002. DOI: 10.1051/e3sconf/202127312002

Beck, N., & Katz, J. N. (1995). What to do (and not to do) with time-series cross-section data. *The American Political Science Review*, *89*(3), 634–647. DOI: 10.2307/2082979

Behjati, S. (2017). Critical remarks about environmentalism implication by Iranian SMEs. *European Journal of Sustainable Development*, *6*(3), 209–219. DOI: 10.14207/ejsd.2017.v6n3p209

Ben Youssef, A., Dahmani, M., & Mabrouki, M. (2024). Decoupling carbon emissions and economic growth in Tunisia: pathways to sustainable development. Handbook on Energy and Economic Growth, 103–127. DOI: 10.4337/9781802204803.00012

Benzaouia, M., Hajji, B., Mellit, A., & Rabhi, A. (2023). Fuzzy-IoT smart irrigation system for precision scheduling and monitoring. *Computers and Electronics in Agriculture*, *215*, 108407. DOI: 10.1016/j.compag.2023.108407

Bergek, A., Hekkert, M., & Jacobsson, S. (2013). Functions in innovation systems: A framework for analysing energy system dynamics and identifying goals for system-building activities by entrepreneurs and policymakers. *Technology Analysis and Strategic Management*, *25*(10), 1026–1041.

Bhandari, M., Tiwari, G., & Dhakal, M. (2025). Enhancing transparency and accountability in sustainable finance through blockchain technology: A systematic review of the literature. *Journal of Intelligent Management Decision*, *4*(1), 23–43.

Bibri, S. E., Huang, J., Jagatheesaperumal, S. K., & Krogstie, J. (2024). The synergistic interplay of artificial intelligence and digital twin in environmentally planning sustainable smart cities: A comprehensive systematic review. *Environmental Science and Ecotechnology*, *20*, 100433. DOI: 10.1016/j.ese.2024.100433 PMID: 38831974

Bindu, S., Sridharan, P., Swain, R. K., & Das, C. P. (2022). Causal linkage between remittances and financial development: Evidence from the BRICS (Brazil, Russia, India, China, and South Africa). *Journal of East-West Business*, *28*(2), 117–149. DOI: 10.1080/10669868.2021.1976348

Binns, A. (2018). Ethical implications of artificial intelligence in environmental management. *Technology Ethics Review, 25*(4), 36–48.

Birişik, N. (2019). *Global and National Agricultural and Food Policies: Facts, Issues, and Proposed Solutions*. Dh Basın Yayın Matbacılık.

BloombergNEF (BNEF). (2023). *Battery price survey 2023*. Retrieved from https://about.bnef.com

Bocken, N. M. P., Short, S. W., Rana, P., & Evans, S. (2014). A literature and practice review to develop sustainable business model archetypes. *Journal of Cleaner Production, 65*, 42–56. DOI: 10.1016/j.jclepro.2013.11.039

Bošković, D., Husremović, D., Muslic, M., & Kapo, A. (2023). Teachers and Students as Promoters or Repressors of Sustainable Education: Navigating the Blended Learning Landscape. *Sustainability (Basel), 15*(24), 16812. DOI: 10.3390/su152416812

Boulaid, F., & Handoko, H. (2024). Cultivating Ecological Awareness Through the Integration of Eco-Linguistic Themes in English for Specific Purposes: A Case Study of Higher Schools of Technology. *Gramatika Stkip Pgri Sumatera Barat, 10*(1). Advance online publication. DOI: 10.22202/jg.2024.v10i1.8062

Bower, J. L., & Christensen, C. M. (1995). Disruptive technologies: Catching the wave. *Harvard Business Review, 73*(1), 43–53.

Bozatli, O., & Akca, H. (2023). The effects of environmental taxes, renewable energy consumption and environmental technology on the ecological footprint: Evidence from advanced panel data analysis. *Journal of Environmental Management, 345*, 118857. DOI: 10.1016/j.jenvman.2023.118857 PMID: 37657289

Brahmia, S. Y., & Mannai, S. (2024). Environmental Degradation in Gulf Cooperation Council: Role of ICT Development, Trade, FDI, and Energy Use. *Sustainability (Basel), 17*(1), 54. DOI: 10.3390/su17010054

Brien, E. O., & Hamburg, I. (2014). Supporting Sustainable Strategies for SMEs through Training, Cooperation and Mentoring. *Higher Education Studies, 4*(2), 61–69. DOI: 10.5539/hes.v4n2p61

Brundtland, G. H. (1987). *Report of the World Commission on Environment and Development: Our Common Future*. United Nations.

Büchi, G., Cugno, M., & Castagnoli, R. (2020). Smart Factory Performance and Industry 4.0. *Technological Forecasting and Social Change, 150*, 150. DOI: 10.1016/j.techfore.2019.119790

Burlea-Schiopoiu, A., & Mihai, L. S. (2019). An integrated framework on the sustainability of SMEs. *Sustainability (Basel), 11*(21), 6026. DOI: 10.3390/su11216026

Bwambale, E., Naangmenyele, Z., Iradukunda, P., Agboka, K.M., Houessou-Dossou, E.A.Y., Akansake, D.A., Bisa, M.E., Hamadou, A.Z.H., Hakizayezu, J., Onofua, O.E., Chikabvumbwa, S.R., Tomasiello, S. 2022. Towards precision irrigation management: A review of GIS, remote sensing and emerging technologies. 9 (1), 2100573. https://doi.org/DOI: 10.1080/23311916.2022.2100573

Çakmak, S. (2019). Meet the AI-Supported Autonomous Agricultural Robot: Mamut. Retrieved from https://bilimdili.com/teknoloji/yapayzeka/yapay-zeka-destekli-otonom-tarim-robotu-ile-tanisin-mamut/ (Accessed: 29.01.2023).

Caldera, H. T. S., Desha, C., & Dawes, L. (2019). Evaluating the enablers and barriers for successful implementation of sustainable business practice in 'lean' SMEs. *Journal of Cleaner Production, 218*, 575–590. DOI: 10.1016/j.jclepro.2019.01.239

Camilleri, M. A. (2024). Artificial intelligence governance: Ethical considerations and implications for social responsibility. *Expert Systems: International Journal of Knowledge Engineering and Neural Networks, 41*(7), e13406. DOI: 10.1111/exsy.13406

Cantele, S., & Zardini, A. (2018). Is sustainability a competitive advantage for small businesses? An empirical analysis of possible mediators in the sustainability–financial performance relationship. *Journal of Cleaner Production, 182*, 166–176. DOI: 10.1016/j.jclepro.2018.02.016

Cao, J., Bhuvaneswari, G., Arumugam, T., & Aravind, B. R. (2023). The Digital Edge: Examining the Relationship Between Digital Competency and Language Learning Outcomes. *Frontiers in Psychology, 14*, 1187909. Advance online publication. DOI: 10.3389/fpsyg.2023.1187909 PMID: 37397293

Celik, A., Kostekci, A., & Alola, A. A. (2024). Carbon neutrality implication of material productivity, total factor productivity and renewable energy uptake in the Nordics. *Ecological Indicators, 160*, 111813. DOI: 10.1016/j.ecolind.2024.111813

Çetin, Ç., Yaman, N., Sabah, L., Ayday, E., & Ayday, C. (2013). Application Possibilities of Cloud Computing Technology in Remote Sensing and Geographic Information Systems. *VII. Technical Symposium of Turkey Photogrammetry and Remote Sensing Association (TUFUAB'2013)*, 23–25.

Chakraborty, P. P. (2024). The role of technology in enhancing sustainable tourism practices. In *Advances in Hospitality, tourism and the Services Industry (AHTSI) book series* (pp. 195–230). DOI: 10.4018/979-8-3693-5903-7.ch011

Charfeddine, L., Hussain, B., & Kahia, M. (2024). Analysis of the impact of information and communication technology, digitalization, renewable energy and financial development on environmental sustainability. *Renewable & Sustainable Energy Reviews*, *201*, 114609. DOI: 10.1016/j.rser.2024.114609

Chaudhry, S., & Sidhu, G. P. S. (2022). Climate change regulated abiotic stress mechanisms in plants: A comprehensive review. *Plant Cell Reports*, *41*(1), 1–31. DOI: 10.1007/s00299-021-02759-5 PMID: 34351488

Chen, F., Wang, Q., Umar, M., & Zheng, L. (2023). Towards sustainable resource management: The role of governance, natural resource rent and energy productivity. *Resources Policy*, 85(A), 104026. DOI: 10.1016/j.resourpol.2023.104026

Cheng, G., Zhao, C., Iqbal, N., Gülmez, Ö., Işik, H., & Kirikkaleli, D. (2021). Does energy productivity and public-private investment in energy achieve carbon neutrality target of China? *Journal of Environmental Management*, *298*, 113464. DOI: 10.1016/j.jenvman.2021.113464 PMID: 34385111

Cheng, N., Stambula, S., Wang, D., Banis, M. N., Liu, J., Riese, A., Xiao, B., Li, R., Sham, T.-K., Liu, L., Botton, G. A., & Sun, X. (2016). Platinum single-atom and cluster catalysis of the hydrogen evolution reaction. *Nature Communications*, *7*(1), 13638. Advance online publication. DOI: 10.1038/ncomms13638 PMID: 27901129

Chen, M. R. A. (2020). Effects of experiencing authentic contexts on English speaking performances, anxiety and motivation of EFL students with different cognitive styles. *Interactive Learning Environments*. Advance online publication. DOI: 10.1080/10494820.2020.1734626

Chen, P. (2022). Relationship Between the Digital Economy, Resource Allocation and Corporate Carbon Emission Intensity: New Evidence From Listed Chinese Companies. *Environmental Research Communications*, *4*(7), 75005. DOI: 10.1088/2515-7620/ac7ea3

Chen, S., Luria, S., & Shaub, K. (2022). Bridging the Gap Between Socio-Cognitive Theory and Practical Second Language Instruction. *Journal of Student Research*, *11*(3). Advance online publication. DOI: 10.47611/jsrhs.v11i3.3304

Chen, Y., Subhan, M., Ahmad, G., Adil, M., & Zamir, M. N. (2024). Unveiling the linkages among digital technology, economic growth, and carbon emissions: A resource management perspective. *Resources Policy*, *91*, 104868. DOI: 10.1016/j.resourpol.2024.104868

Chen, Y., & Wang, L. (2022). Scaling up sustainable investment through blockchain-based project bonds. *Development Policy Review*, *40*(6), e12582.

Chen, Y., Zhao, J., & Li, W. (2021). Artificial intelligence and air pollution: A review of applications, challenges, and future directions. *Environmental Science & Technology*, *55*(9), 6169–6182.

Chien, F., Anwar, A., Hsu, C. C., Sharif, A., Razzaq, A., & Sinha, A. (2021). The role of information and communication technology in encountering environmental degradation: Proposing an SDG framework for the BRICS countries. *Technology in Society*, *65*, 101587. DOI: 10.1016/j.techsoc.2021.101587

Chien, F., Sadiq, M., Nawaz, M. A., Hussain, M. S., Tran, T. D., & Thanh, T. L. (2021). A step toward reducing air pollution in top Asian economies: The role of green energy, eco-innovation, and environmental taxes. *Journal of Environmental Management*, *297*, 113420. DOI: 10.1016/j.jenvman.2021.113420 PMID: 34333309

Chirilu , A., & Costea, A. (2023). The effect of FDI on environmental degradation in Romania: Testing the pollution Haven hypothesis. *Sustainability (Basel)*, *15*(13), 10733. DOI: 10.3390/su151310733

Christensen, C. M. (2015). *The innovator's dilemma: when new technologies cause great firms to fail*. Harvard Business Review Press.

Christensen, C. M., Raynor, M. E., & McDonald, R. (2015). What is disruptive innovation? *Harvard Business Review*, *93*(12), 44–53. PMID: 17183796

Ciasullo, M. V., & Troisi, O. (2013). Sustainable value creation in SMEs: A case study. *The TQM Journal*, *25*(1), 44–61. DOI: 10.1108/17542731311286423

Ciroth, A., Frischknecht, R., & Bickel, P. (2019). Life cycle assessment and the environment: A key tool for sustainability. *Environmental Science & Technology*, *53*(8), 4472–4480.

Cohen, J. E., & Regan, M. (2018). Privacy and the challenges of new technology. *Environmental Ethics*, *40*(2), 151–173.

Copenhagen Solutions Lab. (2020). *Copenhagen: Leading the way to carbon neutrality*. https://www.solutionslab.dk/copenhagen-carbon-neutral

Corwin, D. L. (2024). Dealing with the impact of climate change-induced drought on the management of soil salinity under irrigated agriculture. *Advances in Agronomy*, *184*, 67–124. DOI: 10.1016/bs.agron.2023.12.001

Creswell, J. W. (2017). *30 Essential Skills for Qualitative Researchers* (Özcan, H., Trans.). Anı Publishing.

CSRE,. (2024). Research project the UAE As A Sustainable Tourism Destination, Stage1, The Center for Sustainability Through Research and Education- CSRE.

da Xu, L., Xu, E. L., & Li, L. (2018). Industry 4.0: State of the Art and Future Trends. *International Journal of Production Research, 56*(8), 2941–2962. DOI: 10.1080/00207543.2018.1444806

Dabić, M., Stojčić, N., Simić, M., Potocan, V., Slavković, M., & Nedelko, Z. (2021). Intellectual agility and innovation in micro and small businesses: The mediating role of entrepreneurial leadership. *Journal of Business Research, 123*, 683–695. DOI: 10.1016/j.jbusres.2020.10.013

Dadashi, Z., Mahmoudi, A., & Rashidi, S. (2022). Capacity and strategies of energy production from renewable sources in Arab countries until 2030: A review from renewable energy potentials to environmental issues. *Environmental Science and Pollution Research International, 29*(32), 47837–47866. DOI: 10.1007/s11356-022-20544-z PMID: 35522406

Dagnino, D., Guerrina, M., Minuto, L., Mariotti, M. G., Médail, F., & Casazza, G. (2020). Climate change and the future of endemic flora in the South Western Alps: Relationships between niche properties and extinction risk. *Regional Environmental Change, 20*(4), 1–12. DOI: 10.1007/s10113-020-01708-4

Dahmani, M., Mabrouki, M., & Ragni, L. (2021). Decoupling Analysis of Greenhouse Gas Emissions from Economic Growth: A Case Study of Tunisia. *Energies, 14*(22), 7550. DOI: 10.3390/en14227550

Dalenogare, L. S., Benitez, G. B., Ayala, N. F., & Frank, A. G. (2018). The Expected Contribution of Industry 4.0 Technologies for Industrial Performance. *International Journal of Production Economics, 204*, 383–394. DOI: 10.1016/j.ijpe.2018.08.019

Dalipi, F., Kastrati, Z., & Öberg, T. (2023). "The Impact of Artificial Intelligence on Tourism Sustainability: A Systematic Mapping Review," *International Conference on Computational Intelligence and Knowledge Economy (ICCIKE)*, Dubai, United Arab Emirates, 2023, pp. 119-125, DOI: 10.1109/ICCIKE58312.2023.10131818

Danciu, V. (2018). The Changing Focus of Green Marketing: From Ecological to Sustainable Marketing (III). *Romanian Economic Journal, 21*(68), 121–144.

Dangelico, R. M., Pujari, D., & Pontrandolfo, P. (2017). Green product innovation in manufacturing firms: A sustainability-oriented dynamic capability perspective. *Business Strategy and the Environment, 26*(4), 490–506. DOI: 10.1002/bse.1932

Danish. (2019). Effects of information and communication technology real income, and CO2 emissions: The experience of countries along Belt and Road. *Telematics and Informatics*, 45. 101300. DOI: 10.1016/j.tele.2019.101300

Dantas, T. E. T., de-Souza, E. D., Destro, I. R., Hammes, G., Rodriguez, C. M. T., & Soares, S. R. (2021). How the combination of Circular Economy and Industry 4.0 can contribute towards achieving the Sustainable Development Goals. *Sustainable Production and Consumption*, *26*, 213–227. DOI: 10.1016/j.spc.2020.10.005

Das, K. (2019). The role and impact of ICT in improving the quality of education: An overview. *International Journal of Innovative Studies in Sociology and Humanities*, *4*(6), 97–103.

David, A., Rao, P. S., Srivastava, D., Barthwal, A., Thomas, A., Thomas, T., & Thomas, C. A. (2024). Big Data, AI, and Geoinformatics for Sustainable Agriculture: An Introduction. In *Geoinformatics* (pp. 131–161). Apple Academic Press. DOI: 10.1201/9781003498452-10

Davutoğlu, N. A. (2020). Analysis of Fundamental and Systematic Differences Between the Third and Fourth Industrial Revolutions Using a Deterministic Approach. *Management and Political Sciences Review*, *2*(1), 176–194.

De Andrade, I. M., & Tumelero, C. (2022). Increasing customer service efficiency through artificial intelligence chatbot. *Revista De Gestão*, *29*(3), 238–251. DOI: 10.1108/REGE-07-2021-0120

De Castro, A. I., Jurado-Expósito, M., Peña-Barragán, J. M., & López-Granados, F. (2012). Airborne Multi-Spectral Imagery for Mapping Cruciferous Weeds in Cereal and Legume Crops. *Precision Agriculture*, *13*(3), 302–321. DOI: 10.1007/s11119-011-9247-0

De Haan, D. (2020). William Hazledine: Pioneering Ironmaster. *Industrial Archaeology Review*, *42*(2), 171. DOI: 10.1080/03090728.2020.1822648

Dede, C., Richards, J., & Pungowiyi, E. (2016). The role of education in advancing sustainability through disruptive technologies. *The Journal of Environmental Education*, *47*(2), 122–136.

Dehghan, Z., & Shahnazi, R. (2018). Energy consumption, carbon dioxide emissions, information and communications technology, and gross domestic product in Iranian economic sectors: A panel causality analysis. *Energy*, *169*, 1064–1078. Advance online publication. DOI: 10.1016/j.energy.2018.11.062

Depren, S. K., Kartal, M. T., Çelikdemir, N. Ç., & Depren, Ö. (2022). Energy consumption and environmental degradation nexus: A systematic review and meta-analysis of fossil fuel and renewable energy consumption. *Ecological Informatics*, *70*, 101747. DOI: 10.1016/j.ecoinf.2022.101747

Derya, H. (2018). Industrial Revolutions and Industry 4.0. *GÜ İslâhiye Faculty of Economics and Administrative Sciences International E-Journal*, *2*(2), 1–20.

Despommier, D. (2013). *The vertical farm: Feeding the world in the 21st century*. Thomas Dunne Books.

Dhahri, S., Omri, A., & Mirza, N. (2024). Information technology and financial development for achieving sustainable development goals. *Research in International Business and Finance*, *67*, 102156. DOI: 10.1016/j.ribaf.2023.102156

Dias, A., Santinha, G., Rodrigues, M., Queirós, A., Rodrigues, C., & Rocha, N. P. (2021). Smart cities and accessible tourism: A systematic review. In Eusébio, C., Teixeira, L., & Carneiro, M. J. (Eds.), *ICT tools and applications for accessible tourism* (pp. 96–114). IGI Global. DOI: 10.4018/978-1-7998-6428-8.ch005

Dissanayake, H., Perera, N., Abeykoon, S., Samson, D., Jayathilaka, R., Jayasinghe, M., & Yapa, S. (2023). Nexus between carbon emissions, energy consumption, and economic growth: Evidence from global economies. *PLoS One*, *18*(6), e0287579. DOI: 10.1371/journal.pone.0287579 PMID: 37352276

Dogan, E., & Turkekul, B. (2016). CO2 emissions, real output, energy consumption, trade, urbanization and financial development: Testing the EKC hypothesis for the USA. *Environmental Science and Pollution Research International*, *23*(2), 1203–1213. DOI: 10.1007/s11356-015-5323-8 PMID: 26351068

Doğan, Z., Arslan, S., & Berkman, A. (2015). Economic Development and Issues in Turkey's Agricultural Sector: A Historical Perspective. *Niğde University Journal of Economics and Administrative Sciences*, *8*(1), 29–41.

Dong, H., Xue, M., Xiao, Y., & Liu, Y. (2021). Do carbon emissions impact the health of residents? Considering China's industrialization and urbanization. *The Science of the Total Environment*, *758*, 143688. DOI: 10.1016/j.scitotenv.2020.143688 PMID: 33338785

Du, L., Razzaq, A., & Waqas, M. (2022). The impact of COVID-19 on small-and medium-sized enterprises (SMEs): Empirical evidence for green economic implications. *Environmental Science and Pollution Research International*, *30*(1), 1540. DOI: 10.1007/s11356-022-22221-7 PMID: 35917071

Duman, B., & Özsoy, K. (2019). Smart Agriculture from an Industry 4.0 Perspective. *4th International Congress on 3D Printing (Additive Manufacturing) Technologies and Digital Industry*, 540–555.

Dumitrescu, E. I., & Hurlin, C. (2012). Testing for Granger non-causality in heterogeneous panels. *Economic Modelling*, *29*(4), 1450–1460. DOI: 10.1016/j.econmod.2012.02.014

Durai, S., Manoharan, G., & Ashtikar, S. P. (2024). Harnessing artificial intelligence: Pioneering sustainable solutions for a greener future. In *Social and Ethical Implications of AI in Finance for Sustainability* (pp. 89–117). IGI Global. DOI: 10.4018/979-8-3693-2881-1.ch003

Duran, D. C., Artene, A. E., Gogan, L. M., & Duran, V. (2015). The objectives of sustainable development - ways to achieve welfare. *Procedia Economics and Finance*, *26*, 812–817. DOI: 10.1016/S2212-5671(15)00852-7

Dusenge, M. E., Duarte, A. G., & Way, D. A. (2019). Plant carbon metabolism and climate change: Elevated CO2 and temperature impacts on photosynthesis, photorespiration and respiration. *The New Phytologist*, *221*(1), 32–49. DOI: 10.1111/nph.15283 PMID: 29983005

Düzkaya, H. (2016). The Fourth Wave of the Industrial Revolution and Education: Is Turkey Ready for the Fourth Industrial Revolution? *Journal of Education and Humanities*, *7*(13), 49–88.

Eden, C. A., Chisom, O. N., & Adeniyi, I. S.Chima Abimbola EdenOnyebuchi Nneamaka ChisomIdowu Sulaimon Adeniyi. (2023). Education Policy and Social Change: Examining the Impact of Reform Initiatives on Equity and Access. *International Journal of Science and Research Archive*, *11*(2), 139–146. DOI: 10.30574/ijsra.2024.11.2.0372

Elfaki, K. E., & Ahmed, E. M. (2024). Digital technology adoption and globalization innovation implications on Asian Pacific green sustainable economic growth. *Journal of Open Innovation*, *10*(1), 100221. DOI: 10.1016/j.joitmc.2024.100221

El-Khadir, I., Ktaoui, S., Mouniane, Y., Chriqui, A., Mabrouki, J., Ameziane, H., & Hmouni, D. (2024). Improved Salt Stress Tolerance of Salvia Officinalis Grown in the Presence of a Halophytic Plant *Spergularia Maritima*: Analysis of Morpho-Physiological Parameters. In Mabrouki, J., & Azrour, M. (Eds.), *Integrated Solutions for Smart and Sustainable Environmental Conservation. Studies in Systems, Decision and Control* (Vol. 527). Springer., DOI: 10.1007/978-3-031-55787-3_2

El-Khadir, I., Mouniane, Y., Chriqui, A., Kouighat, M., & Hmouni, D. (2025). Effects of Salt Stress on Origanum majorana and Improvement via Association with Halophytes Plants: *Spergularia salina* and *Aptenia cordifolia.Acta Universitatis Agriculturae et Silviculturae Mendelianae Brunensis*, *73*(1), 27–35. DOI: 10.11118/actaun.2025.003

Ellen MacArthur Foundation. (2015). *Towards the circular economy: Economic and business rationale for an accelerated transition*. Ellen MacArthur Foundation.

Elpisah, E. (2023). Towards Inclusive Growth: Community-Centered Management Strategies for SMEs. *Golden Ratio of Community Services and Dedication*, *3*(1), 29–39. DOI: 10.52970/grcsd.v3i1.606

El-Tabakh, M. A. M., El-Samea, M. A., Roby, Y. A., Mohamed, Y. A., Harb, H. E., Saleh, A. M., Abd El-Aziz, M. H., & El-Abeid, S. E. (2024). Digital Agriculture for Enhancing Yield, Nutrition, and Biological Stress Resistance. In Priyadarshan, P. M., Jain, S. M., Penna, S., & Al-Khayri, J. M. (Eds.), *Digital Agriculture*. Springer., DOI: 10.1007/978-3-031-43548-5_15

Enaifoghe, A., & Vezi-Magigaba, M. F. (2023). Conceptualizing the role of entrepreneurship and SME in fostering South Africa's local economic development. *International Journal of Research in Business and Social Science (2147-4478)*, *12*(4), 96-105.

Enaifoghe, A. (2024). Enhancing Small and Medium Enterprises (SMEs) in a Globalized and Innovative Economy: Challenges and Opportunities. *International Journal of Business. Economics and Social Development*, *5*(2), 130–138.

Ercan, Ş., Öztep, R., Güler, D., & Saner, G. (2019). Agriculture 4.0 and Its Applicability in Turkey. *Agricultural Economics Journal*, *25*(2), 259–265.

Eregha, P. B., Nathaniel, S. P., & Vo, X. V. (2024). COP28 Roadmap: Do ICT, education, and renewable energy consumption matter for environmental quality? Quantile evidence for emerging economies. *Journal of Cleaner Production*, *143744*, 143744. Advance online publication. DOI: 10.1016/j.jclepro.2024.143744

Ertaş, B. (2020). A sustainable future with Agriculture 4.0. *Icontech International Journal*, *4*(1), 1–12.

Ervin, D., Wu, J., Khanna, M., Jones, C., & Wirkkala, T. (2013). Motivations and barriers to corporate environmental management. *Business Strategy and the Environment*, *22*(6), 390–409. DOI: 10.1002/bse.1752

Esmer, Y., & Şaylan, O. (2019). The Link Between Industry 4.0 and Organizational Transformation. *Bilge International Journal of Social Researches*, *3*(1), 1–8.

Espinel, R., Herrera-Franco, G., Rivadeneira García, J. L., & Escandón-Panchana, P. (2024). Artificial Intelligence in Agricultural Mapping. *Agriculture*, *14*(7), 1071. DOI: 10.3390/agriculture14071071

Eyieyien, O. G., Idemudia, C., Paul, P. O., & Ijomah, T. I. (2024). The Impact of ICT Projects on Community Development and Promoting Social Inclusion.

Fagal, A. J. B. (2021). British Responses to the US Steam Frigate Fulton the First. *The Mariner's Mirror*, *107*(2), 188–201. DOI: 10.1080/00253359.2021.1898208

FAO. (2017). *E-Agriculture in Action*. Bangkok: Food and Agriculture Organization of the United Nations and International Telecommunication Union. https://www.fao.org/newsroom/detail/New-partnership-between-ITU-and-FAO-to-bolster-ICT-innovation-in-agriculture/en (Accessed: 03.12.2022).

Fei, S., Chen, Z., Li, L., Ma, Y., & Xiao, Y. (2023). Bayesian model averaging to improve the yield prediction in wheat breeding trials. *Agricultural and Forest Meteorology*, *328*, 109237. DOI: 10.1016/j.agrformet.2022.109237

Fenyi, K., & Sapaty, G. A. (2021). Ideological Underpinnings in President John Dramani Mahama's 2013 State of the Nation Address. *Linguistics Initiative*, *1*(1), 1–17. DOI: 10.53696/27753719.17

Fernie, A. R., Bachem, C. W. B., Helariutta, Y., Neuhaus, H. E., Prat, S., Ruan, Y. L., Stitt, M., Sweetlove, L. J., Tegeder, M., Wahl, V., Sonnewald, S., & Sonnewald, U. (2020). Synchronization of developmental, molecular and metabolic aspects of source–sink interactions. *Nature Plants*, *6*(2), 55–66. DOI: 10.1038/s41477-020-0590-x PMID: 32042154

Ferrag, M. A., Shu, L., Djallel, H., & Choo, K. K. R. (2021). Deep Learning-Based Intrusion Detection for Distributed Denial of Service Attack in Agriculture 4.0. *Electronics (Basel)*, *10*(11), 1257. DOI: 10.3390/electronics10111257

Foley, J. A., Ramankutty, N., Brauman, K. A., Cassidy, E. S., Gerber, J. S., Johnston, M., Mueller, N. D., O'Connell, C., Ray, D. K., West, P. C., Balzer, C., Bennett, E. M., Carpenter, S. R., Hill, J., Monfreda, C., Polasky, S., Rockström, J., Sheehan, J., Siebert, S., & Zaks, D. P. (2011). Solutions for a cultivated planet. *Nature*, *478*(7369), 337–342. DOI: 10.1038/nature10452 PMID: 21993620

Fonseca, L., Oliveira, E., Pereira, T., & Sá, J. C. (2024). Leveraging ChatGPT for Sustainability: A Framework for SMEs to Align with UN Sustainable Development Goals and tackle sustainable development challenges. *Management & Marketing*, *19*(3).

Fowler, S. J., & Hope, C. (2007). Incorporating sustainable business practices into company strategy. *Business Strategy and the Environment*, *16*(1), 26–38. DOI: 10.1002/bse.462

Franco, M., & Rodrigues, M. (2021). Sustainable practices in SMEs: Reducing the ecological footprint. *The Journal of Business Strategy, 42*(2), 137–142. DOI: 10.1108/JBS-07-2019-0136

Freeman, R. E. (1984). *Strategic Management: A Stakeholder Approach.* Pitman.

Gaines, L., Nelson, P., & Birky, A. (2014). Recycling of lithium-ion batteries. *Nature Materials, 13*(9), 809–818.

Geels, F. W., Sovacool, B. K., Schwanen, T., & Sorrell, S. (2017). Sociotechnical transitions for deep decarbonization. *Science, 357*(6357), 1242–1244. DOI: 10.1126/science.aao3760 PMID: 28935795

Geissdoerfer, M., Savaget, P., Bocken, N. M. P., & Hultink, E. J. (2017). The circular economy – A new sustainability paradigm? *Journal of Cleaner Production, 143*, 757–768. DOI: 10.1016/j.jclepro.2016.12.048

Ge, N., Wang, E., & Li, Y. (2023). Foreign Language Education for Sustainable Development in China: A Case Study of German Language Education. *Sustainability (Basel), 15*(8), 6340. DOI: 10.3390/su15086340

Ghana Statistical Service. (2020) Population and Housing Census Report. Regional Office, Accra. https://statsghana.gov.gh

Ghobakhloo, M. (2020). Industry 4.0, digitization, and opportunities for sustainability. *Journal of Cleaner Production, 252*, 119869. DOI: 10.1016/j.jclepro.2019.119869

Ghobakhloo, M., Fathi, M., Iranmanesh, M., Vilkas, M., Grybauskas, A., & Amran, A. (2024). Generative artificial intelligence in manufacturing: Opportunities for actualizing Industry 5.0 sustainability goals. *Journal of Manufacturing Technology Management, 35*(9), 94–121. DOI: 10.1108/JMTM-12-2023-0530

Ghorbal, S., Farhani, S., & Youssef, S. B. (2021). Do renewable energy and national patents impact the environmental sustainability of Tunisia? *Environmental Science and Pollution Research International, 29*(17), 25248–25262. DOI: 10.1007/s11356-021-17628-7 PMID: 34839440

Ghosh, A., & Venkatesh, P. (2021). Public-private partnerships in renewable energy: Lessons from India's solar revolution. *Energy Policy, 149*, 112013.

Ghosh, P., Jha, A., & Sharma, R. (2020). Managing carbon footprint for a sustainable supply chain: A systematic literature review. *Modern Supply Chain Research and Applications, 2*(3), 123–141. DOI: 10.1108/MSCRA-06-2020-0016

Gikunda, K. (2024). Harnessing artificial intelligence for sustainable agricultural development in Africa: opportunities, challenges, and impact. *arXiv preprint arXiv:2401.06171.*

Gill, F. L., Viswanathan, K. K., & Karim, M. Z. A. (2018). The critical review of the pollution haven hypothesis. *International Journal of Energy Economics and Policy*, *8*(1), 167–174.

Glesne, C. (2012). *Introduction to qualitative research* (Ersoy, A., & Yalçınoğlu, P., Trans.). Anı Publishing.

Global Carbon Budget. (2022), "Per capita CO 2 emissions, 1750-2022", available at: https:// ourworldindata.org/co2-and-greenhouse-gas-emissions

Glover, J., Champion, D., Daniels, K., & Dainty, A. (2014). An Institutional Theory perspective on sustainable practices across the dairy supply chain. *International Journal of Production Economics*, *152*, 152. DOI: 10.1016/j.ijpe.2013.12.027

Gökçe, G., Goncu, S., & Bozkurt, S. (2020). Industry 4.0 and animal husbandry. *International Anatolian Journal of Agricultural Engineering Sciences*, *2*(3), 21–26.

Gopinath, G. R., Sattanathan, M., Muthukannan, M., Pandian, R. S., & Manokar, A. M. (2021). Experimental study of solar dryer with thermal energy storage system for drying of agro products. *Research Square*. DOI: 10.21203/rs.3.rs-673328/v1

Gregersen, E. (2020). History of Technology Timeline. https://www.britannica.com/story/history-of-technology-timeline (Accessed: 09.11.2022).

Gubanova, A., Clasen, M., & Theuvsen, L. (2017). How do digital agricultural startups achieve critical mass? A qualitative analysis. In *37. GIL-Jahrestagung 2017-Fokus: Digitale Transformation–Wege in eine zukunftsfähige Landwirtschaft, Lecture Notes in Informatics,* 61-64). Gesellschaft für Informatik eV.

Guimarães, F. F., & Júnior, C. A. H. (2021). Digital Resources and English as an Additional Language in Higher Education: Possibilities for Internationalization. *Ilha do Desterro*, *74*(3). Advance online publication. DOI: 10.5007/2175-8026.2021.e80233

Gule, L. (2021). The effects of FDI and economic growth on CO2 emissions in SADC. *PEOPLE: International Journal of Social Sciences*, *7*(1), 75–87. DOI: 10.20319/pijss.2021.71.7587

Gunay, B., & Yavuz, A. (2020). The impact of the Internet of Things (IoT) in water management: A review. *Environmental Science and Pollution Research International*, *27*(19), 23882–23899.

Gunina, N. A., Mordovina, T. V, & Shelenkova, I. (2021). Integrating Sustainability Issues Into English Language Courses at University. *E3s Web of Conferences, 295*, 5006. DOI: 10.1051/e3sconf/202129505006

Gupta, U., Kim, Y. G., Lee, S., Tse, J., Lee, H.-H. S., Wei, G.-Y., Brooks, D., & Wu, C.-J. (2021). Chasing carbon: The elusive environmental footprint of computing. DOI: 10.1109/HPCA51647.2021.00076

Gupta, A. A. (2023). Diversity, Dignity, Equity, and Inclusion in the Age of Division, Discord and Disunion: Stereotyping, Sexist, Hegemony in Education. *International Education Studies, 16*(1), 110. DOI: 10.5539/ies.v16n1p110

Gyawali, Y. P. (2023). Exploring Dimensions of Teacher Education in Nepal: Perspectives, Challenges, and Reform Opportunities. *Journal of General Education and Humanities, 2*(4), 285–298. DOI: 10.58421/gehu.v2i4.128

Hameed, A. (2023). Environmental Concerns and English Language Teaching in Saudi Context: Perceptions and Practices. *Journal of Language Teaching and Research, 14*(4), 1127–1137. DOI: 10.17507/jltr.1404.31

Han, H., Liu, Z., Li, J., & Zeng, Z. (2024). Challenges in remote sensing based climate and crop monitoring: Navigating the complexities using AI. *Journal of Cloud Computing (Heidelberg, Germany), 13*(1), 34. DOI: 10.1186/s13677-023-00583-8

Han, J., Fang, S., Wang, X., Zhuo, W., Yu, Y., Peng, X., & Zhang, Y. (2024). The impact of intra-annual temperature fluctuations on agricultural temperature extreme events and attribution analysis in mainland China. *The Science of the Total Environment, 949*, 174904. DOI: 10.1016/j.scitotenv.2024.174904 PMID: 39059648

Hao, L., Umar, M., Khan, Z., & Ali, W. (2021). Green growth and low carbon emission in G7 countries: How critical the network of environmental taxes, renewable energy and human capital is? *The Science of the Total Environment, 752*, 141853. DOI: 10.1016/j.scitotenv.2020.141853 PMID: 32889278

Hart, S. L. (1995). A natural-resource-based view of the firm. *Academy of Management Review, 20*(4), 986–1014. DOI: 10.2307/258963

Hassink, J., Agricola, H., Veen, E. J., Pijpker, R., de Bruin, S. R., Meulen, H. A. B., & Plug, L. B. (2020). The Care Farming Sector in The Netherlands: A Reflection on Its Developments and Promising Innovations. *Sustainability (Basel), 12*(9), 3811. DOI: 10.3390/su12093811

Hatamleh, A. I., & Hatamleh, I. M. (2024, January). The Importance of Using ICT in Improving the Quality of Education. In *2024 ASU International Conference in Emerging Technologies for Sustainability and Intelligent Systems (ICETSIS)* (pp. 95-99). IEEE. DOI: 10.1109/ICETSIS61505.2024.10459655

Hati, A. J., & Singh, R. R. (2021). Smart Indoor Farms: Leveraging Technological Advancements for a Sustainable Agricultural Revolution. *AgriEngineering*, *3*(4), 728–767. DOI: 10.3390/agriengineering3040047

Hawkins, T. R., Gunter, E. W., & Franke, M. (2013). Environmental impacts of electric vehicles: A comparison of different powertrain technologies. *Environmental Science & Technology*, *47*(14), 8066–8073.

Heenkenda, H. M. J. C. B., Xu, F., Kulathunga, K. M. M. C. B., & Senevirathne, W. A. R. (2022). The role of innovation capability in enhancing sustainability in SMEs: An emerging economy perspective. *Sustainability (Basel)*, *14*(17), 10832. DOI: 10.3390/su141710832

Heinicke, A. (2018). Performance measurement systems in small and medium-sized enterprises and family firms: A systematic literature review. *Journal of Management Control*, *28*(4), 457–502. DOI: 10.1007/s00187-017-0254-9

He, J., Zhao, Y., Zhang, B., Chen, L., & Ma, X. (2025). The impact of blockchain technology on green investment decisions for a sustainable supply chain with an overconfident manufacturer. *Sustainability*, *17*(1), 284.

Hektaş. (2021). *What is Smart Agriculture?*https://hektas.com.tr/akilli-tarim-nedir/ (Accessed: 07.12.2022)

Hendratmoko, S. (2023). Developing Green Sustainable Entrepreneurship for Competitive Advantages through Green Management. *Indonesian Journal of Business Analytics*, *3*(6), 2163–2176. DOI: 10.55927/ijba.v3i6.5972

Higón, D. A., Gholami, R., & Shirazi, F. (2017). ICT and environmental sustainability: A global perspective. *Telematics and Informatics*, *34*(4), 85–95. DOI: 10.1016/j.tele.2017.01.001

Hill, C. B., & Li, C. (2022). Genetic Improvement of Heat Stress Tolerance in Cereal Crops. *Agronomy (Basel)*, *12*(5), 1205. DOI: 10.3390/agronomy12051205

Hong, Z., & Xiao, K. (2024). Digital economy structuring for sustainable development: The role of blockchain and artificial intelligence in improving supply chain and reducing negative environmental impacts. *Scientific Reports*, *14*(1), 3912. DOI: 10.1038/s41598-024-53760-3 PMID: 38365912

Hoogendoorn, B., Guerra, D., & Van Der Zwan, P. (2015). What drives environmental practices of SMEs? *Small Business Economics*, *44*(4), 59–781. DOI: 10.1007/s11187-014-9618-9

Hopkins, M. S., Kruschwitz, N., Haanaes, K., Kong, M. T., Arthur, D., & Reeves, M. (2011). Sustainability: The 'embracers' seize advantage. *MIT Sloan Management Review*, *52*(3), 3.

Hossain, S. F. A., Ahsan, F. T., Nadi, A. H., Ahmed, M., & Neyamah, H. (2022). Exploring the Role of Technology Application in Tourism Events, Festivals and Fairs in the United Arab Emirates: Strategies in the Post-Pandemic Period. In Hassan, A. (Ed.), *Technology Application in Tourism Fairs, Festivals and Events in Asia*. Springer., DOI: 10.1007/978-981-16-8070-0_19

Hsu, C., Zhang, Y. Ch. P., Aqdas, R., Chupradit, S., & Nawaz, A. (2021). A step towards sustainable environment in China: The role of eco-innovation renewable energy and environmental taxes. *Journal of Environmental Management*, *299*, 113609. DOI: 10.1016/j.jenvman.2021.113609 PMID: 34467863

Huang, J., Li, X., Wang, Y., & Lei, H. (2021). The effect of energy patents on China's carbon emissions: Evidence from the STIRPAT model. *Technological Forecasting and Social Change*, *173*, 121110. DOI: 10.1016/j.techfore.2021.121110

Huang, J., Pang, C., Yang, W., Zeng, X., Zhang, J., & Huang, C. (2022). A Deep Learning Neural Network for the Residential Energy Consumption Prediction. *IEEJ Transactions on Electrical and Electronic Engineering*, *17*(4), 575–582. DOI: 10.1002/tee.23543

Huang, Y., Chen, F., Wei, H., Xiang, J., Xu, Z., & Akram, R. (2022). The impacts of FDI inflows on carbon emissions: Economic development and regulatory quality as moderators. *Frontiers in Energy Research*, *9*, 820596. DOI: 10.3389/fenrg.2021.820596

Huang, Y., Haseeb, M., Usman, M., & Ozturk, I. (2022). Dynamic association between ICT, renewable energy, economic complexity and ecological footprint: Is there any difference between E-7 (developing) and G-7 (developed) countries? *Technology in Society*, *68*, 101853. DOI: 10.1016/j.techsoc.2021.101853

Hur, E., & Cassidy, T. (2019). Perceptions and attitudes towards sustainable fashion design: Challenges and opportunities for implementing sustainability in fashion. *International Journal of Fashion Design, Technology and Education*, *12*(2), 208–217. DOI: 10.1080/17543266.2019.1572789

Hussein, B. (2021). Addressing Collaboration Challenges in Project-Based Learning: The Student's Perspective. *Education Sciences, 11*(8), 434. DOI: 10.3390/educsci11080434

IBM. (2021). *Smarter cities: Transforming urban sustainability with AI and data analytics.* https://www.ibm.com/smarter-cities

IDC. (2024). Worldwide Spending on Digital Transformation. Accessed from https://www.idc.com/getdoc.jsp?containerId=prUS52305724

Indriastuti, N., Prilliandani, N. M. I., Sutadji, I. M., Setiyaningsih, T. A., & Gunawan, A. (2024). *Opportunities and Challenges: Implementation of Sustainable Business Practices in MSME's. Proceeding*. Islamic University of Kalimantan.

Inglesi-Lotz, R., & Dogan, E. (2018). The role of renewable versus non-renewable energy to the level of CO2 emissions a panel analysis of sub- Saharan Africa's Big 10 electricity generators. *Renewable Energy, 123*, 36–43. DOI: 10.1016/j.renene.2018.02.041

Intergovernmental Panel on Climate Change (IPCC). (2021). *Climate change 2021: The physical science basis.* Cambridge University Press.

International Renewable Energy Agency (IRENA). (2021). *Renewable energy statistics 2021.* IRENA.

International Renewable Energy Agency (IRENA). (2023). *Innovations in digital energy solutions.* Retrieved from https://www.irena.org

Iqbal, A., Tang, X., & Rasool, S. F. (2023). Investigating the nexus between CO2 emissions, renewable energy consumption, FDI, exports and economic growth: Evidence from BRICS countries. *Environment, Development and Sustainability, 25*(3), 2234–2263. DOI: 10.1007/s10668-022-02128-6

IRENA. (2021). *Renewable energy and jobs – Annual review 2021.* International Renewable Energy Agency. https://www.irena.org/Publications/2021/Jun/Renewable-energy-and-jobs-Annual-review-2021

Islam, A., Wahab, S. A., & Tehseen, S. (2024). Rethinking holistic sustainable growth of Malaysian SMEs: A University Helix-Quadruple Bottom Line perspective. *International Journal of Innovation Science*. Advance online publication. DOI: 10.1108/IJIS-07-2023-0158

Jahanger, A., Usman, M., & Ahmad, P. (2022). Investigating the effects of natural resources and institutional quality on CO2 emissions during globalization mode in developing countries. *International Journal of Environmental Science and Technology, 20*(9), 9663–9682. Advance online publication. DOI: 10.1007/s13762-022-04638-2

Jambeck, J. R., Geyer, R., Wilcox, C., Siegler, T. R., Perryman, M., Andrady, A., Narayan, R., & Law, K. L. (2015). Plastic waste inputs from land into the ocean. *Science, 347*(6223), 768–771. DOI: 10.1126/science.1260352 PMID: 25678662

Jenkins, H. (2009). A 'business opportunity' model of corporate social responsibility for small- and medium-sized enterprises. *Business Ethics (Oxford, England), 18*(1), 21–36. DOI: 10.1111/j.1467-8608.2009.01546.x

Jeong, S., Ko, J., Ban, J., Shin, T., & Yeom, J. (2024). Deep learning-enhanced remote sensing-integrated crop modeling for rice yield prediction. *Ecological Informatics, 84*, 102886. DOI: 10.1016/j.ecoinf.2024.102886

Jiang, P., Akhter, S., Azizi, Z., Gheisari, A., & Kumar, T. (2022). Exploring the Role of Content and Language Integrated Learning Approach in Developing Transversal Skills in University Students With Respect to the Mediating Role of Emotional Intelligence. *Frontiers in Psychology, 13*, 988494. Advance online publication. DOI: 10.3389/fpsyg.2022.988494 PMID: 36275209

Jose, S. V., Sandhya, H., & Varghese, B. (2023). Artificial intelligence towards smart green transportation. In *Advances in marketing, customer relationship management, and e-services book series* (pp. 102–124). DOI: 10.4018/978-1-6684-8140-0.ch005

Joshua, A., Xusheng, Q., Emmanuel, B. G., Sakoane, K., & Appiah, M. (2024). Unveiling the dynamic nexuses between foreign investment, trade openness, energy consumption, and CO2 emissions in South Africa: A vector error correction model approach. *Energy & Environment*, 0958305X241266529.

Journeault, M., Perron, A., & Vallières, L. (2021). The collaborative roles of stakeholders in supporting the adoption of sustainability in SMEs. *Journal of Environmental Management, 287*, 112349. DOI: 10.1016/j.jenvman.2021.112349 PMID: 33735683

Kagermann, H. (2017). Using Industry 4.0 Opportunities. In *Handbook Industry 4.0*, Vol. 4, 237–248, Springer Vieweg.

Kahouli, B., Hamdi, B., Nafla, A., & Chabaane, N. (2022). Investigating the relationship between ICT, green energy, total factor productivity, and ecological footprint: Empirical evidence from Saudi Arabia. *Energy Strategy Reviews, 42*, 100871. DOI: 10.1016/j.esr.2022.100871

Karakuş, G., Yılmaztekin, F., Hamamcı, S., Bulut, B., & Ünver, R. (2022). Digital Transformation in Airports within the Scope of Aviation 4.0: A Study at Ankara Esenboğa Airport, *Proceedings Book of the 16th International Congress of Social Sciences with Current Research*, 796-813.

Kashem, M. A., Shamsuddoha, M., Nasir, T., & Chowdhury, A. A. (2023). The role of artificial intelligence and blockchain technologies in sustainable tourism in the Middle East. *Worldwide Hospitality and Tourism Themes, 15*(2), 178–191. DOI: 10.1108/WHATT-10-2022-0116

Kassahun, S. K., Assefa, B., Henzler, K., Weißert, J., Oteng-Ababio, M., Admassu, M., Mohammed-Amin, I., & Tesfahun, G. (2025). Firsthand report on solid waste management practice in the major town of Addis Ababa-Adama Economic Corridor, Ethiopia. *Heliyon, 11*(2), e41699. DOI: 10.1016/j.heliyon.2025.e41699 PMID: 39882454

Kassinis, G., Oikonomou, V., & Alexopoulos, C. (2022). Emerging technologies for urban air quality management. *Environmental Science & Policy*, *132*, 25–35.

Katten, G. (2021). Issuing green bonds on the Algorand blockchain. *arXiv* preprint arXiv:2108.10344.

Kelly, D. (2016). Perceptions of Taylorism and a Marxist scientific manager. *Journal of Management History*, *22*(3), 298–319. DOI: 10.1108/JMH-11-2015-0205

Kganyago, M., Adjorlolo, C., Mhangara, P., Tsoeleng, L. (2024). Optical remote sensing of crop biophysical and biochemical parameters: An overview of advances in sensor technologies and machine learning algorithms for precision agriculture. 218, 108730. https://doi.org/DOI: 10.1016/j.compag.2024.108730

Khaled, M. W. B., Dahmani, M., & Ben Youssef, A. (2024). Overcoming Barriers to Energy Transition in the MENA Region: New Institutional Dynamics. DOI: 10.20944/preprints202402.1440.v1

Khan, F., Zahir, S., Rahman, H. U., Raza, A., & Noor, S. (2024). Exploring the role of ICT and education in reducing environmental degradation among Asian countries. In *Natural Resources Forum*. Oxford, UK: Blackwell Publishing Ltd. DOI: 10.1111/1477-8947.12548

Khan, A., & Wu, X. (2022). Digital economy and environmental sustainability: Do information communication and technology (ICT) and economic complexity matter? *International Journal of Environmental Research and Public Health*, *19*(19), 12301. DOI: 10.3390/ijerph191912301 PMID: 36231602

Khan, F. N., Sana, A., & Arif, U. (2020). Information and communication technology (ICT) and environmental sustainability: A panel data analysis. *Environmental Science and Pollution Research International, 27*(29), 36718–36731. DOI: 10.1007/s11356-020-09704-1 PMID: 32566985

Khan, Z., Ali, S., Umar, M., Kirikkaleli, D., & Jiao, Z. (2020). Consumption-based carbon emissions and international trade in G7 countries: The role of environmental innovation and Renewable energy. *The Science of the Total Environment, 730*, 138945. DOI: 10.1016/j.scitotenv.2020.138945 PMID: 32416502

Kharche, S. S., Gaikwad, V. L., & Kaware, D. M. (2020). IoT-based smart irrigation system for water conservation. *Materials Today: Proceedings, 28*(2), 455–460.

Khoshnava, S. M., Rostami, R., Zin, R. M., Štreimikienė, D., Yousefpour, A., Strielkowski, W., & Mardani, A. (2019). Aligning the Criteria of Green Economy (GE) and Sustainable Development Goals (SDGs) to Implement Sustainable Development. *Sustainability (Basel), 11*(17), 4615. DOI: 10.3390/su11174615

Khumalo, P. (2014). Leadership will remain a key factor in attaining global environmental sustainability: What are the stakes? *Journal of Human Ecology (Delhi, India), 46*(2), 187–194. DOI: 10.1080/09709274.2014.11906719

Kiesnere, A. L., & Baumgartner, R. J. (2019). Sustainability management in practice: Organizational change for sustainability in smaller large-sized companies in Austria. *Sustainability (Basel), 11*(3), 572. DOI: 10.3390/su11030572

Kılavuz, E., & Erdem, İ. (2019). Agriculture 4.0 applications in the world and the transformation of Turkish agriculture. *Social Sciences, 14*(4), 133–157.

Kirmikil, M., & Ertaş, B. (2020). Sustainable Future with Agriculture 4.0. *Icontech International Journal of Surveys, Engineering. Technology, 4*(1), 1–12.

Koca, D. (2020). Historical background of industrial revolutions and their impact on workforce skills. *OPUS International Journal of Society Researches, 16*(31), 4531–4558.

Koenker, R. (2004). Quantile regression for longitudinal data. *Journal of Multivariate Analysis, 91*(1), 74–89. DOI: 10.1016/j.jmva.2004.05.006

Kolawole, A. S., & Iyiola, A. O. (2023). Environmental Pollution: Threats, Impact on Biodiversity, and Protection Strategies. In Izah, S. C., & Ogwu, M. C. (Eds.), *Sustainable Utilization and Conservation of Africa's Biological Resources and Environment. Sustainable Development and Biodiversity* (Vol. 32). Springer., DOI: 10.1007/978-981-19-6974-4_14

Kongbuamai, N., Bui, Q., Adedoyin, F., & Victor Bekun, F. (2022). Developing environmental policy framework for sustainable development in Next-11 countries: The impacts of information and communication technology and urbanization on the ecological footprint. *Environment, Development and Sustainability*. Advance online publication. DOI: 10.1007/s10668-022-02528-8

Kosyak, A., & Popov, S. (2020). Analyzing Consumer Preferences for Green Products and Environmental Impact. *Journal of Energy and Environmental Policy Options, 3*(4), 134–140.

Kouhizadeh, M., Saberi, S., & Sarkis, J. (2021). Blockchain technology and the sustainable supply chain: Theoretically exploring adoption barriers. *International Journal of Production Economics, 217*, 164–176. DOI: 10.1016/j.ijpe.2020.107831

Kovács, I., & Husti, I. (2018). The role of digitalization in the agricultural 4.0–how to connect the industry 4.0 to agriculture? *Hungarian Agricultural Engineering*, (33), 38–42. DOI: 10.17676/HAE.2018.33.38

Krishnendu, J., & Patra, B. (2025). Environmental impacts of economic growth: A STIRPAT analysis using machine learning algorithms. *Sustainable Futures : An Applied Journal of Technology, Environment and Society, 9*, 100404. DOI: 10.1016/j.sftr.2024.100404

Kumar, R., & Kumar, S. (2017). Different technologies and components of renewable energy sources based power system: A Review. *International Journal of Trend in Scientific Research and Development*, DOI: 10.31142/ijtsrd165

Kumari, M., Guleria, S. and Kumar, S. (2024). Sustainability in tourism and hospitality: Artificial intelligence role in eco-friendly practices in Indian hotels. *Journal of tourism theory and research*, 10(2):73-83. DOI: 10.24288/jttr.1523976

Kuratko, D. F., Fisher, G., & Audretsch, D. B. (2021). Unraveling the entrepreneurial mindset. *Small Business Economics, 57*(4), 1681–1691. DOI: 10.1007/s11187-020-00372-6

Kurnaz, Z. (2016). *Political Fear and the Construction of Academic Life in Turkey (Published doctoral thesis),* Hacettepe University, Ankara, Türkiye.

Kusumaryoko, P. (2024). Exploring the Impact of Green Entrepreneurship on the Competitive Advantage of SMEs: A Sustainable Management Approach. *The Journal of Academic Science, 1*(7), 805–816.

Kwakwa, P. A. (2023). Climate change mitigation role of renewable energy consumption: Does institutional quality matter in the case of reducing Africa's carbon dioxide emissions? *Journal of Environmental Management*, *342*, 118234. DOI: 10.1016/j.jenvman.2023.118234 PMID: 37327731

Kwakwa, P. A., Adjei-Mantey, K., & Adusah-Poku, F. (2023). The effect of transport services and ICTs on carbon dioxide emissions in South Africa. *Environmental Science and Pollution Research International*, *30*(4), 10457–10468. DOI: 10.1007/s11356-022-22863-7 PMID: 36071365

Kwakwa, P. A., Adzawla, W., Alhassan, H., & Oteng-Abayie, E. F. (2023). The effects of urbanization, ICT, fertilizer usage, and foreign direct investment on carbon dioxide emissions in Ghana. *Environmental Science and Pollution Research International*, *30*(9), 23982–23996. DOI: 10.1007/s11356-022-23765-4 PMID: 36331739

Kwakwa, P. A., Dankwah, J. B., Adu Boahen, E., & Hammond, P. (2023). Financial development in South Africa: The role of natural resources, IT infrastructure, and government size. *Cogent Economics & Finance*, *11*(2), 2281844. DOI: 10.1080/23322039.2023.2281844

Lange, S., Pohl, J., & Santarius, T. (2020). Digitalization and Energy Consumption. Does ICT Reduce Energy Demand? *Ecological Economics*, *176*, 106760. DOI: 10.1016/j.ecolecon.2020.106760

Latif, Z., Xin, W., Khan, D., Iqbal, K., Pathan, Z. H., Salam, S., & Jan, N. (2017). ICT and sustainable development in South Asian countries. *Human Systems Management*, *36*(4), 353–362. DOI: 10.3233/HSM-17166

Leal Filho, W., Mbah, M. F., Dinis, M. A. P., Trevisan, L. V., de Lange, D., Mishra, A., Rebelatto, B., Ben Hassen, T., & Aina, Y. A. (2024). The role of artificial intelligence in the implementation of the UN Sustainable Development Goal 11: Fostering sustainable cities and communities. *Cities (London, England)*, *150*, 105021. DOI: 10.1016/j.cities.2024.105021

Lee, H. S., Moseykin, Y. N., & Chernikov, S. (2021). Sustainable relationship between FDI, R&D, and CO2 emissions in emerging markets: An empirical analysis of BRICS countries. *Russian Journal of Economics*, *7*(4), 297–312. DOI: 10.32609/j.ruje.7.77285

Lee, J. W., & Brahmasrene, T. (2014). ICT, CO2 emissions and economic growth: Evidence from a panel of ASEAN. *Global Economic Review*, *43*(2), 93–109. Advance online publication. DOI: 10.1080/1226508X.2014.917803

Li, D., Du, P., & He, H. (2022) Artificial Intelligence-Based Sustainable Development of Smart Heritage Tourism, Wireless Communications and Mobile Computing 2022, 13 p. DOI: DOI: 10.1155/2022/5441170

Liang, L., Wang, Z., & Li, J. (2019). The effect of urbanization on environmental pollution in rapidly developing urban agglomerations. *Journal of Cleaner Production, 237*, 117649. DOI: 10.1016/j.jclepro.2019.117649

Li, J., Irfan, M., Samad, S., Ali, B., Zhang, Y., Bădulescu, D., & Bădulescu, A. (2023). The relationship between energy consumption, CO2 emissions, economic growth, and health indicators. *International Journal of Environmental Research and Public Health, 20*(3), 2325. DOI: 10.3390/ijerph20032325 PMID: 36767690

Li, L., Fu-quan, H., Chen, S., Pan, L., Zeng, W., & Wu, X. (2020). Exploring the Curriculum Development in Content and Language Integrated Learning: A Systematic Review. *International Journal of Evaluation and Research in Education (Ijere), 9*(4), 1102. DOI: 10.11591/ijere.v9i4.20705

Liu, H., Wong, W. K., Cong, P., Nassani, A. A., Haffar, M., & Abu-Rumman, A. (2023). Linkage among urbanization, energy consumption, economic growth and carbon Emissions: Panel data analysis for China using ARDL model. *Fuel, 332*(2), 126122. DOI: 10.1016/j.fuel.2022.126122

Lodhi, S. K., Gill, A. Y., & Hussain, H. K. (2024). Green innovations: Artificial intelligence and sustainable materials in production. *BULLET: Jurnal Multidisiplin Ilmu, 3*(4), 492–507.

Loewen, S., Crowther, D., Isbell, D. R., Kim, K. M., Maloney, J., Miller, Z. F., & Rawal, H. (2019). Mobile-Assisted Language Learning: A Duolingo Case Study. *ReCALL, 31*(3), 293–311. DOI: 10.1017/S0958344019000065

Lone, U. M., Darzi, M. A., & Islam, K. U. (2023). Macroeconomic variables and stock market performance: A PMG/ARDL approach for BRICS economies. *Macroeconomics and Finance in Emerging Market Economies, 16*(2), 300–325. DOI: 10.1080/17520843.2021.1983704

Lopez, A., Mai, T., Lantz, E., Harrison-Atlas, D., Williams, T., & Maclaurin, G. (2021). Land use and turbine technology influences on wind potential in the United States. *Energy, 223*, 120044. DOI: 10.1016/j.energy.2021.120044

Louati, A., Louati, H., Alharbi, M., Kariri, E., Khawaji, T., Almubaddil, Y., & Aldwsary, S. (2024). Machine learning and Artificial intelligence for a Sustainable tourism: A case study on Saudi Arabia. *Information (Basel), 15*(9), 516. DOI: 10.3390/info15090516

Lourenço, V. M., Ogutu, J. O., Rodrigues, R. A. P., Posekany, A., & Piepho, H. (2024). Genomic prediction using machine learning: A comparison of the performance of regularized regression, ensemble, instance-based and deep learning methods on synthetic and empirical data. *BMC Genomics*, *25*(1), 152. DOI: 10.1186/s12864-023-09933-x PMID: 38326768

Lv, T., Hu, H., Xie, H., Zhang, X., Wang, L., & Shen, X. (2023). An empirical relationship between urbanization and carbon emissions in an ecological civilization demonstration area of China based on the STIRPAT model. *Environment, Development and Sustainability*, *25*(3), 2465–2486. DOI: 10.1007/s10668-022-02144-6

Machado, J. A., & Silva, J. S. (2019). Quantiles via moments. *Journal of Econometrics*, *213*(1), 145–173. DOI: 10.1016/j.jeconom.2019.04.009

Mahmood, H., Asadov, A., Tanveer, M., Furqan, M., & Yu, Z. (2022). Impact of oil price, economic growth and urbanization on CO2 emissions in GCC countries: Asymmetry analysis. *Sustainability (Basel)*, *14*(8), 4562. DOI: 10.3390/su14084562

Majeed, M. T. (2018). Information and communication technology (ICT) and environmental sustainability in developed and developing countries. *Pakistan Journal of Commerce and Social Sciences*, *12*(3), 758–783.

Majid, G. M., Tussyadiah, I., & Kim, Y. R. (2024). Exploring the potential of chatbots in extending tourists' sustainable travel practices. *Journal of Travel Research*, 00472875241247316. Advance online publication. DOI: 10.1177/00472875241247316

Malesios, C., De, D., Moursellas, A., Dey, P. K., & Evangelinos, K. (2021). Sustainability performance analysis of small and medium-sized enterprises: Criteria, methods and framework. *Socio-Economic Planning Sciences*, *75*, 100993. DOI: 10.1016/j.seps.2020.100993

Manteaw, B. O. (2020). Education and Learning in Sustainable Development: Foregrounding an Emergent Discourse. *Discourse and Communication for Sustainable Education*, *11*(2), 5–19. DOI: 10.2478/dcse-2020-0015

Maqsood, A., Gul, S., Noureen, N., & Yaswi, A. (2024). Dynamics of Perceived Stress, Stress Appraisal, and Coping Strategies in an Evolving Educational Landscape. *Behavioral Sciences (Basel, Switzerland)*, *14*(7), 532. DOI: 10.3390/bs14070532 PMID: 39062355

Markard, J., Raven, R., & Truffer, B. (2012). Sustainability transitions: An emerging field of research and its prospects. *Research Policy*, *41*(6), 955–967. DOI: 10.1016/j.respol.2012.02.013

Martini, B., Bellisario, D., & Coletti, P. (2024). Human-centered and sustainable artificial intelligence in industry 5.0: Challenges and perspectives. *Sustainability (Basel), 16*(13), 5448. DOI: 10.3390/su16135448

Mazza, B. (2023). A Theoretical model of Strategic Communication for the Sustainable development of sport tourism. *Sustainability (Basel), 15*(9), 7039. DOI: 10.3390/su15097039

Mell, P., & Grance, T. (2011). The NIST definition of cloud computing (SP 800-145). *NIST Spec. Publ, 145*, 7.

Mengelkamp, E., Gärttner, J., Rock, K., & Weinhardt, C. (2018). Designing microgrid energy markets: A case study: The Brooklyn Microgrid. *Applied Energy, 210*, 870–880. DOI: 10.1016/j.apenergy.2017.06.054

Mensah, C. N., Long, X., Boamah, K. B., Bediako, I. A., Dauda, L., & Salman, M. (2018). The effect of innovation on CO2 emissions of OCED countries from 1990 to 2014. *Environmental Science and Pollution Research International, 25*(29), 29678–29698. DOI: 10.1007/s11356-018-2968-0 PMID: 30144011

Merriam, S. B., & Bierema, L. L. (2013). *Adult learning: Linking theory and practice*. John Wiley & Sons.

Mete, D. E. (2018). Incorporating Environmental Education in English Language Teaching Through Bloom's Revised Taxonomy. *Selçuk Üniversitesi Edebiyat Fakültesi Dergisi, 40*(40), 33–44. DOI: 10.21497/sefad.514847

Metris, D., Kraugusteeliana, K., Amory, J. D. S., Mustafa, F., & Risdwiyanto, A. (2024). The Influence of Community Support on the Sustainability of MSMEs in the Digital Era. [TIRAKAT]. *Jurnal Terobosan Peduli Masyarakat, 1*(1), 21–29. DOI: 10.61100/j.tirakat.v1i1.102

Meyer, E., & Vilsmaier, U. (2020). Economistic Discourses of Sustainability: Determining Moments and the Question of Alternatives. *Sustainability in Debate, 11*(1), 98–124. DOI: 10.18472/SustDeb.v11n1.2020.26663

Mishra, D. (2024). Strategic Communication for AI-Driven Sustainability Initiatives: Bridging Technology, Policy, and Public Engagement in Achieving SDGs. In Riswandi, B., Singh, B., Kaunert, C., & Vig, K. (Eds.), *AI Applications for Clean Energy and Sustainability* (pp. 187–211). IGI Global., DOI: 10.4018/979-8-3693-6567-0.ch010

Mission Innovation. (2021). *Mission Innovation: Accelerating clean energy innovation*. https://www.mission-innovation.net

Mngumi, F., Huang, L., Geng, X., & Ayub, B. (2024). Financial efficiency and CO2 emission in BRICS: Does digital economy development matter? *Heliyon*, *10*(2), e24321. Advance online publication. DOI: 10.1016/j.heliyon.2024.e24321 PMID: 38304825

Mohamed, A. M. (2023). Investigating the Benefits of Multimodal Project-Based Learning in Teaching English to International Students. *International Journal of Educational Innovation and Research*, *2*(2), 114–129. DOI: 10.31949/ijeir.v2i2.5085

Mohammed, I., & Bunyaminu, A. (2021). Major obstacles facing business enterprises in an emerging economy: The case of Ghana using the World Bank Enterprise Survey. *Journal of Small Business and Enterprise Development*, *28*(3), 475–487. DOI: 10.1108/JSBED-04-2020-0110

Mohammed, K. S., Serret, V., Jabeur, S. B., & Nobanee, H. (2024). The role of artificial intelligence and fintech in promoting eco-friendly investments and non-greenwashing practices in the US market. *Journal of Environmental Management*, *359*, 120977. DOI: 10.1016/j.jenvman.2024.120977 PMID: 38678903

Moşteanu, N. R. (2023). Navigating the intersection of sustainability and intelligent automation: Opportunities and challenges for entrepreneurs. *Human Resources Management and Services*, *5*(1).

Moursi, M., Wehn, N., & Hammoud, B. (2025). Smart environmental monitoring of marine pollution using edge AI. *arXiv* preprint arXiv:2504.21759.

Moyer, J. D., & Hughes, B. B. (2012). ICTs: Do they contribute to increased carbon emissions? *Technological Forecasting and Social Change*, *79*(5), 919–931. DOI: 10.1016/j.techfore.2011.12.005

Msomi, T. S., & Olarewaju, O. M. (2024). Navigating the Green Frontier and Sustainable Innovation in the Dynamic Business Landscape. In *Waste Management and Life Cycle Assessment for Sustainable Business Practice*, 241-268. DOI: 10.4018/979-8-3693-2595-7.ch011

Mudd, G. M., & Weng, Z. (2012). The environmental costs of cobalt mining in the Democratic Republic of Congo. *Environmental Science & Technology*, *46*(8), 4648–4655.

Mugano, G., & Dorasamy, N. (2024). *SMEs Perspective in Africa*. Springer Books. DOI: 10.1007/978-3-031-69103-4

Muhammad, M., Waheed, A., Wahab, A., Majeed, M., Nazim, M., Liu, Y. H., Li, L., & Li, W.-J. (2024). Soil salinity and drought tolerance: An evaluation of plant growth, productivity, microbial diversity, and amelioration strategies. *Plant Stress*, *11*, 100319. DOI: 10.1016/j.stress.2023.100319

Mumuni, S., & Mwimba, T. (2023). Modeling the impact of green energy consumption and natural resources rents on economic growth in Africa: An analysis of dynamic panel ARDL and the feasible generalized least squares estimators. *Cogent Economics & Finance*, *11*(1), 2161774. DOI: 10.1080/23322039.2022.2161774

Naderi, N., & Tian, Y. (2022). Leveraging blockchain technology and tokenizing green assets to fill the green finance gap. *Energy Research Letters*, *3*(3), 1–9.

Naseem, S., Hu, X., Shi, J., Mohsin, M., & Jamil, K. (2023). Exploring the optical impact of information communication technology and economic growth on CO2 emission in BRICS countries. *Optik (Stuttgart)*, *273*, 170339. DOI: 10.1016/j.ijleo.2022.170339

Nehra, A., Kalwan, G., Gill, R., Nehra, K., Agarwala, N., Jain, P.K., Naeem, M., Tuteja, N., Pudake, R.N., Gill, S.S. (2024). Status of impact of abiotic stresses on global agriculture. Nanotechnology for Abiotic Stress Tolerance and Management in Crop Plants. 1-21. https://doi.org/DOI: 10.1016/B978-0-443-18500-7.00001-6

Nguyen, V. N., Tarełko, W., Sharma, P., El-Shafay, A. S., Chen, W. H., Nguyen, P. Q. P., Nguyen, X. P., & Hoang, A. T. (2024). Potential of explainable artificial intelligence in advancing renewable energy: Challenges and prospects. *Energy & Fuels*, *38*(3), 1692–1712. DOI: 10.1021/acs.energyfuels.3c04343

Nilius, H., Tsouka, S., Nagler, M., & Masoodi, M. (2024). Machine learning applications in precision medicine: Overcoming challenges and unlocking potential. *Trends in Analytical Chemistry*, *179*, 117872. DOI: 10.1016/j.trac.2024.117872

Novitasari, N., Wahyuningsih, N., & Agustina, H. N. (2022). Improving Students' Speaking Skills Through Clil in Tourist Guiding Online Class. *Celtic a Journal of Culture English Language Teaching Literature and Linguistics*, *9*(1), 53–68. DOI: 10.22219/celtic.v9i1.20903

Nurmala, I., Irianto, S., Franchisca, S., Amsa, H., & Susanti, R. (2023). Technology-Enhanced Language Learning: A Meta-Analysis Study on English Language Teaching Tools. *Journal of Education*, *6*(1), 2188–2195. DOI: 10.31004/joe.v6i1.3221

Nyika, F., Muzekenyi, M., Akbar, K., Moodley, M., & Nzimande, S. (2024). Economic inclusion of rural small businesses in policy formulation: Strategies for sustainable development in Africa. *International Journal of Development and Sustainability, 13*(1), 51–67.

Obobisa, E. S., & Ahakwa, I. (2024). Stimulating the adoption of green technology innovation, clean energy resources, green finance, and environmental taxes: The way to achieve net zero CO2 emissions in Europe? *Technological Forecasting and Social Change, 205*, 123489. DOI: 10.1016/j.techfore.2024.123489

OECD. (2022). *Australia in Financing SMEs and Entrepreneurs 2022: An OECD Scoreboard*. OECD Publishing.

Ogbeibu, S., Emelifeonwu, J., Pereira, V., Oseghale, R., Gaskin, J., Sivarajah, U., & Gunasekaran, A. (2024). Demystifying the roles of organisational smart technology, artificial intelligence, robotics and algorithms capability: A strategy for green human resource management and environmental sustainability. *Business Strategy and the Environment, 33*(2), 369–388. DOI: 10.1002/bse.3495

Ogidi, O. I., & Akpan, U. M. (2022). Aquatic biodiversity loss: impacts of pollution and anthropogenic activities and strategies for conservation. In Chibueze Izah, S. (Ed.), *Biodiversity in Africa: Potentials, Threats and Conservation. Sustainable Development and Biodiversity* (Vol. 29). Springer., DOI: 10.1007/978-981-19-3326-4_16

Okeke, A. (2024). Evaluating sustainable practices and supply chain management effectiveness in African small and medium-sized enterprises (SMEs). *Journal of Sustainability Research, 6*(2), e240033.

Olawade, D. B., Wada, O. Z., David-Olawade, A. C., Fapohunda, O., Ige, A. O., & Ling, J. (2024). Artificial intelligence potential for net zero sustainability: Current evidence and prospects. *Next sustainability, 4*, 100041.

Olawade, D. B., Fapohunda, O., Wada, O. Z., Usman, S. O., Ige, A. O., Ajisafe, O., & Oladapo, B. I. (2024). *Smart waste management: A paradigm shift enabled by artificial intelligence*. Waste Management Bulletin.

Ollo-López, A., & Aramendia-Muneta, M. E. (2011). ICT impact on competitiveness, innovation and environment. *Telematics and Informatics, 29*(2), 204–210. DOI: 10.1016/j.tele.2011.08.002

Omowole, B. M., Olufemi-Philips, A. Q., Ofodili, O. C., Eyo-Udo, N. L., & Ewim, S. E.Bamidele Micheal OmowoleAmarachi Queen Olufemi-PhillipsOnyeka Chrisanctus OfodileNsisong Louis Eyo-UdoSomto Emmanuel Ewim. (2024). Conceptualizing green business practices in SMEs for sustainable development. *International Journal of Management & Entrepreneurship Research*, *6*(11), 3778–3805. DOI: 10.51594/ijmer.v6i11.1719

Omri, E., Chtourou, N., & Bazin, D. (2015). Solar thermal energy for sustainable development in Tunisia: The case of the PROSOL project. *Renewable & Sustainable Energy Reviews*, *41*, 1312–1323. DOI: 10.1016/j.rser.2014.09.023

Onu, P., & Mbohwa, C. (2021). Industry 4.0 opportunities in manufacturing SMEs: Sustainability outlook. *Materials Today: Proceedings*, *44*, 1925–1930. DOI: 10.1016/j.matpr.2020.12.095

Onwe, J. C., Bandyopadhyay, A., Hamid, I., Rej, S., & Hossain, E. (2023). Environment sustainability through energy transition and globalization in G7 countries: What role does environmental tax play? *Renewable Energy*, *218*, 119302. DOI: 10.1016/j.renene.2023.119302

Oppong, N. B. (2022). Sustainable development goals and small and medium enterprises: A comparative study of emerging economies and sub-Saharan Africa. *Global Business Review*, 1-14, 09721509221087848.

Oral, O., & Çakır, M. (2017). Internet of Things Concept and Creation of a Sample Prototype. *Mehmet Akif Ersoy University Journal of the Institute of Science*, *1*(Special Issue), 172–177.

Orhan, A. (2019). Industry 4.0 Revolution or Movement? *Iğdır University Social Sciences Journal*, (18), 557–576.

Osei, A. (2024). Investing for a greener future: Unpacking the dual impact of FDI on carbon emissions and the transformative role of renewable energy. *SN Business & Economics*, *5*(1), 14. DOI: 10.1007/s43546-024-00779-3

Oudghiri, M., Yamani, B., Benlemlih, N., El Aammouri, S., Abid, N., Brhadda, N., Bouhassoun, S., Ziri, R., Chriqui, A., Aoujil, F. Z., El Bakkali, M., Mouniane, Y., & Ibriz, M. (2025). Effect of Heavy Metals on the Morphological and Physiological Responses of the Torro Plus Variant of *Zea mays.Journal of Environment and Earth Science*, *7*(4), 165–179. DOI: 10.30564/jees.v7i4.8208

Özden, A. T. (2022). The World from 1.0 to 5.0: Web, Marketing, Industry and Society. *Journal of Business in The Digital Age*, *5*(1), 29–44.

Özdoğan, O. (2017). Industry 4.0: The fourth industrial revolution and the keys to industrial transformation. *Compass*, *327*, 1–26.

Özsoylu, A. F. (2017). Industry 4.0. *Çukurova University Faculty of Economics and Administrative Sciences Journal*, 21(1), 41-64.

Pakdemirli, B., Birişik, N., Aslan, İ., Sönmez, B., & Gezici, M. (2021). Use of Digital Technologies in Turkish Agriculture and Agriculture 4.0 in the Agricultural-Food Chain. *Soil Water Journal*, *10*(1), 78–87.

Parks, G. A. (1967). Aqueous surface chemistry of oxides and complex oxide minerals: Isoelectric point and zero point of charge.

Patton, M. Q. (2014). *Qualitative Research and Evaluation Methods* (Bütün, M., & Demir, S. B., Trans.). Pegem Academy.

Pellow, D. N. (2018). Wind energy and wildlife: An overview of the environmental impacts. *Environmental Research Letters*, *13*(3), 035003.

Pesaran, M. H. (2004). General diagnostic tests for cross section dependence in panels. Cambridge Working Papers in Economics, 0435, Faculty of Economics, University of Cambridge.

Pesaran, M. (2007). A simple panel unit root test in the presence of cross-section dependence. *Journal of Applied Econometrics*, *22*(2), 265–312. DOI: 10.1002/jae.951

Pesaran, M. H., Shin, Y., & Smith, R. P. (1999). Pooled mean group estimation of dynamic heterogeneous panels. *Journal of the American Statistical Association*, *94*(446), 621–634. DOI: 10.1080/01621459.1999.10474156

Pesaran, M. H., Ullah, A., & Yamagata, T. (2008). A bias-adjusted Lm test of error cross-section independence. *The Econometrics Journal*, *11*(1), 105–127. DOI: 10.1111/j.1368-423X.2007.00227.x

Pesaran, M. H., & Yamagata, T. (2008). Testing slope homogeneity in large panels. *Journal of Econometrics*, *142*(1), 50–93. DOI: 10.1016/j.jeconom.2007.05.010

Pham, C., Do, T., Doan, T., Nguyen Thi, X. H., & Pham, T. (2021). The impact of sustainability practices on financial performance: Empirical evidence from Sweden. *Cogent Business & Management*, *8*(1), 1912526. DOI: 10.1080/23311975.2021.1912526

Phuong, T. T. H. (2020). Gamified Learning: Are Vietnamese EFL Learners Ready Yet? *International Journal of Emerging Technologies in Learning*, *15*(24), 242. DOI: 10.3991/ijet.v15i24.16667

Pizzichini, L. (2024). Green tourism. In *Edward Elgar Publishing eBooks* (pp. 220–242). DOI: 10.4337/9781789904833.00023

Porter, M. E., & Kramer, M. R. (2011). Creating shared value. *Harvard Business Review, 89*(1/2), 62–77.

Porto, M. (2018). Intercultural Citizenship in Foreign Language Education: An Opportunity to Broaden CLIL's Theoretical Outlook and Pedagogy. *International Journal of Bilingual Education and Bilingualism*, •••, 1–21. DOI: 10.1080/13670050.2018.1526886

Pu, Z., Liu, J., & Yang, M. (2024). The effect of digital technology on residential and non-residential carbon emission. *International Review of Economics & Finance*, 95, 103495. DOI: 10.1016/j.iref.2024.103495

Qaim, M. (2020). The economics of genetically modified crops. *Annual Review of Resource Economics, 12*, 109–132.

Radanliev, P. (2024). Cyber diplomacy: defining the opportunities for cybersecurity and risks from Artificial Intelligence, IoT, Blockchains, and Quantum Computing. *Journal of Cyber Security Technology*, 1-51.

Radjabov, O. O., & Serrano, M. D. T. (2024). Analysis of tourism Development indicators and their Relationship with communication at Cultural Heritage Sites. HEDIT 2024 - International Congress for Heritage Digital Technologies and Tourism Management.

Rafique, M. Z., Fareed, Z., Ferraz, D., Ikram, M., & Huang, S. (2022). Exploring the heterogenous impacts of environmental taxes on environmental footprints: An empirical assessment from developed economies. *Energy*, 238(A), 121753. DOI: 10.1016/j.energy.2021.121753

Raheem, I., Tiwari, A., & Lorente, D. (2020). The role of ICT and financial development in CO2 emissions and economic growth. *Environmental Science and Pollution Research International, 27*(2), 1912–1922. Advance online publication. DOI: 10.1007/s11356-019-06590-0 PMID: 31760620

Rahim, H. U., Ali, W., Uddin, M., Ahmad, S., Khan, K., Bibi, H., & Alatalo, J. M. (2024). Abiotic stresses in soils, their effects on plants, and mitigation strategies: A literature review. *Chemistry and Ecology*. Advance online publication. DOI: 10.1080/02757540.2024.2439830

Raifu, I. A., Okunoye, I. A., & Aminu, A. (2024). The effect of ICT on financial sector development in Africa: Does regulatory quality matter? *Information Technology for Development, 30*(3), 424–451. DOI: 10.1080/02681102.2023.2233458

Raihan, A., Paul, N. A., Rahman, M. S., Islam, S., Paul, P., & Karmakar, S. (2024). Artificial Intelligence (AI) for Environmental Sustainability: A concise review of technology innovations in energy, transportation, biodiversity, and water management. *Journal of Technology Innovations and Energy*, *3*(2), 64–73. DOI: 10.56556/jtie.v3i2.953

Rane, N., Choudhary, S., & Rane, J. (2023). Sustainable Tourism Development Using Leading-edge Artificial Intelligence (AI), Blockchain, Internet of Things (IoT), Augmented Reality (AR) and Virtual Reality (VR) Technologies (October 31, 2023). Available at *SSRN*: https://ssrn.com/abstract=4642605 or DOI: 10.2139/ssrn.4642605

Rani, T., Amjad, M. A., Asghar, N., & Rehman, H. U. (2022). Revisiting the environmental impact of financial development on economic growth and carbon emissions: Evidence from South Asian economies. *Clean Technologies and Environmental Policy*, *24*(9), 2957–2965. DOI: 10.1007/s10098-022-02360-8 PMID: 35874941

Rauf, M., Kumar, L., Zulkifli, S. A., & Jamil, A. (2024). Aspects of artificial intelligence in future electric vehicle technology for sustainable environmental impact. *Environmental Challenges*, *14*, 100854. DOI: 10.1016/j.envc.2024.100854

Reem, M., Rasoolimanesh, S. M., & Wijesinghe Sara, N. R. (2022). Sustainability indicators in hotels: A systematic literature review. *Asia-Pacific Journal of Innovation in Hospitality and Tourism*, *11*(1), 149–178.

Regan, Á., Green, S., & Maher, P. (2018). Smart Farming in Ireland: Anticipating positive and negative impacts through a qualitative study of risk and benefit perceptions amongst expert actors in the Irish agri-food sector. In *Proceedings of the 13th European International Farm Systems Association Symposium,* Chania, Greece 1-5.

Regona, M., Yigitcanlar, T., Hon, C., & Teo, M. (2024). Artificial intelligence and sustainable development goals: Systematic literature review of the construction industry. *Sustainable Cities and Society*, *108*, 105499. DOI: 10.1016/j.scs.2024.105499

Reis, S. C., Linck, A. J. M., Figueiredo, M. F., & Pfeifer, D. L. (2023). Gamification Into the Design of the E-3d Online Course. *Frontiers in Education*, *8*, 1152999. Advance online publication. DOI: 10.3389/feduc.2023.1152999

Remeikiene, R., & Vebraite, V. (2021). The role of ICT development in boosting economic growth in transition economies. *Journal of International Students*, *14*(4).

REN21. (2022). *Renewables 2022 global status report*. Retrieved from https://www.ren21.net

Revell, A., Stokes, D., & Chen, H. (2009). Small businesses and the environment: Turning over a new leaf? *Business Strategy and the Environment, 19*(5), 273–288. DOI: 10.1002/bse.628

Ribeiro, J. P. S. (2024). *The role of artificial intelligence in achieving environmental sustainability for the health and technological industry* (Master's thesis).

Rockström, J., Steffen, W., Noone, K., Persson, Å., Chapin, F. S.III, Lambin, E., Lenton, T. M., Scheffer, M., Folke, C., Schellnhuber, H. J., Nykvist, B., de Wit, C. A., Hughes, T., van der Leeuw, S., Rodhe, H., Sörlin, S., Snyder, P. K., Costanza, R., Svedin, U., & Foley, J. A. (2009). A safe operating space for humanity. *Nature, 461*(7263), 472–475. DOI: 10.1038/461472a PMID: 19779433

Rodrik, D., & Stiglitz, J. (2024). A new growth strategy for developing nations. *Universidad Harvard [en línea]*https://drodrik. scholar. harvard. edu/publications/new-growth-strategy-developing-nations

Rogers, E. M. (2003). *Diffusion of innovations* (5th ed.). Free Press.

Rohde, F., Wagner, J., Meyer, A., Reinhard, P., Voss, M., Petschow, U., & Mollen, A. (2024). Broadening the perspective for sustainable artificial intelligence: Sustainability criteria and indicators for Artificial Intelligence systems. *Current Opinion in Environmental Sustainability, 66*, 101411. DOI: 10.1016/j.cosust.2023.101411

Rolando, B., & Mulyono, H. (2024). Managing Risks In Fintech: Applications And Challenges Of Artificial Intelligence-Based Risk Management. [ECBIS]. *Economics and Business Journal, 2*(3), 249–268. DOI: 10.47353/ecbis.v2i3.127

Rolnick, D., Donti, P. L., Kaack, L. H., Kochanski, K., Lacoste, A., Sankaran, K., Ross, A. S., Milojevic-Dupont, N., Jaques, N., Waldman-Brown, A., Luccioni, A. S., Maharaj, T., Sherwin, E. D., Mukkavilli, S. K., Kording, K. P., Gomes, C. P., Ng, A. Y., Hassabis, D., Platt, J. C., & Bengio, Y. (2019). Tackling climate change with machine learning. *ACM Computing Surveys, 55*(2), 1–36. DOI: 10.1145/3485128

Rubaj, P. (2023). Emerging markets as key drivers of the global economy. *European Research Studies, XXVI*(4), 431–445. DOI: 10.35808/ersj/3294

Russo-Tait, T. (2022). Science Faculty Conceptions of Equity and Their Association to Teaching Practices. *Science Education, 107*(2), 427–458. DOI: 10.1002/sce.21781

Ruyffelaert, A. (2022). Raising Concepts and Awareness of Sustainability and the Environment in Higher Education Through French Foreign Language Teaching: A Multidisciplinary Didactic Proposal. *Language Learning in Higher Education, 12*(2), 617–626. DOI: 10.1515/cercles-2022-2063

Saadaoui, H., & Chtourou, N. (2022). Do Institutional Quality, Financial Development, and Economic Growth Improve Renewable Energy Transition? Some Evidence from Tunisia. *Journal of the Knowledge Economy*, *14*(3), 2927–2958. DOI: 10.1007/s13132-022-00999-8

Saadaoui, H., & Omri, E. (2023). Towards a gradual transition to renewable energies in Tunisia: Do foreign direct investments and information and communication technologies matter? *Energy Nexus*, *12*, 100252. DOI: 10.1016/j.nexus.2023.100252

Saadaoui, H., Omri, E., & Chtourou, N. (2024). The transition to renewable energies in Tunisia: The asymmetric impacts of technological innovation, government stability, and democracy. *Energy*, *293*, 130686. DOI: 10.1016/j.energy.2024.130686

Sabancı, K., Ünlerşen, M. F., & Dilay, Y. (2016). Determination of Classification Parameters of Apple Varieties Grown in Karaman Region Using Image Processing Techniques. *Journal of Agricultural Machinery Science*, *12*(2), 133–139.

Sabari, N. A. S., & Hashim, H. (2023). Sustaining Education With Digital Storytelling in the English Language Teaching and Learning: A Systematic Review. *International Journal of Academic Research in Business & Social Sciences*, *13*(4). Advance online publication. DOI: 10.6007/IJARBSS/v13-i4/16860

Saberi, S., Kouhizadeh, M., Sarkis, J., & Shen, L. (2019). Blockchain technology and its relationships to sustainable supply chain management. *International Journal of Production Research*, *57*(7), 2117–2135. DOI: 10.1080/00207543.2018.1533261

Sabir, R. M., Mehmood, K., Sarwar, A., Safdar, M., Muhammad, N. E., Gul, N., Athar, F., Majeed, M. D., Sattar, J., Khan, Z., Fatima, M., Hussain, M. A., & Akram, H. M. B. (2024). Remote Sensing and Precision Agriculture: A Sustainable Future. In Kanga, S., Singh, S. K., Shevkani, K., Pathak, V., & Sajan, B. (Eds.), *Transforming Agricultural Management for a Sustainable Future. World Sustainability Series*. Springer., DOI: 10.1007/978-3-031-63430-7_4

Sahoo, M., & Sethi, N. (2021). The intermittent effects of renewable energy on ecological footprint: Evidence from developing countries. *Environmental Science and Pollution Research International*, *28*(40), 56401–56417. DOI: 10.1007/s11356-021-14600-3 PMID: 34053045

Salem, H. S., Pudza, M. Y., & Yihdego, Y. (2023). Harnessing the energy transition from total dependence on fossil to renewable energy in the Arabian Gulf region, considering population, climate change impacts, ecological and carbon footprints, and United Nations' Sustainable Development Goals. *Sustainable Earth Reviews*, *6*(1), 10. Advance online publication. DOI: 10.1186/s42055-023-00057-4

Sallam, M. A. (2025). Determinants of Economic Complexity: Evidence from BRICS countries, Egypt and Turkey using Panel ARDL Approach. *Faculty of Mediterranean Business Studies Tivat. Montenegro, 21*(2), 107–121.

Salman, M., Wang, G., Qin, L., & He, X. (2024). G20 roadmap for carbon neutrality: The role of Paris agreement, artificial intelligence, and energy transition in changing geopolitical landscape. *Journal of Environmental Management, 367*, 122080. DOI: 10.1016/j.jenvman.2024.122080 PMID: 39111003

Sankarasubramanian, P. (2025). Enhancing precision in agriculture: A smart predictive model for optimal sensor selection through IoT integration. *Smart Agricultural Technology, 10*, 100749. DOI: 10.1016/j.atech.2024.100749

Santos, R. C., & Martinho, J. L. (2020). An Industry 4.0 maturity model proposal. *Journal of Manufacturing Technology Management, 31*(5), 1023–1043. DOI: 10.1108/JMTM-09-2018-0284

Sarango-Lalangui, P., Álvarez-García, J., & del Río-Rama, M. D. L. C. (2018). Sustainable practices in small and medium-sized enterprises in Ecuador. *Sustainability (Basel), 10*(6), 2105. DOI: 10.3390/su10062105

Satornino, C. B., Du, S., & Grewal, D. (2024). Using artificial intelligence to advance sustainable development in industrial markets: A complex adaptive systems perspective. *Industrial Marketing Management, 116*, 145–157. DOI: 10.1016/j.indmarman.2023.11.011

Saunders, M. Lewis, P., & Thornhill, A. (2012). Research methods for business students. Published by Pearson Education, prentice Hall New York USA.

Saunders, B., Sim, J., Kingstone, T., Baker, S., Waterfield, J., Bartlam, B., Burroughs, H., & Jinks, C. (2018). Saturation in qualitative research: Exploring its conceptualization and operationalization. *Quality & Quantity, 52*(4), 1893–1907. DOI: 10.1007/s11135-017-0574-8 PMID: 29937585

Sawal, J. N., & Anjum, R. (2023). BRICS the Major Emerging Economies: Prospects for a Multi-Polar World. *Global International Relations Review, VI*(II), 72–82. DOI: 10.31703/girr.2023(VI-II).08

Saxena, R., Joshi, A., Joshi, S., Borkotoky, S., Singh, K., Rai, P. K., Mueed, Z., & Sharma, R. (2023). The role of artificial intelligence strategies to mitigate abiotic stress and climate change in crop production. In *Visualization Techniques for Climate Change with Machine Learning and Artificial Intelligence* (pp. 273–293). Elsevier., DOI: 10.1016/B978-0-323-99714-0.00006-6

Scott, W. (2004). Institutional carriers: Reviewing modes of transporting ideas over time and space and considering their consequences. *Industrial and Corporate Change*, *12*(4), 879–894. DOI: 10.1093/icc/12.4.879

Sekarini, S., & Selvabaskar, S. (2024). AI-Powered Branding: Enhancing Consumer Experience in Emerging Markets. In *Integrating AI-Driven Technologies into Service Marketing* (pp. 19–48). IGI Global. DOI: 10.4018/979-8-3693-7122-0.ch002

Shad, M. K., Lai, F. W., Fatt, C. L., Klemeš, J. J., & Bokhari, A. (2019). Integrating sustainability reporting into enterprise risk management and its relationship with business performance: A conceptual framework. *Journal of Cleaner Production*, *208*, 415–425. DOI: 10.1016/j.jclepro.2018.10.120

Shafi, U., Mumtaz, R., García-Nieto, J., Hassan, S. A., Zaidi, S. A. R., & Iqbal, N. (2019). Precision Agriculture Techniques and Practices: From Considerations to Applications. *Sensors (Basel)*, *19*(17), 3796. DOI: 10.3390/s19173796 PMID: 31480709

Shahbaz, M., Shahzad, S. J. H., & Mahalik, M. K. (2017). Is globalization detrimental to CO2 emissions in Japan? New threshold analysis. *Environmental Modeling and Assessment*. Advance online publication. DOI: 10.1007/s10666-017-9584-0

Shahhosseini, M., Hu, G., Khaki, S., & Archontoulis, S. V. (2021). Corn Yield Prediction With Ensemble CNN-DNN. *Frontiers in Plant Science*, *12*, 709008. DOI: 10.3389/fpls.2021.709008 PMID: 34408763

Shah, I. H., Manzoor, M. A., Jinhui, W., Li, X., Hameed, M. K., Rehaman, A., Li, P., Zhang, Y., Niu, Q., & Chang, L. (2024). Comprehensive review: Effects of climate change and greenhouse gases emission relevance to environmental stress on horticultural crops and management. *Journal of Environmental Management*, *351*, 119978. DOI: 10.1016/j.jenvman.2023.119978 PMID: 38169258

Shahzad, U., Fareed, Z., Shahzad, F., & Shahzad, K. (2021). Investigating the nexus between economic complexity, energy consumption and ecological footprint for the United States: New insights from quantile methods. *Journal of Cleaner Production*, *279*, 123806. DOI: 10.1016/j.jclepro.2020.123806

Shanmugavel, D., Rusyn, I., Solorza-Feria, O., & Kamaraj, S. K. (2023). Sustainable SMART fertilizers in agriculture systems: A review on fundamentals to in-field applications. *The Science of the Total Environment*, *904*, 166729. DOI: 10.1016/j.scitotenv.2023.166729 PMID: 37678530

Shao, M., Fan, J., Huang, Z., & Chen, M. (2022). The Impact of Information and Communication Technologies (ICTs) on Health Outcomes: A Mediating Effect Analysis Based on Cross-National Panel Data. *Journal of Environmental and Public Health, 2022*(1), 2225723. DOI: 10.1155/2022/2225723 PMID: 35990542

Sharma, K., Sharma, C., Sharma, S., & Asenso, E. (2022). Broadening the Research Pathways in Smart Agriculture: Predictive Analysis Using Semiautomatic Information Modeling. *Journal of Sensors, 2022*, 1–19. DOI: 10.1155/2022/5442865

Sharma, N., Bhattacharjee, S., Garg, R. D., Sharma, K., & Salim, M. (2024). Sustainable management and agriculture resource technology system using remote sensing descriptors and IoT. *Geomatica., 76*(2), 100040. DOI: 10.1016/j.geomat.2024.100040

Shuford, J. (2024). Interdisciplinary perspectives: fusing artificial intelligence with environmental science for sustainable solutions. *Journal of Artificial Intelligence General science (JAIGS) ISSN: 3006-4023, 1*(1), 106-123.

Shykun, A. (2023). Content and Language Integrated Learning: Enhancing Language Acquisition and Content Understanding. *International Science Journal of Education & Linguistics, 2*(4), 39–44. DOI: 10.46299/j.isjel.20230204.05

Sidek, S., Rosli, M. M., Khadri, N. A. M., Hasbolah, H., Manshar, M., & Abidin, N. M. F. N. Z. (2020). Fortifying small business performance sustainability in the era of Ir 4.0: E-marketing as a catalyst of competitive advantages and business performance. *Journal of Critical Reviews, 7*(13), 2143–2155.

Sinha, M., Fukey, L. N., & Sinha, A. (2021). Artificial Intelligence and Internet of Things readiness: inclination for hotels to support a sustainable environment. In *Elsevier eBooks* (pp. 327–353). DOI: 10.1016/B978-0-323-85769-7.00015-X

Smidt, H. J., & Jokonya, O. (2022). Factors affecting digital technology adoption by small-scale farmers in agriculture value chains (AVCs) in South Africa. *Information Technology for Development, 28*(3), 558–584. DOI: 10.1080/02681102.2021.1975256

Smith, H., Discetti, R., Bellucci, M., & Acuti, D. (2022). SMEs engagement with the Sustainable Development Goals: A power perspective. *Journal of Business Research, 149*, 112–122. DOI: 10.1016/j.jbusres.2022.05.021

Sommer, C. (2017). *Drivers and constraints for adopting sustainability standards in small and medium-sized enterprises (SMEs)* (No. 21/2017). Discussion Paper.

Somoye, O. A., Ozdeser, H., Seraj, M., & Turuc, F. (2023). The determinants of CO2 emissions in Brazil: The application of the STIRPAT model. *Energy Sources. Part A, Recovery, Utilization, and Environmental Effects, 45*(4), 10843–10854. DOI: 10.1080/15567036.2023.2251921

Song, J., & Tang, M. (2020). Ecological Discourse Analysis From the Perspective of Systemic Functional Linguistics. *Destech Transactions on Social Science Education and Human Science, icesd*. DOI: 10.12783/dtssehs/icesd2020/34131

Song, Y., Wang, C., Linderholm, H. W., Fu, Y., Cai, W., Xu, J., Zhuang, L., Wu, M., Shi, Y., Wang, G., & Chen, D. (2022). The negative impact of increasing temperatures on rice yields in southern China. *The Science of the Total Environment, 820*, 153262. DOI: 10.1016/j.scitotenv.2022.153262 PMID: 35065105

Sovacool, B. K., Schwanen, T., & Sorrell, S. (2019). Transforming energy systems: A critical analysis of global clean energy collaboration. *Global Environmental Change*, *59*, 101987.

Statista (2024). Information technology (IT) worldwide spending from 2005 to 2024. Accessed from https://www.statista.com/statistics/203935/overall-it-spending-worldwide/

Streimikiene, D., & Kyriakopoulos, G. L. (2025). Climate Change Mitigation in Visegrad Countries. *Faculty of Mediterranean Business Studies Tivat. Montenegro*, *21*(2), 233–246.

Su, F., Zou, D., Xie, H., & Wang, F. L. (2021). A Comparative Review of Mobile and Non-Mobile Games for Language Learning. *SAGE Open*, *11*(4), 21582440211067247. Advance online publication. DOI: 10.1177/21582440211067247

Sun, Y. (2023). Income inequality, carbon emissions, and green development efficiency. *Environmental Science and Pollution Research International*, *30*(8), 21081–21091. DOI: 10.1007/s11356-022-23583-8 PMID: 36264465

Sunyani Municipal Assembly. (2019). Medium-Term Development Plan 2018-2021 Sunyani, Ghana.

Surji, A. L., & Ulker, V. (2021). Undergraduate Foreign Language Learners Perspective on Project Based Learning. *International Journal of Social Sciences and Educational Studies*, *8*(4). Advance online publication. DOI: 10.23918/ijsses.v8i4p254

Sushma, Y., Lakshmi, C. H. J., Rajesh, K., Hemanth, V., & Sowmyarao, V. 2024. IoT Based Soil Nutrient Monitoring and Analysis System. In: *International Conference on IoT Based Control Networks and Intelligent Systems (ICICNIS)*, 348-353. IEEE. DOI: 10.1109/ICICNIS64247.2024.10823219

Syathroh, I. L., Kareviati, E., Lestari, A. P., & Fitria, N. (2021). Exploring the Potentials of Technology Integration for Teaching Language Skills: A Literature Review. [Professional Journal of English Education]. *Project*, *4*(3), 488. DOI: 10.22460/project.v4i3.p488-496

Sy, M. (2016). Impact of sustainability practices on the firms' performance. *Asia Pac. Bus. Econ. Perspect*, *4*, 4–21.

Tagnin, L., & Ríordáin, M. N. (2021). Building Science Through Questions in Content and Language Integrated Learning (CLIL) Classrooms. *International Journal of STEM Education*, *8*(1), 34. Advance online publication. DOI: 10.1186/s40594-021-00293-0

Tanış, H., Akçay, A., Yilmaz, N., Yiğit, M. F., & Tüzün, H. (2022). How Usable Is Coursera? A Usability Analysis Through Eye-Tracking and Authentic Tasks. *Participatory Educational Research*, *9*(4), 379–395. DOI: 10.17275/per.22.96.9.4

Tariq, M. U. (2024). Leveraging artificial intelligence for a sustainable and climate-neutral economy in Asia. In *Strengthening sustainable digitalization of Asian economy and society* (pp. 1–21). IGI Global. DOI: 10.4018/979-8-3693-1942-0.ch001

Tariq, M. U. (2025). Empowering Learning Through Networked and Connected Education: Case Studies in Digital Engagement. In Gierhart, A. (Ed.), *Cases on Enhancing P-16 Student Engagement With Digital Technologies* (pp. 169–198). IGI Global Scientific Publishing., DOI: 10.4018/979-8-3693-5633-3.ch007

Tariq, M. U. (2025). Harnessing Emotional Intelligence for Strategic Leadership: A Contemporary Approach. In Sayed, Z. (Ed.), *Emotionally Intelligent Methods for Meaningful Leadership* (pp. 197–222). IGI Global Scientific Publishing., DOI: 10.4018/979-8-3693-7372-9.ch009

Tariq, M. U. (2025). Sustainability of Quality Processes in Higher Education: Strategies for Continuous Improvement. In Saeed, S., Rana, O., & Dhanaraj, R. (Eds.), *Higher Education and Quality Assurance Practices* (pp. 305–334). IGI Global Scientific Publishing., DOI: 10.4018/979-8-3693-6765-0.ch011

Tariq, M. U., & Sergio, R. P. (2025). Nurturing Digital Citizenship in Society 5.0 Through AI and Computational Intelligence Education. In Pandey, R., Srivastava, N., Prasad, R., Prasad, J., & Garcia, M. (Eds.), *Open AI and Computational Intelligence for Society 5.0* (pp. 59–84). IGI Global Scientific Publishing., DOI: 10.4018/979-8-3693-4326-5.ch003

Tedeneke, A. (2023). Exploring the Practices and Contextual Factors in Teaching Writing Skills in EFL Classrooms: A Case Study. *Erjssh*, *9*(2), 83–101. DOI: 10.4314/erjssh.v9i2.6

Terzi, H., & Pata, U. (2020). Is the pollution haven hypothesis (PHH) valid for Turkey? *Panoeconomicus*, *67*(1), 93–109. DOI: 10.2298/PAN161229016T

Thomas, R., O'Hare, G., & Coyle, D. (2023). Understanding technology acceptance in smart agriculture: A systematic review of empirical research in crop production. *Technological Forecasting and Social Change*, *189*, 122374. DOI: 10.1016/j.techfore.2023.122374

Thorlakson, T., De Zegher, J. F., & Lambin, E. F. (2018). Companies' contribution to sustainability through global supply chains. *Proceedings of the National Academy of Sciences of the United States of America*, *115*(9), 2072–2077. DOI: 10.1073/pnas.1716695115 PMID: 29440420

Tiago, F., Gil, A., Stemberger, S., & Borges-Tiago, T. (2020). Digital sustainability communication in tourism. *Journal of Innovation & Knowledge*, *6*(1), 27–34. DOI: 10.1016/j.jik.2019.12.002

Tilman, D., Cassman, K. G., Matson, P. A., Carpenter, S. R., & Dixon, J. (2011). Agricultural sustainability and intensive production practices. *Nature*, *418*(6898), 671–677. DOI: 10.1038/nature01014 PMID: 12167873

Tomšič, N., Bojnec, Š., & Simčič, B. (2015). Corporate sustainability and economic performance in small and medium sized enterprises. *Journal of Cleaner Production*, *108*, 603–612. DOI: 10.1016/j.jclepro.2015.08.106

Tong, H., & Nikoloski, Z. (2021). Machine learning approaches for crop improvement: Leveraging phenotypic and genotypic big data. *Journal of Plant Physiology*, *257*, 153354. DOI: 10.1016/j.jplph.2020.153354 PMID: 33385619

Toromade, A. S., & Chiekezie, N. R.Adekunle Stephen ToromadeNjideka Rita Chiekezie. (2024). Driving sustainable business practices in SMEs: Innovative approaches for environmental and economic synergy. *Int. J. Manag. Entrep. Res*, *6*(8), 2637–2647. DOI: 10.51594/ijmer.v6i8.1411

Trabelsi, E. (2024). Transition to sustainable environment and economic growth in Tunisia: An ARDL approach. *World Development Sustainability*, *4*, 100165. DOI: 10.1016/j.wds.2024.100165

Tran, T. T. D., & Smith, A. D. (2017). Evaluation of renewable energy technologies and their potential for technical integration and cost-effective use within the U.S. energy sector. https://ideas.repec.org/a/eee/rensus/v80y2017icp1372-1388.html

Trivelli, L., Apicella, A., Chiarello, F., Rana, R., Fantoni, G., & Tarabella, A. (2019). From precision agriculture to Industry 4.0: Unveiling technological connections in the agrifood sector. *British Food Journal*, *121*(8), 1730–1743. DOI: 10.1108/BFJ-11-2018-0747

Türker, U., Akdemir, B., Topakcı, M., Tekin, B., Aydın, İ. Ü. A., Özoğul, Ö., & Evrenosoğlu, M. (2015). Developments in Precision Agriculture Technologies, *Proceedings Book of the 8th Turkish Agricultural Engineering Technical Congress-1*, 295

Türkseven, S., Kızmaz, M. Z., Tekin, A. B., Urkan, E., & Serim, A. T. (2016). Digital transformation in agriculture; use of unmanned aerial vehicles. *Journal of Agricultural Machinery Science*, *12*(4), 267–271.

Tussyadiah, I. (2020). a. A review of research into automation in tourism: Launching the annals of tourism research curated collection on artificial intelligence and robotics in tourism. *Annals of Tourism Research*, *81*, 102883. DOI: 10.1016/j.annals.2020.102883

Tussyadiah, I. P., Zach, F. J., & Wang, J. (2020). b. Do travelers trust intelligent service robots? *Annals of Tourism Research*, *81*, 102886. DOI: 10.1016/j.annals.2020.102886

Tyler, B. B., Lahneman, B., Cerrato, D., Cruz, A. D., Beukel, K., Spielmann, N., & Minciullo, M. (2024). Environmental practice adoption in SMEs: The effects of firm proactive orientation and regulatory pressure. *Journal of Small Business Management*, *62*(5), 2211–2246. DOI: 10.1080/00472778.2023.2218435

Ueda, D., Walston, S. L., Fujita, S., Fushimi, Y., Tsuboyama, T., Kamagata, K., Yamada, A., Yanagawa, M., Ito, R., Fujima, N., Kawamura, M., Nakaura, T., Matsui, Y., Tatsugami, F., Fujioka, T., Nozaki, T., Hirata, K., & Naganawa, S. (2024). Climate change and artificial intelligence in healthcare: Review and recommendations towards a sustainable future. *Diagnostic and Interventional Imaging*, *105*(11), 453–459. DOI: 10.1016/j.diii.2024.06.002 PMID: 38918123

Ukoba, K., Olatunji, K. O., Adeoye, E., Jen, T., & Madyira, D. M. (2024). Optimizing renewable energy systems through artificial intelligence: Review and future prospects. *Energy & Environment*, *35*(7), 3833–3879. DOI: 10.1177/0958305X241256293

Ullah, A. S. (2020). What is knowledge in Industry 4.0? *Engineering Reports*, *2*(8), 1–21. DOI: 10.1002/eng2.12217

Ullah, S., Nadeem, M., Ali, K., & Abbas, Q. (2022). Fossil fuel, industrial growth and inward FDI impact on CO2 emissions in Vietnam: Testing the EKC hypothesis. *Management of Environmental Quality*, *33*(2), 222–240. DOI: 10.1108/MEQ-03-2021-0051

Ulucak, R., Danish, , & Khan, S. U.-D. (2020). Does information and communication technology affect CO2 mitigation under the pathway of sustainable development during the mode of globalization? *Sustainable Development (Bradford)*, *28*(4), 857–867. Advance online publication. DOI: 10.1002/sd.2041

United Nations (UN). (2021). *The Sustainable Development Goals report 2021.* United Nations Publications.

United Nations Environment Programme (UNEP). (2020). *Emissions gap report 2020.* UNEP.

Upadhyay, A., Mukhuty, S., Kumar, V., & Kazancoglu, Y. (2021). Blockchain technology and the circular economy: Implications for sustainability and social responsibility. *Technological Forecasting and Social Change, 168*, 120748.

Üstündağ, F. (2022). *Transformation of Agricultural Production and Smart Agriculture (Published master's thesis).* Aydın Adnan Menderes University, Institute of Social Sciences, Aydın, Türkiye.

Uzun, Y., Bilban, M., & Arıkan, H. (2018). Use of Artificial Intelligence in Agriculture and Rural Development, *VI.International KOP Regional Development Symposium, 26*, 27. 1-6

Văidean, V. L., & Achim, M. V. (2022). When more is less: Do information and communication technologies (ICTs) improve health outcomes? An empirical investigation in a non-linear framework. *Socio-Economic Planning Sciences, 80*, 101218. DOI: 10.1016/j.seps.2021.101218

Van der Meer, D., & Tiwari, A. (2017). Smart technologies for water conservation: Opportunities and challenges. *Environmental Management, 60*(5), 775–785.

Van Marrewijk, M., & Werre, M. (2003). Multiple levels of corporate sustainability. *Journal of Business Ethics, 44*(2), 107–119. DOI: 10.1023/A:1023383229086

Veleva, V. (2021). The role of entrepreneurs in advancing sustainable lifestyles: Challenges, impacts, and future opportunities. *Journal of Cleaner Production, 283*, 124658. DOI: 10.1016/j.jclepro.2020.124658 PMID: 33078048

Voigtlaender, S., Pawelczyk, J., Geiger, M., Vaios, E. J., Karschnia, P., Cudkowicz, M., Dietrich, J., Haraldsen, I. R. J. H., Feigin, V., Owolabi, M., White, T. L., Świeboda, P., Farahany, N., Natarajan, V., & Winter, S. F. (2024). Artificial intelligence in neurology: Opportunities, challenges, and policy implications. *Journal of Neurology, 271*(5), 2258–2273. DOI: 10.1007/s00415-024-12220-8 PMID: 38367046

Wach, K. (2015). Entrepreneurial orientation and business internationalisation process: The theoretical foundations of international entrepreneurship. *Entrepreneurial Business and Economics Review, 3*(2), 9–24. DOI: 10.15678/EBER.2015.030202

Wahab, S., Imran, M., Ahmed, B., Rahim, S., & Hassan, T. (2024). Navigating environmental concerns: Unveiling the role of economic growth, trade, resources and institutional quality on greenhouse gas emissions in OECD countries. *Journal of Cleaner Production*, *434*, 139851. DOI: 10.1016/j.jclepro.2023.139851

Wang, J., Xie, C., Huang, Q., & Morrison, A. M. (2020). Smart tourism destination experiences: The mediating impact of arousal levels. *Tourism Management Perspectives*, *35*, 100707. DOI: 10.1016/j.tmp.2020.100707

Wang, Q., Li, Y., & Li, R. (2024). Ecological footprints, carbon emissions, and energy transitions: The impact of artificial intelligence (AI). *Humanities & Social Sciences Communications*, *11*(1), 1–18. DOI: 10.1057/s41599-024-03520-5

Wang, R., Mirza, N., Vasbieva, D. G., Abbas, Q., & Xiong, D. (2020). The nexus of carbon emissions, financial development, renewable energy consumption, and technological innovation: What should be the priorities in light of COP 21 Agreements? *Journal of Environmental Management*, *271*, 111027. DOI: 10.1016/j.jenvman.2020.111027 PMID: 32778307

Wang, T., Jin, H., & Sieverding, H. L. (2023). Factors affecting farmer perceived challenges towards precision agriculture. *Precision Agriculture*, *24*(6), 2456–2478. DOI: 10.1007/s11119-023-10048-2

Wang, Y., Derakhshan, A., & Pan, Z. (2022). Positioning an Agenda on a Loving Pedagogy in Second Language Acquisition: Conceptualization, Practice, and Research. *Frontiers in Psychology*, *13*, 894190. Advance online publication. DOI: 10.3389/fpsyg.2022.894190 PMID: 35668974

Wang, Y., Liao, M., Xu, L., & Malik, A. (2021). The impact of foreign direct investment on China's carbon emissions through energy intensity and emissions trading system. *Energy Economics*, *97*, 105212. DOI: 10.1016/j.eneco.2021.105212

Wassell, B. A., Glynn, C., Carey, B., Sevinç, E., & Baroudi, F. (2023). Equity, Access, and Inclusion in K-12 World Language Education: A System of Failure or Work in Progress? *Journal of Linguistics and Language Teaching*, *15*(1). Advance online publication. DOI: 10.5070/L215159614

Waters, J. (2008). *Sustainable development*. The International Encyclopedia of Communication., DOI: 10.1002/9781405186407.wbiecs122

Webb, K. J., Hodge, T. G., & Thompson, J. H. (2012). Small business sustainability: What is the CPA's role? *International Journal of Business and Social Justice*, *3*(12), 1–7.

Wei, G., Bi, M., Liu, X., Zhang, Z., & He, B. J. (2023). Investigating the impact of multi-dimensional urbanization and FDI on carbon emissions in the belt and road initiative region: Direct and spillover effects. *Journal of Cleaner Production*, *384*, 135608. DOI: 10.1016/j.jclepro.2022.135608

Wei, L. (2023). Artificial Intelligence in Language Instruction: Impact on English Learning Achievement, L2 Motivation, and Self-Regulated Learning. *Frontiers in Psychology*, *14*, 1261955. Advance online publication. DOI: 10.3389/fpsyg.2023.1261955 PMID: 38023040

Wei, Y., Zhang, H., & Zhao, Z. (2024). Digitalization and pollution: Evidence from South Africa. *China Economic Review*, *102266*, 102266. Advance online publication. DOI: 10.1016/j.chieco.2024.102266

Westerlund, J. (2007). Testing for error correction in panel data. *Oxford Bulletin of Economics and Statistics*, *69*(6), 709–748. DOI: 10.1111/j.1468-0084.2007.00477.x

WHO. (2018). Ambient air pollution: A global assessment of exposure and burden of disease. World Health Organization. https://www.who.int/news-room/fact-sheets/detail/ambient-air-pollution

WHO. (2024). Electronic waste (e-waste). Accessed from Information technology (IT) worldwide spending from 2005 to 2024. Accessed from https://www.who.int/news-room/fact-sheets/detail/electronic-waste-(e-waste)

Wilhelm, M. M., Blome, C., Bhakoo, V., & Paulraj, A. (2016). Sustainability in multi-tier supply chains: Understanding the double agency role of the first-tier supplier. *Journal of Operations Management*, *41*(1), 42–60. DOI: 10.1016/j.jom.2015.11.001

Wolfert, S., Ge, L., Verdouw, C., & Bogaardt, M. J. (2017). Big data in smart farming – A review. *Agricultural Systems*, *153*, 69–80. DOI: 10.1016/j.agsy.2017.01.023

World Bank. (2024). *Measuring the Emissions and Energy Footprint of the ICT Sector: Implications for Climate Action.* Washington, D.C. Accessed https://documents.worldbank.org/curated/en/099121223165540890/P17859712a98880541a4b71d57876048abb

World Bank. (2024a). World Development Indicators. Accessed from https://databank.worldbank.org/source/world-development-indicators

World Health Organization (WHO). (2020). *Air pollution and child health: Prescribing clean air.*

Wörsdörfer, M. (2024). Mitigating the adverse effects of AI with the European Union's artificial intelligence act: Hype or hope? *Global Business and Organizational Excellence, 43*(3), 106–126. DOI: 10.1002/joe.22238

Xie, J., Ahmed, Z., Zhang, P., Khan, S., & Alvarado, R. (2023). Financial expansion and CO2 mitigation in top twenty emitters: Investigating the direct and moderating effects of the digital economy. *Gondwana Research, 125*, 1–14. DOI: 10.1016/j.gr.2023.07.013

Xie, P., & Jamaani, F. (2022). Does green innovation, energy productivity and environmental taxes limit carbon emissions in developed economies: Implications for sustainable development. *Structural Change and Economic Dynamics, 63*, 66–78. DOI: 10.1016/j.strueco.2022.09.002

Xing, C., & Ding, Y. (2019). Research on urban eco-environmental problems and sustainable development. *IOP Conference Series. Earth and Environmental Science, 300*(3), 32103. DOI: 10.1088/1755-1315/300/3/032103

Xing, L., Khan, Y. A., Arshed, N., & Iqbal, M. (2023). Investigating the impact of economic growth on environment degradation in developing economies through STIRPAT model approach. *Renewable & Sustainable Energy Reviews, 182*, 113365. DOI: 10.1016/j.rser.2023.113365

Xulu-Gama, N., & Hadebe, S. (2022). Language of Instruction: A Critical Aspect of Epistemological Access to Higher Education in South Africa. *South African Journal of Higher Education, 36*(5). Advance online publication. DOI: 10.20853/36-5-4788

Yadav, M. P., Kumar, S., & Mukherjee, D. (2023). Do green bonds offer a diversification opportunity during COVID-19? Empirical evidence from energy, crypto, and carbon markets. *Environmental Science and Pollution Research International, 30*(1), 7625–7639.

Yakymchuk, A., & Rataj, M. (2024). Development of the green policy strategies of enterprises: A decent work. *Ecology & Environment, 89*(2), 752–752.

Yaman, H., Sungur, O., & Dulupçu, M. A. (2021). Transformation of agriculture and animal husbandry in the world: Technology-based applications and revolutions. *Journal of Agricultural Economics, 27*(2), 123–133.

Yang, L., Karanyathikul, A., & Kotchasit, S. (2024). Effect of Gamification Combined With Mobile Learning on Chinese Language Speaking Skills of Second-Year Foreign Students at Guangdong University of Foreign Studies. *Ijsasr, 4*(6), 563–578. DOI: 10.60027/ijsasr.2024.5054

Yassin, M. M., Abdallah, A. A. J., Altahtamouni, F., Alsmadi, A., & Al-Okaily, M. A. N. A. F. (2025). Achieving Sustainable Excellence: Corporate Governance, Environmental Performance, Innovation, and Financial Success. Faculty of Mediterranean Business Studies Tivat. *Montenegro*, *21*(2), 123–147.

Yilanci, V., & Pata, U. K. (2020). Investigating the EKC hypothesis for China: The role of economic complexity on ecological footprint. *Environmental Science and Pollution Research International*, *27*(26), 32683–32694. DOI: 10.1007/s11356-020-09434-4 PMID: 32514916

Yılmaz, M., & Arslantaş, C. (2022). Industry 4.0 awareness in enterprises and its effects on workforce: Multiple case study on managers' opinions. *Ardahan University Journal of Faculty of Economics and Administrative Sciences*, *4*(2), 139–147.

Yin, Y., Stecke, K. E., & Li, D. (2018). The Evolution of Production Systems from Industry 2.0 Through Industry 4.0. *International Journal of Production Research*, *56*(1–2), 848–861. DOI: 10.1080/00207543.2017.1403664

Yuan, Q., Wang, R., Tang, H., Ma, X., & Zeng, X. (2024). A study on the potential of higher education in reducing carbon intensity. *PLoS One*, *19*(11), e0309546. DOI: 10.1371/journal.pone.0309546 PMID: 39591416

Yue, L., Yan, H., & Han, L. (2024). The carbon emission reduction effect of two-way foreign investment from the perspective of government environmental attention. *Environment, Development and Sustainability*, •••, 1–26. DOI: 10.1007/s10668-024-05679-y

Yüksekbilgili, Z., & Çevik, G. (2018). A current and future-oriented analysis of Turkey's place in the context of Industry 4.0. *Journal of Finance. Economics and Social Research*, *3*(2), 422–436.

Yuldoshboy, S., Karimov, M., & Kuralbaev, J. (2022). The association between co2 and economic growth in central Asian countries: Panel data approach. *Journal of Positive School Psychology*, •••, 5587–5601.

Yu, S., Zhang, Q., Hao, J. L., Ma, W., Sun, Y., Wang, X., & Song, Y. (2023). Development of an extended STIRPAT model to assess the driving factors of household carbon dioxide emissions in China. *Journal of Environmental Management*, *325*, 116502. DOI: 10.1016/j.jenvman.2022.116502 PMID: 36274310

Zafar, M. W., Saud, S., & Hou, F. (2019). The impact of globalization and financial development on environmental quality: Evidence from selected countries in the Organization for Economic Co-operation and Development (OECD). *Environmental Science and Pollution Research International, 26*(13), 13246–13262. DOI: 10.1007/s11356-019-04761-7 PMID: 30900127

Zaidi, S. A., Zafar, M. W., Shahbaz, M., & Hou, F. (2019). Dynamic linkages between globalization, financial development and carbon emissions: Evidence from Asia Pacific Economic Cooperation countries. *Journal of Cleaner Production, 228*, 533–543. Advance online publication. DOI: 10.1016/j.jclepro.2019.04.210

Zandalinas, S. I., & Mittler, R. (2021). Plant responses to multifactorial stress combination. *The New Phytologist, 234*(4), 1161–1167. DOI: 10.1111/nph.18087 PMID: 35278228

Zandalinas, S. I., Sengupta, S., Fritschi, F. B., Azad, R. K., Nechushtai, R., & Mittler, R. (2021). The Impact of Multifactorial Stress Combination on Plant Growth and Survival. *The New Phytologist, 230*(3), 1034–1048. DOI: 10.1111/nph.17232 PMID: 33496342

Zavrazhnyi, K. Y., Kulyk, A. K., Voronenko, V. I., Sokolov, M. A., & Antunes de Abreu, O. (2024). Formation of strategic directions for the use of artificial intelligence in the enterprise to achieve the goals of sustainable development.

Zeghlouli, J., Christophe, G., Guendouz, A., Modafar, C. E., Belkamel, A., Michaud, P., & Delattre, C. (2021). Optimization of Bioethanol Production from Enzymatic Treatment of Argan Pulp Feedstock. In Molecules (Vol. 26, Issue 9, p. 2516). Multidisciplinary Digital Publishing Institute. DOI: 10.3390/molecules26092516

Zengler, T. (2020). Tesla's innovation in sustainable technology. *Journal of Sustainable Business, 32*(1), 42–50.

Zhang, S., & Hasim, Z. (2023). Gamification in EFL/ESL Instruction: A Systematic Review of Empirical Research. *Frontiers in Psychology, 13*, 1030790. Advance online publication. DOI: 10.3389/fpsyg.2022.1030790 PMID: 36687912

Zhang, X., Ibrahim, Z., Khaskheli, M. B., Raza, H., Zhou, F., & Shamsi, I. H. (2024). Integrative Approaches to Abiotic Stress Management in Crops: Combining Bioinformatics Educational Tools and Artificial Intelligence Applications. *Sustainability (Basel), 16*(17), 7651. DOI: 10.3390/su16177651

Zhang, Y., Ren, S., Liu, Y., & Si, S. (2019). A big data analytics architecture for cleaner manufacturing and maintenance processes of complex products. *Journal of Cleaner Production, 228*, 451–467.

Zhang, Y., Wang, L., & Zhang, Z. (2020). Real-time air quality monitoring using IoT and cloud computing. *Sensors (Basel), 20*(24), 6964.

Zhao, L., Zhang, Y., & Zhang, H. (2022). Research on the Impact of Digital Literacy on Farmer Households' Green Cooking Energy Consumption: Evidence From Rural China. *International Journal of Environmental Research and Public Health, 19*(20), 13464. DOI: 10.3390/ijerph192013464 PMID: 36294039

Zhu, J. Z., & Cheung, K. W. (2012). Summary of environment impact of renewable energy resources. *Advanced Materials Research, 1133*, 1133–1136. Advance online publication. . DOI: 10.4028/www.scientific.net/AMR.616-618.1133

Ziaul, I., & Shuwei, W. (2023). Environmental sustainability: A major component of sustainable development. *International Journal of Environmental Sustainability and Social Science, 4*(2), 620–627. DOI: 10.38142/ijesss.v4i2.296

About the Contributors

Ali Ahmadi is currently an Assist Prof at Accounting at the university of Gafsa, Tunisia with an academic experience of 9 years. He holds a master's degree in accounting. He earned his Doctor of Philosophy in financial and accounting methods from Sfax University, Tunisia. He has published research papers in reputed SCOPUS/Web of Science/SCI/ABDC/UGC Care Journals. Assit Prof. Ali has more than 23 publications and 20 National & International seminars and conferences, and 4 book chapters (IGI). He is the editor of books titled Impact of Renewable Energy on Corporate Finance and Economics and Renewable Energy and the Economic Welfare of Society. He is also reviewer in several Scopus/ABDC/ ABS journals. Overall, he is a very dynamic, creative,

Ali Mna is currently an assistant Prof of economy in the Higher institute of business administration of Gafsa Tunisia with an academic experience of more than 20 years. He holds a master's degree in Economic and financial dynamic. He earned his Doctor of Philosophy in economics from Sfax University, Tunisia. He has published research papers in reputed SCOPUS/Web of Science/SCI/ABDC/UGC Care Journals. Dr. Ali has more than 10 publications and 8 National & International seminars and conferences. He is also reviewer in several Scopus/ABDC/ABS journals. He is a member of the core team of editors of book titled Renewable Energy and the Economic Welfare of Society. Overall, he is a very dynamic, creative, and positive individual

***

Babita is an Assistant Professor teaching at CHRIST (Deemed to be University) specializing in Finance & Account where she has a profound knowledge in economic growth corporate finance & international trade. A UGC NET/JRF-qualified scholar, she has done her Ph.D. in Management Studies from Guru Jambheshwar

University of Science and Technology, Hisar, Haryana. Dr. Chaudhary has a diverse academic background in both Finance and Human Resource Management coupled with prior corporate experience of two years. She has published in various international journals such as IFI, ABDC, SAGE, and Inderscience Apart from research, she has presented papers in many national/ international conferences. Dr. Babita's research at present explores the effects of consumer ethnocentrism on purchase behaviors within the context of OTT platforms, considering important consumer behavior domains within a more globalized media environment. Dr. Babita Chaudhary works as an Assistant Professor in School of Business and Management of CHRIST (Deemed to be University) Indore Campus, India., she holds Ph.D., UGC NET-JRF completed her MBA in Human Resource and Finance.

Abhijit Pramod Chandratreya is currently working as Deputy Director in the Department of Post Graduate Research Centre at SCES's Indira Institute of Management, Pune for the past 16 years. He joined IIMP after working in various Industries for 17 years. He graduated in Production Engineering from Birla Vishwakarma Mahavidyalaya (Sardar Patel University), Anand, Gujarat, India. He secured a Master of Business Administration in Human Resource Development from IGNOU. He has been awarded a Ph.D. in the field of Marketing Management from Savitribai Phule Pune University, India. He has presented several papers in National and International Journals, conferences, and Symposiums and has published 15 research papers both in referred and SCOPUS / ABDC-listed Journals. He has published 12 chapters in books released by IGI Global of which 4 are indexed in SCOPUS and three chapters in books released by NOVA SCIENCE of which one is indexed in SCOPUS. He has published three patents. He is a certified trainer of CSCP, CPTP, CLDM, MTF, and CID certified by MiddleEarth HR. His main areas of interest include Marketing, HRM, SCM, Statistics, Business Analytics, and Research Methodologies.

İbrahim Gönen serves as a Mechanical Systems Petty Officer in the Air Forces Command. After graduating from the Air Petty Officer Vocational School in 2005, he graduated from Anadolu University, Department of Business Administration in 2012. In 2023, he completed his master's degree in Konya Necmettin Erbakan University, Department of Industrial Engineering.

Güzide Karakuş is an Assistant Professor at Necmettin Erbakan University, department of Aviation Management. He holds a bachelor's degree in Industrial Engineering from Kocaeli University (1999) and an MBA degree from Gebze Institute of Technology. (2002). She completed her PhD in Production Management and Marketing at Selçuk Universitiy 2014. Prior joining academia, she worked in consulting about quality management systems, CE marking and project management. She works and teaches Innovation Management, Technology Management, Project Management, Quality Management, Sustainable Operations Management, Supply Chain Management, both undergraduate and graduate level.

Vivit Rosmayanti, S.Pd.I., M.Pd was born in Botta on June 26, 1987. She has over 10 years of experience teaching at the University. She obtained a bachelor's degree in English Education from Alauddin State Islamic University. She completed her master's and doctoral degrees in English Education from Makassar State University. She is currently a lecturer in the Post Graduate Program, Department of English Education at Makassar State University, South Sulawesi, Indonesia. She is also actively writing research articles to be published in nationally and internationally accredited journals. She has written in some chapter books, monographs, and reference books. She is interested in teaching methods for teaching English. Her research interests include language teaching methodology, ICT instruction, and ESP teaching. She can be contacted at email: rosmayantivivit@gmail.com

Muhammad Usman Tariq has more than 16+ year's experience in industry and academia. He has authored more than 200+ research articles, 110+ case studies, 130+ book chapters and several books other than 4 patents. He is founder and CEO of The Case HQ, a unique repository for courses, narrative and video case studies. He has been working as a consultant and trainer for industries representing six sigma, quality, health and safety, environmental systems, project management, and information security standards. His work has encompassed sectors in aviation, manufacturing, food, hospitality, education, finance, research, software and transportation. He has diverse and significant experience working with accreditation agencies of ABET, ACBSP, AACSB, WASC, CAA, EFQM and NCEAC. Additionally, Dr. Tariq has operational experience in incubators, research labs, government research projects, private sector startups, program creation and management at various industrial and academic levels. He is Certified Higher Education Teacher from Harvard University, USA, Certified Online Educator from HMBSU, Certified Six Sigma Master Black Belt and has been awarded PFHEA, SFSEDA, SMIEEE, and CMBE.

Index

F

G

I

P